Hacking
Roomba®

Tod E. Kurt

Hacking Roomba®

Published by
Wiley Publishing, Inc.
10475 Crosspoint Boulevard
Indianapolis, IN 46256
www.wiley.com

Copyright © 2007 by Wiley Publishing, Inc., Indianapolis, Indiana

Published simultaneously in Canada

ISBN-13: 978-0-470-07271-4
ISBN-10: 0-470-07271-7

Manufactured in the United States of America

10 9 8 7 6 5 4 3 2 1

Library of Congress Cataloging-in-Publication Data

Kurt, Tod E., 1969–
  Hacking Roomba / Tod E. Kurt.
    p. cm.
  ISBN-13: 978-0-470-07271-4 (paper/website)
  ISBN-10: 0-470-07271-7 (paper/website)
  1. Robots—Programming. 2. Roomba vacuum cleaner. 3. Autonomous robots. 4. Mobile robots. I. Title.
  TJ211.45.K87 2007
  629.8′9251—dc22
                                    2006031031

About the Author

Tod E. Kurt has engineered the hardware and software for robotic camera systems that went to Mars. He was a founding developer and systems architect of Overture Systems, originally GoTo.com, later sold to Yahoo. Now as co-creator of ThingM.com, he's designing sketchable hardware and networked smart objects. He has degrees in electrical engineering from Caltech and physics from Occidental College. He started robotics hacking at the age of twelve when he took apart his BigTrak, RC car, and chemistry set to make an upright programmable robot.

Credits

Executive Editor
Chris Webb

Development Editor
Kelly Talbot

Copy Editor
Michael Koch

Editorial Manager
Mary Beth Wakefield

Production Manager
Tim Tate

**Vice President and
Executive Group Publisher**
Richard Swadley

Vice President and Executive Publisher
Joseph B. Wikert

Compositor
Happenstance Type-O-Rama

Proofreader
Nancy Riddiough

Indexer
Ted Laux

Anniversary Logo Design
Richard Pacifico

Cover Design
Anthony Bunyan

Contents at a Glance

Acknowledgments xi
Introduction xiii

Part I: Interfacing

Chapter 1: Getting Started with Roomba 3
Chapter 2: Interfacing Basics 19
Chapter 3: Building a Roomba Serial Interface Tether 41
Chapter 4: Building a Roomba Bluetooth Interface 65
Chapter 5: Driving Roomba 89
Chapter 6: Reading the Roomba Sensors 109

Part II: Fun Things to Do

Chapter 7: Making RoombaView 131
Chapter 8: Making Roomba Sing 151
Chapter 9: Creating Art with Roomba 167
Chapter 10: Using Roomba as an Input Device 189

Part III: More Complex Interfacing

Chapter 11: Connecting Roomba to the Internet 205
Chapter 12: Going Wireless with Wi-Fi 231
Chapter 13: Giving Roomba a New Brain and Senses 257
Chapter 14: Putting Linux on Roomba 297
Chapter 15: RoombaCam: Adding Eyes to Roomba 333
Chapter 16: Other Projects 365

Appendix A: Soldering and Safety Basics 383
Appendix B: Electrical Diagram Schematics 405
Appendix C: iRobot Roomba Open Interface (ROI) Specification . . . 415

Index 427

Contents

Acknowledgments xi

Introduction xiii

Part I: Interfacing

Chapter 1: Getting Started with Roomba 3
Quick Start 3
What Is Roomba? 3
Which Roomba Cleaners Are Hackable? 5
Internal and External Components 13
OSMO//hacker: Hope for Older Third Generation Roombas 16
Summary 17

Chapter 2: Interfacing Basics 19
What Can Be Done with the ROI? 19
The ROI Connector 21
The ROI Protocol 24
Introducing the RoombaComm API 39
Summary 40

Chapter 3: Building a Roomba Serial Interface Tether 41
Alternatives 41
Parts and Tools 43
The Circuit 44
Building the Serial Tether 48
Connecting to a Computer 56
Commanding Roomba 62
Summary 63

Chapter 4: Building a Roomba Bluetooth Interface 65
Alternatives 65
Why Bluetooth? 66
How Bluetooth Works 67
Parts and Tools 68
The Circuit 70
Building the Bluetooth Adapter 71
Setting Up Bluetooth 78
Testing Bluetooth 83

Using the Adapter 83
Making RoombaComm 84
Summary. 88

Chapter 5: Driving Roomba. 89

The Roomba Motors and Drive Train. 89
The ROI DRIVE Command 92
Simple Tank-Like Motion. 98
Moving in Curves 102
Real-Time Driving 104
Writing Logo-Like Programs 107
Summary. 108

Chapter 6: Reading the Roomba Sensors 109

Roomba Sensors 109
ROI SENSORS Command 115
Parsing Sensor Data 118
Using Sensor Data 121
BumpTurn: Making an Autonomous Roomba. 122
Measuring Distance and Angle 124
Spying on Roomba 126
Summary. 127

Part II: Fun Things to Do

Chapter 7: Making RoombaView 131

About Processing. 131
Using RoombaComm in Processing 136
Designing RoombaView 140
Summary. 150

Chapter 8: Making Roomba Sing 151

Sonic Capabilities of Roomba 151
ROI SONG and PLAY Commands 154
Playing Roomba as a Live Instrument 157
Roomba Ringtones 159
RoombaMidi: Roomba as MIDI Instrument 161
Summary 166

Chapter 9: Creating Art with Roomba 167

Can Robots Create Art? 168
Parts and Tools 168
Adding a Paintbrush to Roomba. 170
What Are Spiral Equations? 178

Drawing Spirals with RoombaSpiro 184
Summary 187

Chapter 10: Using Roomba as an Input Device 189

Ways to Use the Roomba's Sensors. 189
Using Roomba as a Mouse 190
Using Roomba as a Theremin 194
Turning Roomba into an Alarm Clock. 200
Summary 202

Part III: More Complex Interfacing

Chapter 11: Connecting Roomba to the Internet 205

Why Ethernet? 205
What Is Ethernet? 206
Parts and Tools 207
SitePlayer Telnet 208
Lantronix XPort 222
Modifying RoombaComm for the Net. 225
Summary 228

Chapter 12: Going Wireless with Wi-Fi 231

Understanding Wi-Fi 232
Parts and Tools 236
Building the Roomba Wi-Fi Adapter 241
Controlling Roomba through a Web Page 248
Putting It All Together 253
Going Further with LAMP 255
Summary 255

Chapter 13: Giving Roomba a New Brain and Senses 257

Microcontroller vs. Microprocessor 257
Parts and Tools 258
Adding a New Brain with the Basic Stamp 261
Adding a New Roomba Brain with Arduino. 276
Summary 294

Chapter 14: Putting Linux on Roomba 297

Linux on Roomba? 298
Parts and Tools 303
Installing OpenWrt 305
Controlling Roomba in OpenWrt 316
Making It All Truly Wireless. 325
Summary 331

Chapter 15: RoombaCam: Adding Eyes to Roomba **333**

Parts and Tools 334
Upgrading the Brain 334
Controlling Roomba from C 346
Putting It All Together 352
Summary 363

Chapter 16: Other Projects **365**

Autonomous Roombas 365
Roomba Costumes and Personalities 370
Roomba APIs and Applications 372
Warranty-Voiding Hacks 375
Summary 381

Appendix A: Soldering and Safety Basics **383**

Appendix B: Electrical Diagram Schematics **405**

Appendix C: iRobot Roomba Open Interface (ROI) Specification **415**

Index **427**

Acknowledgments

Thanks to my friends and family for being there for me and understanding when I would disappear for days at a time to write this book and commune with the Roombas. Thanks again to them for providing encouragement and many great ideas for fun hacks. Particularly I'm grateful to Ben C, Ben F, Chris L, Liz C, John Joseph, Mike K, Phil T, Paul R, and Preston P for all their great ideas. I owe you all.

I'd like to thank everyone at the iRobot Corporation for providing the ROI specification to the world at large and supplying the assistance I needed for this book. Your company has done an amazing thing by being so open to its users. You're a leader in creating the new conversation between a company and its users.

This book wouldn't be possible without my editors: Kelly Talbot, who kept me going and added to this book in substantial ways, and Chris Webb, who believed in me and the book. Thanks to everyone at Wiley Publishing for doing the hard work of turning bits to atoms.

Thanks to the extended Roomba hacking community, especially those who provided information about their hacks for this book. Also thanks to everyone at Makezine, Roomba Review, and Hackaday for providing a medium for news and discussion about Roomba hacking and hacking in general.

And thanks to you, the reader. Hacking anything, including Roombas, can only get better with the addition of new people, fresh viewpoints, and sharing of ideas. Welcome to the Roomba hacking community and have fun with this book and your Roomba!

Introduction

The iRobot Roomba is perhaps the best example of mobile robotics entering the home. It performs a useful task and is relatively inexpensive, and while it doesn't look like Rosie the Robot on *The Jetsons*, it does have a charm of its own.

The purpose of this book is to introduce robot hacking to people who are interested in programming and modifying their own robot but who don't want to destroy a functioning Roomba. This "reversible hacking" is device modification that can be undone to return the device to its original state. In the case of the Roomba, the ROI connector is the gateway to reversible Roomba hacking. All manner of devices can be plugged into the ROI: a desktop computer, a microcontroller "brain" to replace the original, new sensors and actuators, or maybe just some snazzy running lights. Any of these modifications can be quickly removed, leaving the Roomba in the original state. All hacks presented in this book are reversible hacks. None of the projects in this book will damage your Roomba or even void its warranty. (However, a few potentially warranty-voiding options are included and explained as such, if you are intrepid enough to explore them.)

This book shows how to make the Roomba do more than be a cute vacuum cleaner. With this book, the reader can give the Roomba new or more complex behaviors, connect it to the Internet, control it from a computer, and literally make it dance and sing. This book is a way to learn the basics of robotics and have fun programming a robot without needing to build one. All of the projects can be done without breaking open the Roomba or even voiding its warranty. And like all good hacking books, this one shows how to install Linux on a Roomba. This book is a practical demonstration of several ways to create networked objects, normal devices with intelligence and Internet connectivity.

The History of Hacking the Roomba

Most people who purchased the first Roombas were early adopters of technology and liked the idea of a personal robot to do their bidding. To watch a Roomba roaming around their living room, cleaning up after a mess, was to experience in a small way life in the future.

Unfortunately, the Roomba wasn't very "hackable" by the normal gadgeteer. If you wanted to easily reprogram your Roomba to alter its behavior or make it do tricks, you were out of luck. At the least you had to take the Roomba apart, definitely voiding its warranty. Once inside perhaps you could reverse engineer the small computer (also known as the microcontroller) used as its brain, maybe replace it completely, and hook into the motors and sensors, effectively destroying it for its original purpose. Communities devoted to hacking the Roomba in this low-level way grew and flourished. The hacking section of Roomba Review (http://roombareview.com/hack/) is one of the most famous, and the accompanying forum is still the best place to go to discuss Roomba hacking. Other sites like Hackaday (http://hackaday.com/) and Makezine (http://makezine.com/) routinely featured projects that used stripped-down or heavily modified Roombas. But hacking the Roomba was a difficult and expensive task, only suitable for the most experienced engineers. Recently this has changed.

In December 2005, iRobot Corporation, the maker of the Roomba, recognized the growing hacking community and released documentation describing the Serial Command Interface (SCI) present on third-generation Roombas. In mid-2006 iRobot renamed the SCI to be the Roomba Open Interface (ROI), a name that better fits its role. The ROI allows you to take full control of the Roomba and its behavior. This is no simple remote control interface, but instead a protocol that allows complete sensor readout and full actuator control.

Since the release of the SCI/ROI specification, there has been an explosion of new Roomba hacks. The Roomba hacking community has blossomed to include not just professional hardware engineers, but people from all experience levels, from normal people looking to play with their Roomba in a new way to academics experimenting with low-cost robotics. The ROI turns the Roomba into a true robotics platform. And because these are all reversible hacks, it's easy to try someone else's hacks. No longer do you have to break a Roomba to try something out. To see some of the hacks people are working on and join in discussions about them with others, see the Roomba Review hacking site mentioned above, the accompanying forum at `http://www.roombareview.com/chat/`, and the Roomba hacking wiki at `http://roomba.pbwiki.com/`.

Whom This Book Is For

This book is for those who want to experience the fun of programming a robot without all the problems normally associated with building one.

Most of this book is designed for beginning hackers, those who know a bit of programming and a little bit of electronics. Familiarity but not expertise is assumed with soldering and schematics. No mechanical expertise is required, but if you have it, you can do even more impressive things than what is outlined in this book.

 Note   If your soldering and schematics skills are a little rusty, two useful appendixes are supplied to help get you back up to speed.

Most of the code examples are presented in Java. Java was chosen for its ubiquity and cross-platform capability and is used for all the PC-connected projects. Thus, knowing how to compile and run Java programs is required. The later, more advanced projects are programmed on microcontrollers in either PIC BASIC or AVR GCC.

What This Book Covers

The projects in this book are based around the Roomba and the Roomba Open Interface (ROI). The Roomba's capabilities as presented via the ROI are described and tested. The few capabilities that aren't accessible via the ROI are mentioned briefly. The Roomba's mechanical and electrical internals are also discussed, but since this book is about hacking the Roomba without taking it apart, they're mentioned only briefly.

The ROI protocol is covered in detail, with examples given as each part of the protocol is examined. Practical and fun examples of each ROI command are given. As a way of abstracting the rather low-level commands of the ROI, a code library of software routines is slowly built up to become the RoombaComm API with applications built using it.

Throughout this book, the ideas and practices of Network Objects are developed. As computing and networking become so cheap as to be effectively free, all objects will become network objects. The Roomba robot is already a computing object, a normal everyday device (a vacuum cleaner) that has been imbued with intelligence in the form of a small computer. This book describes several methods of extending the Roomba to become a networked object. These methods are similar to those used for current and future network objects that will exist throughout the home.

How This Book Is Structured

This book is designed mostly for the novice electronics hacker, but it contains several advanced projects toward the end. The book is divided into three parts. Each part is mostly self-contained, depending upon which shortcuts are taken, but knowing the concepts presented in earlier chapters helps in the later ones.

Part I: Interfacing

This part describes the Roomba, its history, and its model variations, to dispel the confusion regarding which Roombas are hackable via the ROI protocol. The ROI protocol is discussed in depth, showing exactly what bytes are sent and received to command the Roomba. To allow a PC to speak to the ROI, two simple hardware interface projects are shown — one wired, one wireless. With those created, a software library is given that provides an easy-to-use abstraction layer on the PC.

Part II: Fun Things to Do

Using the hardware and software infrastructure from the previous part, this part focuses on interesting, or just plain silly, things to do with a computer-controlled Roomba. Make it dance and sing, draw huge artwork on the ground, and create a complete dashboard/remote control PC application called RoombaView.

Part III: More Complex Interfacing

With experience from using a PC to control a Roomba, the focus now becomes making the Roomba a true Internet device and fully autonomous. The first few hacks are Internet versions of the initial interfaces. From there a fully reprogrammable replacement brain is added to the Roomba using microcontrollers like the PIC Basic Stamp or Arduino AVR. This part ends with adding a larger microcontroller board that can run Linux and use a webcam, microphone, or any other sensor imaginable.

Appendixes

If your electronics hacking skills are a little rusty, Appendix A covers the basics on how to solder circuits and work safely with electronics. Appendix B explains how to interpret common schematic circuit diagrams like the ones in many of the projects. Appendix C is a reprint of the ROI specification from iRobot. The ROI is what enables all the hacks in this book, and it is the authority on how the Roomba can be hacked.

What You Need to Use This Book

Of course you will need a Roomba, one with ROI capability. Chapter 1 describes which Roombas have ROI. To run the code, you will need a PC with USB and Java JDK 1.5 installed. Windows, Mac OS X, and Linux computers can all fit this requirement. For Windows and Linux, Java is not installed by default and can be obtained as a free download from http://java.com/. To write and compile programs, you'll need a text editor and knowledge of the command line or experience with a Java IDE. If you're unfamiliar with how to create and compile Java programs, there are many tutorials on the Net. This book assumes basic familiarity with programming and Java. Even so, all code presented in the book is available in ready-to-run form from www.wiley.com/go/extremetech and http://roombahacking.com/.

For projects that have circuits, a soldering iron and other tools are required, as well as basic knowledge of their use. Expect to have on hand a multimeter, wire cutters/strippers, test leads, and so on. Each chapter describes exactly which tools are required. There is an appendix that contains a basic overview on soldering, tool use, and electronics assembly. It also covers how to be safe around these somewhat dangerous tools. There are many good references on the Internet going into more depth on these topics than this book has room for, so some of my favorite electronics "how-to" sites will be listed in that appendix as well.

Many of the circuits presented in this book can be purchased as kits or fully assembled from various suppliers. Notably, RoombaDevTools.com provides fully assembled Roomba interfaces that are functionally identical to the interfaces provided in Chapters 3 and 4.

To build the circuits, various electronic components are required. Only a few components are more than a dollar or two. There are several suppliers for these components: Digikey (http://digikey.com/), Mouser (http://mouser.com/), Jameco (http://jameco.com/), and Radio Shack (http://radioshack.com/) are four of the more popular. Sparkfun (http://sparkfun.com/) is a great source for the specialized components used. Throughout this book, Jameco part numbers will be used when possible for the commonly available parts. Jameco is a great resource that is very popular. They stock almost anything an electronics hobbyist needs, at decent prices; they ship fast; and, most important, they have an easy-to-use web interface. Jameco also sells all the tools needed for the projects in this book.

The later projects assume some experience in microcontroller programming. While this book hasn't the space to go into how to do this, it's not that different from programming on a PC, and controlling the Roomba would be a great excuse to learn about it.

If you'd like to learn about microcontroller programming, a good starting resource is Parallax's *What Is a Microcontroller?* book available as a free PDF download from their website (`http://parallax.com/`) in their Documents/Tutorials section. It's focused on the Basic Stamp microcontroller, but the techniques and concepts are universal. A good repository for other microcontroller info is NYU's ITP program tutorial website (`http://itp.nyu.edu/physcomp/Tutorials/Tutorials`) and Tom Igoe's Physical Computing Site (`http://tigoe.net/pcomp/`).

Conventions Used in This Book

In this book, you'll find several notification icons — Note, Caution, Tip, and Cross-Reference — that point out important information. Here's what the three types of icons look like:

Notes provide you with additional information or resources.

A caution indicates that you should use extreme care to avoid a potential disaster.

A tip is advice that can save you time and energy.

A cross-reference directs you to more information elsewhere in the book.

Code lines are often longer than what will fit across a page. The symbol ↵ indicates that the following code line is actually a continuation of the current line. For example,

```
root@OpenWrt:~# wget http://roombahacking.com/software/openwrt/↵
roombacmd_1.0-1_mipsel.ipk
```

is really one line of code when you type it into your editor.

Code, functions, URLs, and so forth within the text of this book appear in a monospaced font, while content you will type appears either **bold** or monospaced.

What's on the Companion Website

On the companion website at http://roombahacking.com/ and www.wiley.com/go/extremetech, you'll find source code and schematics for all the projects in this book. All the code and schematics are open source. At http://roombahacking.com/, they are improved and added to by the Roomba hacking community. You'll also find additional projects that expand upon the ideas presented in this book. The site also contains mirrors of important documents like the ROI specification and data sheets for useful electrical components. Galleries are available for Roomba hackers to upload and share information about their favorite Roomba hacks. Finally, the site contains links to other Roomba sites, tutorials about electronics assembly and microcontroller programming, and other useful hacking websites.

Interfacing

in this part

Chapter 1
Getting Started with Roomba

Chapter 2
Interfacing Basics

Chapter 3
Building a Roomba
Serial Interface Tether

Chapter 4
Building a Roomba
Bluetooth Interface

Chapter 5
Driving Roomba

Chapter 6
Reading the Roomba Sensors

Getting Started with Roomba

iRobot has produced a dizzying variety of Roomba vacuuming robots since the original Roomba model was introduced in 2002. They now have even the Scooba, a robot that washes floors.

Compared to other robotic vacuum cleaners, the typical Roomba robotic vacuum cleaner is very inexpensive at under $300 for even the most expensive Roombas and $150 for the least expensive. The cheapest new Roombas can be found for around $100 on the Internet. For a vacuum cleaner that's a pretty good price. For a robot that's also a vacuum cleaner, that's an amazing price. And for a robotic vacuum cleaner that's hackable by design?

Quick Start

If you're already familiar with Roomba, know it's compatible with the Roomba Open Interface (ROI), and you'd like to start hacking immediately, skip to Chapter 3 to begin building some hacks. If you're uncertain which Roomba you have, if it is hackable through the ROI, and want to learn the details on the ROI protocol that enables all these hacks, keep reading.

Note All projects in this book will utilize the Roomba Open Interface (ROI). It was previously known as the Roomba Serial Command Interface (SCI) and you'll find many references to the SCI on the Internet. It's exactly the same as the ROI; only the name has changed.

What Is Roomba?

Roomba is an autonomous robotic vacuum cleaner created by iRobot Corporation. To operate, Roomba requires no computer and no technical knowledge from its owner. It only needs a power outlet and occasional cleaning, like any vacuum cleaner.

in this chapter

☑ Uncover how Roomba evolved

☑ Explore which Roomba models are hackable

☑ Examine the components of Roomba

☑ Learn about the OSMO//hacker module

Originally released as just Roomba in 2002, the Roomba design and functionality have evolved over the years. Currently there are five varieties of Roomba available with names like Roomba Discovery and Roomba Red. According to iRobot, with over 2 million units sold, not only is Roomba one of the most successful domestic robots to date, it is also one of the very few robots to have sold over a million units. This accomplishment is the result of a long evolutionary process of robotics design at the iRobot Corporation.

iRobot Corporation

The creators of Roomba have been making robots for over 15 years. iRobot was founded by Rodney Brooks, Colin Angle, and Helen Greiner. These three MIT alumni have been instrumental in guiding robotics research for many years, not only through their research but also through the practical application of their ideas through iRobot.

Subsumption Architecture

Rodney Brooks coined the term *subsumption architecture* in 1986 in his classic paper "Elephants Don't Play Chess." This paper began a shift in artificial intelligence research. At the time it was believed that to create a thinking machine, one needed to start with a symbolic representation of its world from some set of base principles. (For example, a robot butler having a built-in map of a house would be a kind of basic symbol.) This top-down view of cognition is opposite to how nature works: When we enter a new house, we must explore and build up our own unique perception of how to get from place to place. Brooks codified a bottom-up, behavior-based approach to robotics.

In subsumption architecture, increasingly complex behaviors arise from the combination of simple behaviors. The most basic simple behaviors are on the level of reflexes: "avoid an object," "go toward food if hungry," and "move randomly." A slightly less simple behavior that sits on top of the simplest may be "go across the room." The more complex behaviors subsume the less complex ones to accomplish their goal.

Genghis and PackBot

In 1990 while at MIT, Rodney Brooks and iRobot created the Genghis Robot, an insect-like robot with six legs and compound eyes. It was a research platform that bucked the trend in artificial intelligence at the time by using Brook's subsumption architecture. Genghis was designed from an evolutionary perspective instead of the normal high-level cognition perspective of traditional AI. It looked and acted like an insect. This behavior-based robotics architecture would inform the design of all future iRobot robots.

From Genghis, iRobot developed a few other research robots but quickly moved into developing robots for real-world use. iRobot has had great success with their PackBot, a series of ruggedized telepresence (able to withstand harsh outdoor environments and remotely controlled) and autonomous robots for the military and law enforcement. Instead of sending soldiers or a SWAT team into a dangerous area, the PackBot can be pulled from a backpack and thrown into the area. With its onboard video and audio sensors, the area can be inspected without risking a life. The PackBot can withstand 400+ *g*s of force. This makes it much tougher than a human. One *g* is the force you feel every day from gravity. Three *g*s are what most roller coasters make you feel, and at five *g*s you black out. Although the Roomba isn't nearly so rugged, it definitely seems to have inherited some of its cousin's toughness.

Enter Roomba

The Roomba robotic vacuum cleaner is a physical embodiment of Brooks' subsumption architecture. Roomba has no room map or route plan. It has no overall view of what it is doing. Instead it functions much more like an insect: going toward things it likes (dirt, power) and away from things it dislikes (walls, stairs), moving in predefined movement routines while occasionally and randomly jumping out of a predefined routine.

This random walk feature of the Roomba algorithm is perhaps what confuses people the most at first. It will seem to be going along doing the right thing when it suddenly takes off in a different direction to do something else. But for every time it moves from the right place to the wrong place, it has moved from the wrong place to the right place. On average (and if left for a long enough time), Roomba covers the entire area. In terms of time efficiency, Roomba is not the most effective, as it takes several times longer for it to fully cover a region than it would for a person with a normal vacuum cleaner. But whose time is more valuable? Roomba can work while the person does something else.

Which Roomba Cleaners Are Hackable?

There is some confusion as to which Roomba cleaners are easily hackable through the ROI. This is complicated by the fact that iRobot doesn't make obvious the model numbers and firmware versions of the different Roomba cleaners.

All new Roomba cleaners currently have the ROI protocol built-in and ready to use. These are third-generation Roomba cleaners. The two most common Roomba cleaners, Roomba Discovery and Roomba Red, will be used in the examples in this book.

Following is a fairly comprehensive list of Roomba cleaners available in North America. International versions are functionally identical and named the same, with only small modifications to function on different mains voltages.

First Generation

The first generation of Roomba cleaners was astounding in the amount of capability they packed into a small, inexpensive package. This generation did not have any ROI capability. There was only one type of Roomba in the first generation:

- **Roomba:** The original Roomba model, shown in Figure 1-1, was released in 2002 and improved in 2003. It could clean small, medium, or large rooms when instructed through its S, M, and L buttons. It shipped with at least one *virtual wall* (a special battery-powered infrared emitter used to create virtual boundaries) and a plug-in battery charger.

FIGURE 1-1: The original Roomba

Second Generation

The second Roomba generation added what many considered a necessity: a dirt sensor. This generation also featured improvements in battery life and cleaning efficiency. As with the first generation, this generation also did not have ROI functionality. The second generation of Roomba cleaners included two models:

- **Roomba Pro:** This model, shown in Figure 1-2, was released in 2003 as the base model of the new generation. It included the new dirt sensor and could perform spot cleaning.
- **Roomba Pro Elite:** This model, shown in Figure 1-3, was also released in 2003 and was the same as the Roomba Pro model, but colored red, and included both spot cleaning and max cleaning.

FIGURE 1-2: Roomba Pro

FIGURE 1-3: Roomba Pro Elite

Third Generation

The third generation of Roomba cleaners includes a great many more improvements than were made in the first to second generation jump. In addition to a dirt sensor, these models include a home base dock for self-charging, a remote control, a scheduling capability, and, most importantly for hackers, a serial port. This generation introduced ROI functionality as a firmware upgrade in October 2005.

This is the current line of Roombas:

- **Roomba Red:** This model, shown in Figure 1-4, was released in 2004 and improved in 2005. It is the least expensive member of the current Roomba family. It comes with a seven-hour charger instead of a three-hour one and a single dirt sensor. It doesn't have a remote control or a self-charging home base, which are standard with the Discovery model.

FIGURE 1-4: Roomba Red

- **Roomba Sage:** This model, shown in Figure 1-5, was released in 2004 and improved in 2005. It is the next least expensive model. It is the same as the Roomba Red model, except that it comes with a three-hour charger and is light green.

- **Roomba Discovery:** This model, shown in Figure 1-6, was released in 2004 and improved in 2005. It is the one seen in most advertisements. It contains everything the Sage model does, and it also includes the remote control, the self-charging home base, and dual dirt sensors.

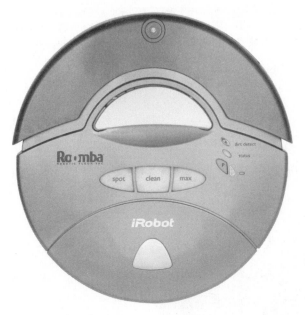

FIGURE 1-5: Roomba Sage

FIGURE 1-6: Roomba Discovery

- **Roomba Discovery SE:** This model, shown in Figure 1-7, was released in 2004 and improved in 2005. It is identical to the Discovery model except for the different colored exterior and the inclusion of a self-charging wall mount in addition to the self-charging home base.

FIGURE 1-7: Roomba Discovery SE

- **Roomba Pink Ribbon Edition:** This model, shown in Figure 1-8, was released in 2005 as a promotional version of Roomba and is functionally the same as the Roomba Sage model. For every Pink Roomba sold, 20 percent of the sale price was donated to the Susan G. Komen Breast Cancer Foundation, with a $45,000 minimum guaranteed donation.

- **Roomba 2.1:** This model, shown in Figure 1-9, was released in 2005 as a special model sold only by the Home Shopping Network. It was the introductory model for a makeover of the third generation. All Roomba robotic vacuum cleaners released since then are 2.1. The 2.1 designator is a blanket term for over 20 enhancements to both software and hardware. The software upgrade (called AWARE robotic intelligence) includes improvements to the cleaning algorithms for better cleaning efficiency and greater room coverage. The hardware improvements are perhaps more numerous and include better battery-charging circuitry, improved brushes and sensors, and a better vacuum design.

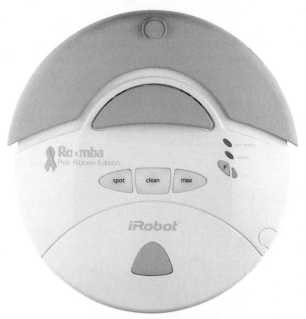

FIGURE 1-8: Roomba Pink Ribbon Edition

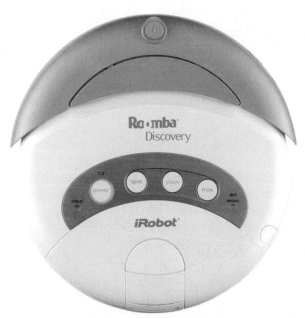

FIGURE 1-9: Roomba 2.1 for the Home Shopping Network

■ **Roomba Scheduler:** This model, shown in Figure 1-10, was released in 2005 and is the same as the Roomba Discovery model, with the inclusion of a special scheduler remote control and a blue exterior. iRobot has also released an improved Scheduler model. This improved model is black and has a Dust Bin Alert feature to let you know when its dust bin is full.

FIGURE 1-10: Roomba Scheduler

What about Scooba?

Scooba is the newest home cleaning robot from iRobot. It is a floor-washing robot. The robot preps the floor by vacuuming loose debris, squirts clean solution, scrubs the floor, and then sucks up the dirty solution leaving a nearly dry floor behind. Although it does vacuum, it's not a general purpose vacuum cleaner like Roomba (for example, it doesn't work on carpet). The cleaning solution, which has been nicknamed Scooba juice, is a special non-bleach formula that is safe for sealed hardwood floors.

Scooba apparently contains an ROI port and thus would be compatible with the projects presented here, but no tests have been performed with it yet. The ROI specification published by iRobot makes no mention of Scooba-specific commands.

Internal and External Components

Although it's not necessary to know the details of the insides of Roomba to do the projects in this book, it is instructive and neat. Knowing how something works can help you diagnose any problems that are encountered during normal use.

The Underside

To get started on how Roomba is put together, turn it over. Figure 1-11 shows the underside of Roomba with its brushes removed.

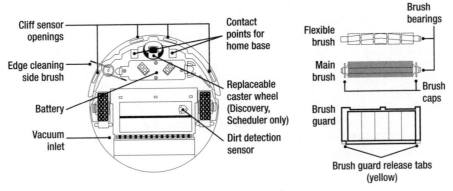

FIGURE 1-11: Bottom view of a typical Roomba vacuum cleaner

Roomba is organized in three sections:

- **Sensor front:** Virtually all of the sensors (bump, wall, cliff, and home base contacts) are up front. In fact, almost all the sensors are mounted on the movable front bumper. This movable bumper both enables a mechanical means to measure contact (the give triggers a switch) and absorbs shock to minimize damage. The Roomba firmware is designed to always travel forward, so it places its most sensitive foot forward, as it were. When programming the Roomba, you can subvert this tendency and make the Roomba drive backward, but doing so makes it difficult for the Roomba to "see" anything.

- **Motor middle:** The main drive motors, vacuum motors, vacuum brushes, side cleaning brush, and battery are all in the center. This gives the Roomba a center-of-mass very close to the center of its body, making it very stable when moving.

- **Vacuum back:** Just like a normal vacuum cleaner, the entire back third contains the vacuum and vacuum bag for holding dirt. The back can be removed when in ROI mode, which slightly unbalances the Roomba and gives it more of a "hot rod" type of movement.

Power

The first consideration for any robotic system is power. Roomba is powered by a custom high-power rechargeable battery pack. This pack provides enough power to run the Roomba for up to 100 minutes of normal operation. It can be re-charged in 3 hours using the 3-hour charger.

Battery Pack Details

Internally this battery pack consists of 12 sub C size nickel metal-hydride (NiMh) cells. Each cell puts out 1.2V so 12 cells wired in series give 14.4 VDC. The newer yellow battery pack uses at least 3000 mAh cells. Some people have taken their packs apart and even found 3200 mAh cells. (The original black Roomba battery pack used 2300 mAh cells.) The mAh differences only affect run time and are otherwise the same. The batteries are good for approximately 1000 charging cycles and do not suffer from any sort of negative memory effect from partial discharge. Do-it-yourselfers can find companies like BatterySpace.com that will sell compatible battery packs using up to 3600 mAh cells. These packs give 20 percent longer run time over the yellow pack and 56 percent longer time over the original pack. Of course, such a hack does void your warranty, but it is a way to save an otherwise old and unused Roomba cleaner.

Note

The main metric of batteries is ampere-hours, which are more commonly referred to as milliamp-hours (mAh). This describes how much current can be drawn from the battery and for how long, and thus how much power a given battery can provide. A 1000 mAh (1 Ah) battery can supply either a 1000 mA (milliampere) circuit for one hour, a 5 mA circuit for 200 hours, or a 2500 mA circuit for 24 minutes. For comparison, a typical LED flashlight might draw 30 mA, while a typical AA battery can provide 1000–1800 mAh. The Roomba batteries have increased in capacity from their original 2300 mAh to at least 3000 mAh, making the current models last 30 percent longer on a charge than previous models.

Available Power

When turned on but sitting idle, the Roomba draws 150 to 250 mA, depending on the Roomba model. During normal operation, a Roomba draws from 1500 mA to 2000 mA of current. This variation in current consumption is due to the variety of floor types: Thick carpets cause more current draw than hard floors. The battery pack can be maximally discharged at a 4 Amp rate, limited by an internal polyswitch (a device that acts somewhat like a fuse that can be reset). Without the polyswitch, a short circuit would damage the battery and the unit.

The full voltage and power available from the pack is available through pins 1 and 2 on the ROI connector. Any projects using power through the ROI can draw as much power as they need. However, drawing too much will shorten the life of the battery, shorten the run time of the unit, and perhaps confuse the system's internal firmware. All projects in this book will draw less than 1 Amp of current and most draw less than 100 mA. A 100 mA project running of Roomba power would shorten the normal Roomba run time by maybe 5 percent.

Motors

The Roomba has five independently controllable electric motors. Two of these, the drive motors, are variable speed through pulse-width modulation (PWM) and run both forward and in reverse. The three motors that run the main brush, side brush, and vacuum have simple on/off controls.

Drive Motors

The two drive wheels can be seen in the previously shown Figure 1-11. They are located on the centerline, right behind the center of gravity. Having the drive wheels behind the center of gravity makes the Roomba lean forward a bit on its front non-rotating caster. The drive motors connected to the wheels can move the Roomba as fast as 500 mm/sec (about 1.64 ft/sec) forward or backward and as slow as 10 mm/sec (about 3/8 in/sec).

The drive motors draw approximately 1000 mA when running at normal speeds, and at their slowest draw about 300 mA.

Vacuum Motors

The three vacuum motors draw about 500 mA when running. The main vacuum motor has about the same amount of suction as a standard hand vacuum. However, due to the design of the main brush motors and the rubber flap around the vacuum inlet, the effective suction is as good as a small upright vacuum.

Sensors

The Roomba navigates mainly by its mechanical bump sensors, infrared wall sensors, and dirt sensors. For detecting dangerous conditions, it also has infrared cliff detectors and wheel-drop sensors.

Bump Sensors

Roomba has two bump sensors on the front, located at the 11 o'clock and 1 o'clock positions. The spring-loaded front bumper moves to trigger one or both of these sensors. Each is implemented as an optical interrupter. An optical interrupter is a simple LED and photodetector pair: the LED shines and the photodetector detects the LED's light. When something (an *interrupter*) is inserted between the LED and photodetector, the photodetector senses the absence of light and changes an electrical signal. The bell that rings when you enter or leave a store is a large example of an optical interrupter. On one side of the door is a focused light source, on the other a detector for that light. You are the interrupter. When you break the light beam, the detector senses that and rings the bell. In the case of Roomba's bump sensor, the interrupter is a small plastic arm connected to the bumper.

Infrared Sensors

There are six infrared sensors on the Roomba, all on the front bumper. Four of these face down and are the cliff sensors, and another faces to the right and is the wall sensor. These five sensors work much like the bump sensors in that there is an LED emitter and a photodetector looking for the LED's light. But unlike the interrupter-based sensor, these are looking for the reflected light of the LEDs. For the cliff sensors, they are looking for light reflected from the floor (meaning the floor's still there). For the wall sensor, it is looking for a wall (to enable it to follow walls). One problem with just having an LED shine and looking for reflection is that the ambient light could trigger false readings. On a bright sunny day you'd find your Roomba prototype not able to find walls and always falling down the stairs. The common way around this is to modulate the light emitted by the LED and then only look for light that's been encoded in that way. For most robotics applications, including the Roomba, this is done by turning on and off the LED 40,000 times a second (40 kHz).

The last infrared sensor is the remote control/virtual wall/docking station sensor that is visible as the small round clear plastic button at the 12-o'clock position on the bumper. This sensor works just like any other remote control sensor for consumer electronics. It has an interesting 360-degree lens that enables it to see from any orientation.

Internal Sensors

The most commonly used internal sensors are the wheel-drop sensors. All three wheels have a microswitch that detects when the wheel has extended down. In the case of Roomba, these wheel drops are equivalent to cliff detection since they are indicative that the Roomba is in some dire situation and should abort its current algorithm.

The dirt sensor is a small metal disk (or two) under the main brush and appears to be a capacitive touch sensor. Capacitive sensors are used in those touch lamps that can be controlled by simply placing a finger on a metal surface of the lamp. Although the touch lamp sensor only provides an on/off result, the dirt sensors provide an analog value.

The last set of internal sensors is the various power measurement sensors. Because power is so important in a robotic system, there are many battery and charge sensors. There is an estimated capacity and present capacity (charge) of the battery. Both of these are analog values with units of mAh. You also have analog values for voltage, temperature, and charge/discharge current of the battery. The latter is useful for determining in real time how much extra power your project is using. In Chapter 6 you'll learn how to read these values, allowing you to dynamically adjust how much power the Roomba and your project are using to maximize run time.

OSMO//hacker: Hope for Older Third Generation Roombas

The ROI functionality wasn't built into the third generation Roomba models when they first came out in 2004. Only around October 2005 (around the time of Roomba 2.1) did iRobot start including ROI. With an amazing degree of savvy regarding the gadget-using population, iRobot has released a firmware updater module called OSMO//hacker, shown in Figure 1-12, that revs up the software inside the Roomba to include ROI.

FIGURE 1-12: The OSMO//hacker

This is a one-time use device that plugs into the Roomba to be upgraded. The OSMO//hacker upgrades the Roomba and from that point on, the module is no longer needed.

There are two variations of this $30 device, and you must inspect your Roomba's serial number to determine which variation you need. If you have one of these older third generation Roomba models, visit http://irobot.com/hacker for details on how to determine which OSMO//hacker module is right for you.

Summary

iRobot has created an astounding variety of Roomba vacuuming robots over the years, and hopefully this chapter assuages the confusion as to which Roomba models are hackable. Even if you feel a little reluctant about hacking a brand new Roomba and decide to buy a used one on eBay or from your local classified ads, it would be a shame to get one that's not hackable.

Regardless of what type of Roomba you have, the next time you run it, see if you can determine what basic impulses are competing to create the complex actions it performs. Seeing a real device implement subsumption architecture is fascinating. From looking at how the Roomba is built and its capabilities, you may have ideas on how to improve it or add on to it.

Interfacing Basics

All projects in this book utilize the Roomba Open Interface (ROI), previously known as the Roomba Serial Command Interface (SCI). Although you'll find many references to the SCI on the Internet, know that the SCI is the same as the ROI; only the name has changed. All third-generation Roomba models produced since 2004 are compatible with the ROI. Roombas produced after October 2005 have ROI built-in, while older third-generation Roombas will need the OSMO//hacker updater as described in Chapter 1.

Originally, the ROI appears to have been a diagnostic port used by iRobot to test the Roomba robotic vacuum cleaner before shipment and as a way to release firmware upgrades if bugs were ever discovered. The OSMO//hacker device that enables ROI for older Roombas is one of these firmware updaters.

The ROI protocol is fairly simple, but a few factors, such as variable command length and reading sensors, complicate using it. Even so, the ROI is a basic serial protocol similar to the type that is spoken between a computer and a modem. It is much simpler than Ethernet or Wi-Fi.

The full ROI specification released by iRobot is located at `http://irobot .com/developers`. The ROI specification is also available in Appendix C and on this book's web site at `http://roombahacking.com/docs/`. This chapter is an excellent guidebook to the official ROI specification.

in this chapter

- ☑ What's possible with ROI (and what's not)

- ☑ ROI connector

- ☑ ROI protocol

- ☑ Introducing the RoombaComm API

What Can Be Done with the ROI?

The ROI offers an almost complete view of the Roomba's internals. It abstracts certain functions, making them easier to use. Much of the low-level hard work dealing with motors and sensors has been taken care of inside the Roomba itself, so users of the ROI don't have to deal with it. However, some of these abstractions can also make it difficult to accomplish certain types of hacks, as you will see at the end of this section.

Sensing

The Roomba contains many sensors to observe the real world and the ROI allows access to all of them. They include:

- Bump sensors (left, right)
- Cliff sensors (left, right, left-front, right-front)
- Wall sensor

- Dirt sensors (left, right)
- Wheel drop sensors (left, right, caster)
- Button press sensors (power, spot, clean, max)
- Infrared sensor (virtual wall, home base, and remote control functions)

Control

The Roomba also contains several actuators and annunciators that can be controlled through the ROI:

- Drive-wheel motors
- Vacuum motor
- Main brush motor
- Side brush motor
- Status LEDs
- Piezoelectric beeper

Internal State

Additionally, the ROI makes available certain internal states of Roomba:

- Battery state and charge level
- Motor over-current
- Distance traveled
- Angle turned

What You Cannot Do

The ROI is simply an interface into the existing microcontroller program running in the Roomba. It doesn't bypass it. You cannot get direct access to the Roomba hardware. In general this isn't a bad thing. Some of the sensor data is constructed or massaged by this program to be easier to use. For example, the infrared detector on the top of the Roomba is a single sensor that responds to the virtual wall unit and remote control, but the ROI provides different sensor values for those functions. Roomba is parsing the infrared bit stream emitted by those devices and presenting the result as multiple binary values. It is not possible to parse custom infrared bit streams, so detecting commands from other remote controls cannot be done. Most disappointingly, it doesn't provide a sensor interface to the charging dock beacon of the home base beyond telling Roomba to go into "force-seeking dock" mode.

Beyond access to those data massaging routines, the ROI doesn't provide any access to the various cleaning algorithms used by the Roomba. But that doesn't mean new ones can't be created and commanded through the ROI.

The hardware design of the Roomba itself prevents some other potential hacking ideas. For example, although the main drive motors are driven by pulse-width modulation (PWM), allowing for varying speeds, the vacuum motors are not. Thus the ROI has only simple on/off commands for the vacuum motors.

The ROI Connector

The ROI connector is a Mini DIN 7-pin jack. "DIN" is a standard connector format; "mini" is a smaller variation of the original DIN format. The Mini DIN standard is incredibly common in consumer electronics. S-Video cables use a 4-pin Mini DIN connector, and old PC PS/2 keyboard and mouse connecters are 6-pin Mini DIN.

Figure 2-1 shows what the ROI connector looks like on the Roomba, and Table 2-1 lists the available signals.

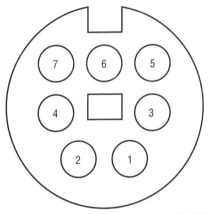

FIGURE 2-1: Roomba ROI connector Mini DIN
7-pin socket pin-out

Table 2-1   Roomba ROI Connector Signal Names

Pin	Name	Description
1	Vpwr	Roomba battery + (unregulated)
2	Vpwr	Roomba battery + (unregulated)
3	RXD	0–5V serial input to Roomba
4	TXD	0–5V serial output from Roomba
5	DD	Device detect (active low), used to wake up Roomba from sleep mode
6	GND	Roomba battery - (ground)
7	GND	Roomba battery - (ground)

The physical interface allows two-way serial communication through 0–5V binary levels. This serial communication is identical to normal RS-232-style PC serial port communication, except that it is at different voltage levels. RS-232 uses +12V to represent a zero bit and -12V to represent a one bit. This is sometimes called *negative logic* because it is opposite of what might logically be expected (which would be using a positive value to represent a one bit). The ROI uses 0–5V *positive logic*, where 0V indicates a zero bit and 5V indicates a one bit. This is the same standard used in most microcontrollers like the Basic Stamp and Arduino (see Chapter 13), allowing direct connection of these devices to the Roomba.

Caution

Do not directly connect a PC's RS-232 port to the Roomba ROI port. A converter circuit must be inserted between the two.

The available signals are:

- **Vpwr:** A direct unregulated tap off the Roomba's main battery. This is normally around +16 VDC but fluctuates as the battery is charged and discharged.

- **RXD:** Serial data into Roomba. This is a 5V signal referenced to GND. This will normally be connected to the TX line of an external microcontroller or transceiver. Must be used to send commands and data to the Roomba.

- **TXD:** Serial data out of the Roomba. This is a 5V signal referenced to GND. This will normally be connected to the RX line of an external microcontroller or transceiver.

- **DD:** Device detection into Roomba. This is a 5V signal that, when held to GND for at least 500 ms, will wake Roomba from sleep. It's an optional signal and most of this book will not use it and assume Roomba is already powered on.

- **GND:** The ground reference for all of the above signals. Must be used. This is also the negative terminal of the Roomba main battery.

For basic communication with Roomba, only three connections are absolutely required: RXD, TXD, and GND. If only commands are to be sent and sensor data isn't going to be received, then even the TXD connection can be left off, leaving only a two-wire connection.

Note

In all the hardware projects in this book, the DD line will be included if possible, but the software presented never assumes it is connected. Therefore, be sure to turn on the Roomba with the Power button before trying any of the software.

Alternatives to the 7-pin Mini DIN

The 7-pin variant of the Mini DIN connector is fairly uncommon and hard to find. The parts distributors that do carry it often charge a premium for it.

It turns out the 8-pin Mini DIN cables used for old Macintosh serial cables are mechanically and electrically compatible with the 7-pin variety. Because of the wide availability of the 8-pin Mini DIN cables and jacks, they are much cheaper than 7-pin and they will be used as the ROI connection components in the projects in this book unless otherwise noted.

Figure 2-2 illustrates what an 8-pin Mini DIN connector looks like.

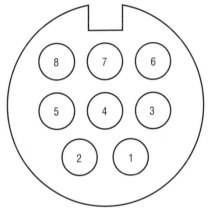

FIGURE 2-2: Mini DIN 8-pin socket pin-out

It looks very similar, which is good. All the pins in the 7-pin jack line up with an 8-pin connector. The extra pin in the 8-pin cable goes into the hole meant for the plastic guide pin in a normal 7-pin connector. Due to the extra pin, all pin numbers after pin 3 are offset by one, which can lead to confusion when wiring up cables. Table 2-2 demonstrates the signals for each pin.

Table 2-2   Roomba ROI 8-Pin Mini DIN Connector Signal Names

Pin	Name	Description
1	Vpwr	Roomba battery + (unregulated)
2	Vpwr	Roomba battery + (unregulated)
3	RXD	0–5V Serial input to Roomba
4	n/c	not connected
5	TXD	0–5V Serial output from Roomba
6	DD	Device detect (active low), used to wake up Roomba from sleep mode
7	GND	Roomba battery - (ground)
8	GND	Roomba battery - (ground)

Locating the Roomba ROI Connector

The ROI connector is located on the top edge of the Roomba, at about the four-o'clock position when looking down at the Roomba. See Figure 2-3. This location is the same for all models of Roomba.

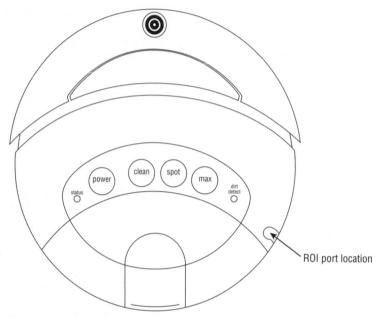

FIGURE 2-3: Location of the ROI port

The port is covered by a small plastic hood, which can be popped off easily with a small flat-blade screwdriver. Don't worry about breaking anything. The hood can be quickly snapped back on.

The ROI Protocol

The ROI protocol is rather rudimentary, as protocols go. The protocol is a simple byte-oriented binary UART-like serial protocol, operating at 57,600 bps, 8 bits, no parity, one stop-bit (often represented as 57600, 8N1). The interaction method is command-response, and Roomba never sends data unless commanded. In fact, there is only one command that elicits a response from Roomba. Since there is no flow-control, data responses, if present, can be ignored. This allows very simple devices with only serial output to be connected.

The commands and the responses are binary, not text. Thus communication is not as simple as connecting to Roomba with a terminal program and typing commands. Determining how to test the connection after a physical interface has been built will be covered in Chapter 3.

When using the ROI, Roomba exists in one of several modes (or states). These modes represent both how Roomba behaves and how it responds to subsequent ROI commands. Actions by Roomba can also change the mode. Several of the ROI commands are dedicated to selecting the appropriate mode/state because certain commands only work in certain states. The mode change commands are single bytes with no arguments and are invoked by just sending the command byte for the desired mode. Other commands have arguments, like how fast to drive or which vacuum motors to turn on. Those arguments are sent immediately after the command byte as data bytes with a particular format. The format varies among commands and can either be a single byte, a 16-bit value represented as two bytes, several binary values (bits) encapsulated in a single byte, or some combination thereof. One of the challenges of using the ROI is knowing all the data type variations. Only one command causes Roomba to return data, and the block of bytes it returns must also be carved up into bits, bytes, and 16-bit words.

Roomba Modes

When using the ROI, Roomba can exist in one of five states (see Figure 2-4). The states are:

- **Sleep (Off):** Roomba responds to no commands over the ROI, but can be woken up and put into the on mode through either the Power button or by toggling the DD hardware line.

- **On:** Roomba is awake and is awaiting a START command over the ROI. In this mode Roomba is able to operate normally via its buttons or remote control. The only way out of this mode through the ROI is through the START command.

- **Passive:** Roomba has received the START command. In this mode, sensors can be read and songs defined, but no control of the robot is possible through the ROI. The Roomba buttons work as normal. This is the mode to use to spy on Roomba as it goes about its business. The usual path from this mode is to send the CONTROL command to enter safe mode.

- **Safe:** Roomba has received the CONTROL command from passive mode or the SAFE command from full mode. Everything that could be done in passive mode is still possible, but now Roomba can be controlled. The buttons on Roomba no longer change the robot's behavior, but instead their states are reflected in the Roomba sensors data. All commands are now available, but a built-in safety feature exists to help you not kill your Roomba. This safety feature is triggered if Roomba detects any of the following:

    - A cliff is encountered while moving forward (or moving backward with a small turning radius).

    - Any wheel drops.

    - The charger is plugged in and powered.

- Triggering this safety feature will stop all Roomba motors and send Roomba back to passive mode. Another way back to passive mode is to send one of the SPOT, CLEAN, or MAX virtual button commands. Sending the POWER virtual button command will put the robot in passive mode and then put it to sleep.

Note One way to quickly stop Roomba if your code creates a Roomba gone haywire is to run in safe mode and just lift the robot a little. This triggers the safety feature.

- **Full:** If Roomba receives a FULL command while in safe mode, it will switch to this mode. This mode is the same as safe mode except that the safety feature is turned off. To get out of this mode, send the SAFE command. Sending one of the SPOT, CLEAN, or MAX virtual button commands will put Roomba in the passive mode. Sending the POWER virtual button command will put Roomba in passive mode and then put it to sleep.

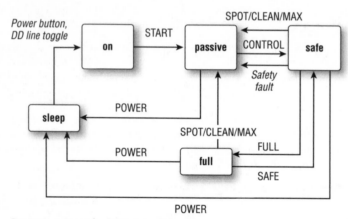

FIGURE 2-4: Roomba ROI state diagram

The Roomba changes from one mode to the next depending on either ROI commands or external events. In Figure 2-4, the ROI commands are listed in CAPITALS and the external events are in *italics*.

Cross-Reference Controlling the modes is discussed in the section "ROI Mode Commands" later in this chapter.

Note Allow at least 20 milliseconds between sending commands that change the ROI mode (CONTROL, SAFE, FULL, and the virtual button commands SPOT, CLEAN, and MAX). Any ROI commands issued in the 20 millisecond window will be ignored or only partially received.

Note There is no way to determine which state the robot is in. If you are unsure, explicitly set its state to what you want.

If the Roomba battery is removed or when the battery is too low for Roomba to operate reliably, the system will switch to the off state.

One undocumented event that has happened a few times to me is an over-current event. If the Roomba battery is low and the robot attempts to do something that uses a lot of current (like drive across thick carpet), it appears to trigger an over-current sensor that switches Roomba out of safe or even full mode and into passive mode.

ROI Command Structure

All ROI commands start with a command opcode, a single byte with a value between 128 (0x80) and 143 (0x8F), inclusive. Only 16 opcodes have been published, but these 16 allow almost complete control. There may be more, unpublished opcodes.

In addition to the command opcode byte, an ROI command may include one or more data bytes that are arguments to the command. Many commands, like the START command, consist of only the command opcode byte and no data bytes. The PLAY command is an example of a command that takes one extra command byte, a data byte with the number 1–16 of the song to play. All commands except the SONG command have a fixed number of data bytes. The SONG command has a variable number of data bytes, depending on the length of the song, N. The number of data bytes is determined by the formula 2+2N. Therefore, if the song to be sent is one note long (N=1), the number of data bytes to send is 4. For a 10-note song (N=10), 22 data bytes are sent.

To send a complete ROI command, send the appropriate command byte and the appropriate data bytes, if any, from the serial port of your controlling computer and to the RXD pin of the ROI port.

Table 2-3 shows the opcodes and the number of data bytes for each command. Hexadecimal representation of the opcode values is useful since each hexadecimal digit represents exactly four bits, so two hex digits exactly represent 8 bits, or one byte.

Table 2-3 Command Opcodes and Data Bytes

Command	Opcode	Hexadecimal Values	Number of Data Bytes
START	128	0x80	0
BAUD	129	0x81	1
CONTROL	130	0x82	0
SAFE	131	0x83	0
FULL	132	0x84	0
POWER	133	0x85	0

Continued

Table 2-3   *Continued*

Command	Opcode	Hexadecimal Values	Number of Data Bytes
SPOT	134	0x86	0
CLEAN	135	0x87	0
MAX	136	0x88	0
DOCK	143	0x8F	0
DRIVE	137	0x89	4
MOTORS	138	0x8A	1
LEDS	139	0x8B	3
SONG	140	0x8C	2+2N
PLAY	141	0x8D	1
SENSORS	142	0x8E	1

Since a command can take a varying number of arguments, a buggy program sending ROI commands may run into the problem where it's sending bytes in the wrong sequence, with Roomba interpreting a data byte as a command byte or vice-versa. The only way around this is to either power off-and-on the robot or to send the same single-byte command (like START) enough times to ensure that Roomba really did interpret the byte as a START command. In practice, especially when using an API library like RoombaComm, this isn't a big concern unless the serial connection is severed in the middle of a command sequence.

The largest possible command is a full SONG command at 35 bytes long. This is a rare command usually. The second largest command is the DRIVE command at five bytes (one byte for the command byte itself and four data bytes).

Cross-Reference   To better understand how to use bits and bytes, see the sidebar "Setting and Clearing Bits" later in this chapter.

ROI Mode Commands

These commands are the ones that alter the operating mode of the Roomba. The START command is required for any Roomba hacking through the ROI and the CONTROL command is required for any meaningful hacking. The following list provides an overview of each of the mode commands:

- START: The START command starts the ROI. Roomba must be in the on mode through either the Power button or toggling the DD line. If the robot is in an unknown state or you need to reset it, send the START command a few times to put Roomba in a known state. This command puts Roomba in the passive mode.

- BAUD: The BAUD command sets the baud rate (the speed of data transferal) in bits per second (bps). The data byte is the baud code (0–11) of the baud rate to use. See the ROI specification in Appendix C or at http://irobot.com/developers for getting the right baud code for a baud rate. The default rate is 57,600 bps, which has a baud code of 10. This speed is well-supported and fast enough for most every use. Changing the baud rate can lead to the age-old problem of never being sure at what speed each side is transferring data. In general do not change the baud rate unless 57,600 bps will not work for your application. All the projects in this book that use a PC to talk to Roomba use 57,600 bps.

- CONTROL: The CONTROL command enables control of Roomba. It is almost always sent immediately after the START command. This command puts the robot in the safe mode.

- SAFE: The SAFE command returns Roomba to the safe mode if it isn't already in it. This is only relevant if Roomba is in the full mode.

- FULL: The FULL command puts the robot in the full mode if it isn't already in it. Roomba must be in the safe mode for this command to work.

- POWER: This is a virtual button-press command. It is equivalent to pushing the Power button on the Roomba. This command puts Roomba in the sleep mode.

- SPOT: This is a virtual button-press command. It is equivalent to pushing the Spot button on Roomba. It starts the Spot cleaning algorithm.

- CLEAN: This is a virtual button-press command. It is equivalent to pushing the Clean button. It starts the Clean cleaning algorithm.

- MAX: This is a virtual button-press command. It is equivalent to pushing the Max button. It starts the Max cleaning algorithm.

- DOCK: This turns on the force-seeking dock algorithm of the robot. It is equivalent to pressing both Spot and Clean buttons simultaneously. Roomba stops whatever it was doing and tries to find the charging dock.

Drive Motors

The DRIVE command controls the drive wheels. It takes four bytes of data: two for velocity and two for direction, each treated as 16-bit signed values using two's complement. (Two's complement is a way of representing negative numbers in computers, accomplished by counting backward from the maximum value in the range. For example, -5 in a 16-bit range is $2^{\wedge}16-5 = 65536-5 = 65531 = 0xFFFB$.)

The velocity is simply the speed, forward or back, at which Roomba should move. The direction is a bit trickier. It's mostly the turn radius, but has some special case values, as seen below. (The values in parentheses are the hexadecimal representation of the values shown.)

- **Velocity:** from -500 to 500 mm/s (0xFE0C to 0x01F4)
- **Direction:** either a radius or some special purpose values
  - **Turn on a radius:** -2000 to 2000 mm (0xF830 to 0x07D0)

- **Straight:** 32768 (0x8000)
- **Spin counter-clockwise:** 1 (0x0001)
- **Spin clockwise:** -1 (0xFFFF)

The turn radius is the radius in millimeters of an imaginary circle upon whose edge Roomba would drive along. So, larger radii make Roomba drive straighter, and shorter radii make Roomba turn more quickly. When going forward, a positive radius turns Roomba to the right, and a negative radius turns Roomba to the left.

In addition to the radius values, there are three special values for direction. The direction value of 32768 is the special value to make the Roomba drive straight. And instead of using -1 and +1 to represent a radius of -1mm and +1mm, they're used to mean spin clockwise or counter-clockwise.

Taken together, the velocity and direction describe what direction a Roomba should go and how fast. The hexadecimal representation above makes it easy to turn these values into a byte sequence to send to the Roomba. For instance, to drive straight at 500 mm/s, the complete DRIVE command would be: 0x89,0x01,0xF4,0x80,0x00. Or, to spin clockwise at -500 mm/s, the complete command would be: 0x89,0xFE,0x0C,0xFF,0xFF.

It's interesting that iRobot chose to expose the drive functionality as a combination of velocity and direction instead of the presumably simpler velocity of each wheel. Apparently it was presented this way, because this is how the robot's algorithms use the drive wheels. The typical spiral the Roomba does is certainly easier with the above abstraction: Set the drive speed once and then slowly increment the radius value over time.

In Chapter 5 you'll learn more about DRIVE command and create code to abstract out these special values.

Cleaning Motors

The MOTORS command controls Roomba's three cleaning motors. Unlike the drive motors, only on/off control is available for them. Each motor's state is represented by a bit in the single data byte (hexadecimal bit values in parentheses):

- **Main brush:** bit 2 (0x04)
- **Vacuum:** bit 1 (0x02)
- **Side brush:** bit 0 (0x01)

To turn on a motor, set the bit. To turn off the motor, clear the bit, as explained in the sidebar "Setting and Clearing Bits." All other bits of the data byte are not used and should be set to zero.

For example, to turn all cleaning motors off, the full ROI command is: 0x8A,0x00. To turn them all on, the full command is: 0x8A,0x07.

Setting and Clearing Bits

Many parts of the ROI represent multiple binary values within a single byte. For command data bytes, these bits are like mini-commands within the larger command, telling Roomba to do one small thing or another. In the MOTORS command, for instance, you can turn on or off any combination of the three cleaning motors.

Bytes are divided into 8 binary digits called bits. Each bit has a position in a byte and value equal to $2^{position}$. Bit positions are numbered from 0 to 7. The corresponding bit values are 1,2,4,8,16,32,64,128, more commonly represented in hexadecimal as 0x01,0x02,0x04,0x08,0x10,0x20,0x40,0x80. Sometimes a set of bits is called a bit field. A byte is a bit field of 8 bits.

The act of turning on a bit (and thus turning on a motor in the case of the MOTORS command) is called *setting* a bit. Turning off a bit is called *clearing* a bit. Most of the time, you can start with all bits cleared and then set the ones you want. In practice this means starting with a value of zero and then joining together the bit values of the bits you want to set.

For example, in the MOTORS command if you want to turn on only the main brush (bit 2, bit value 0x04) and the side brush (bit 0, bit value 0x01), you can do this:

```
databyte = 0x04 | 0x01
```

The vertical bar (|) is the logical OR operator and works sort of like addition in this context. You can also build up a byte incrementally using the "|=" operator:

```
databyte = 0
databyte |= 0x04
databate |= 0x01
```

The result is the same as before, but now you can add code to easily pick which statement to run. Setting bits is similar but a little more complex, by using the logical NOT and logical AND operators on a bit's value. For example, to clear the side brush bit (and thus turn it off), you would do this:

```
databyte &= ~0x01
```

This type of complex setting and clearing of bits isn't used much in the ROI. Usually the very first method of setting bits is used. For completeness, the general definitions in a C-like language for setting and clearing a bit in a byte given the bit's position are as follows:

```
#define setbit(byte,bit) (byte) |= (1 << (bit))
#define clrbit(byte,bit) (byte) &= ~(1 << (bit))
```

ROI Indicator Commands

Roomba has two types of indicators it uses to let humans know what it is up to: LED lights on its control panel and a piezoelectric beeper to play status melodies. These melodies are particularly useful when the robot gets stuck somewhere hidden. It will forlornly play its "uh-oh" melody in an attempt to get attention.

Lights

The LEDS command is one of the more complex ones. The first data byte of the command is a bit field for turning on/off the LEDs:

- **Status (green):** bit 5 (0x20)
- **Status (red):** bit 4 (0x10)
- **Spot (green):** bit 3 (0x08)
- **Clean (green):** bit 2 (0x04)
- **Max (green):** bit 1 (0x02)
- **Dirt detect (blue):** bit 0 (0x01)

To turn on an LED, set the corresponding bit; to turn it off, clear the corresponding bit. Notice how the Status light has both a red LED and a green LED. Both can be used at the same time to create an amber light. Bits 6 and 7 of the first data byte are not used and should be set to zero.

The second and third data bytes of the LEDS command represent the color (byte 2) and intensity (byte 3) of the Power LED.

- **Power color:** 0=green, 255=red, and values in the middle are mix of those two colors.
- **Power intensity:** 0=off, 255=full on, and intermediate values are intermediate intensities.

So to create a medium amber light, the color and intensity data bytes would be 128,128 (0x80,0x80). Or, a full LEDS command to turn the Status and Spot light green and have the power light be medium amber would be: 0x8B,0x28,0x80,0x80. To turn all the LEDS off, the full command is: 0x8B,0x00,0x00,0x00.

When charging, Roomba sets the Power color to be the percentage charged, and pulses the Power intensity about once per second to indicate charging.

Music

There are two types of music commands: SONG and PLAY. The SONG command defines a song to be played with the PLAY command. Roomba can remember up to 16 songs, and each song can be up to 16 notes long. Each note in a song is specified by two bytes: a note number for pitch, and a note number for duration. To specify a musical rest, use a zero for pitch. There is no way to change note loudness.

This command is the most complex one in the ROI as it has a varying number of data bytes, dependent on the length of the song being defined. The first data byte specifies which song is being defined. The second data byte specifies how many notes are in the song. Every pair of data bytes after that defines a note: note number and note duration.

The shortest song is one note long. To define a one-note song would take four data bytes. The longest song is 16 notes long and to define that song would take 32 data bytes. This makes for the largest possible command in the ROI at 35 bytes (1 command + 1 byte song num + 1 byte song length + 32 bytes song data).

Note numbers are very much like note numbers in MIDI. A note number of 36 (0x24) defines the note C in the first octave (denoted C1). Numbers increase sequentially (for example, C#1 =37, D1=38, and so on). The familiar middle C (C3) is note number 60 (0x3C). The standard A440 pitch (the first A above middle C) is note number 69 (0x45). The smallest note number is 32 (0x1f) corresponding to G0, and the highest pitch is note number 127 (0x7f) corresponding to G8. Roomba has an eight-octave range, better than any opera star.

Note durations are specified in units of 1/64ths of a second. Thus, a half-second-long note would have a duration value of 32 (0x20), and a three-second note would have a duration value of 192 (0xC0). The longest duration that can be specified is $255 \times 1/64 = 3.984$ seconds.

For example, a complete SONG command to program into song slot #3 a one-note song that plays a C3 for three seconds would be: 0x8C, 0x03, 0x01, 0x3C, 0xC0.

After a song has been defined with the SONG command, it can be triggered to play with the PLAY command. The PLAY command takes one data byte, the song number to play. The song plays immediately and does not repeat. To play the above defined song, the full ROI command would be: 0x8D, 0x03.

ROI Sensors

Of all the ROI commands, only one returns data. There is no need to get sensor data from the Roomba. Roomba will react appropriately to all other ROI commands regardless of whether the SENSORS command is used. In fact, operating Roomba in such an "open-loop" manner may be a preferred mode if external sensors with better performance have been added through an external microcontroller, as you do in Chapter 13.

The SENSORS command takes one data byte as an argument, the packet code. This packet code determines how much and what group of sensor data is being retrieved.

The different groups of sensor data are:

- **Physical sense:** Sensors that detect the environment, like bump, cliff, and wall sensors
- **Buttons and internal sense:** The state of the panel and remote buttons, and the computed distance and angle values
- **Power sense:** The state of the battery and charging systems
- **All:** All of the above data

To choose which group of sensor data to receive, set the packet code data byte to one of the following:

- **0:** All data (26 bytes)
- **1:** Physical sensors (10 bytes)
- **2:** Buttons and internal sensors (6 bytes)
- **3:** Power sensors (10 bytes)

For example, to have Roomba send just the power sensors data packet, the full ROI command is: `0x8E, 0x03`.

In most of the code presented in this book, the full 26-byte sensor packet (group 0) will be fetched.

However, there are also situations where you might want to receive only a portion of the data if the application has critical timing or if receiving the entire 26 bytes takes too long due to a reduced serial port baud rate. For instance, if the ROI is set to 1,200 bps because of a slow microcontroller, then transmitting 26 bytes would take at best (2+26 bytes × 10 bit-times/byte) / 1,200 bit-times/second = 0.233 seconds. A quarter-second is a long time if Roomba is heading toward the stairs. By only getting the physical sensors, the time to know about the stairs is reduced to about 0.10 seconds, perhaps time enough to not drive off the cliff. This constructed example shows one of the problems of systems with high communication time costs. In such a low-speed system, the likely solution would be to move the robot at a slower speed to compensate for the slower sensor update rate. Controlling Roomba with the Basic Stamp, as you'll see in Chapter 13, is an example of a low-speed system that gets smaller data packets.

Physical Sensors

Figure 2-5 shows the arrangement of the 10 bytes that make up the physical sensors packet group.

There are two bytes (16 bits) of digital sensor data, and two bytes of analog data. That is a total of four bytes of data, optimally packed. Then why is this packet 10 bytes?

Although the wheel drop and bump sensors are implemented as bits in a single bit field, all the wall and cliff sensors are given a byte each, even though they are all just binary on/off sensors. This seems strange and rather wasteful, especially considering the aforementioned rationale of dividing up the sensor data into groups.

One possible reason for this distinction might be that the Roomba firmware wants to do the minimum calculation necessary to determine if it detects a cliff or a wall, and perhaps testing an entire byte being non-zero is easier or faster than testing a bit within a byte. This is one of those examples where the ROI could have offered a little more abstraction of the data but didn't.

byte 0	7	6	5	4	3	2	1	0
bump and wheeldrop	n/a	n/a	n/a	wheel drop caster	wheel drop left	wheel drop right	bump left	bump right
byte 1	7	6	5	4	3	2	1	0
wall	n/a	n/a	n/a	n/a	n/a	n/a	n/a	wall
byte 2	7	6	5	4	3	2	1	0
cliff left	n/a	n/a	n/a	n/a	n/a	n/a	n/a	cliff left
byte 3	7	6	5	4	3	2	1	0
cliff front left	n/a	n/a	n/a	n/a	n/a	n/a	n/a	cliff front left
byte 4	7	6	5	4	3	2	1	0
cliff front right	n/a	n/a	n/a	n/a	n/a	n/a	n/a	cliff front right
byte 5	7	6	5	4	3	2	1	0
cliff right	n/a	n/a	n/a	n/a	n/a	n/a	n/a	cliff right
byte 6	7	6	5	4	3	2	1	0
virtual wall	n/a	n/a	n/a	n/a	n/a	n/a	n/a	virtual wall
byte 7	7	6	5	4	3	2	1	0
motor over-currents	n/a	n/a	n/a	drive left	drive right	main brush	vacuum	side brush
byte 8	7	6	5	4	3	2	1	0
dirt detector left	range 0–255							
byte 9	7	6	5	4	3	2	1	0
dirt detector right	range 0–255							

FIGURE 2-5: ROI sensor packet group 1, physical sensors

Tip

To test a single bit in a byte (like the bumps wheeldrops sensor byte), don't use an equality test like `if(bumps==2)` to test for bump left. This is because if more than one bit is set, the value is the combination of those bits. This is similar to the setting and clearing bits issue mentioned earlier. For example if both left and right bump sensors are triggered, then `bumps` equals 3. Instead use a bit mask and bit shifts to get only the bit you want. The proper way to test for bump left would be `if(((bumps&0x02)>>1)==1)`. Such math-heavy bit manipulation is often hard to read, so the RoombaComm API introduced below abstracts these tests out to a set of methods like `bumpLeft()`. In Chapter 6 you'll explore how to create these methods for all sensor bits.

Buttons and Internal Sensors

Figure 2-6 shows the details of the six bytes that make up the buttons and internal sensors packet group.

Buttons and Remote Control Codes

The buttons sensor byte is a bit field with only four of the bits used. Not all Roombas have all buttons. For those Roombas (like Roomba Red with no Max button), the corresponding missing button bits will always read as zero.

byte 10	7	6	5	4	3	2	1	0
remote control	range 0–255							

byte 11	7	6	5	4	3	2	1	0
buttons	n/a	n/a	n/a	n/a	Power	Spot	Clean	Max

byte 12	7	6	5	4	3	2	1	0
distance (high byte)	range -32768 to 32767 mm							

byte 13	7	6	5	4	3	2	1	0
distance (low byte)	range -32768 to 32767 mm							

byte 14	7	6	5	4	3	2	1	0
angle (high byte)	range -32768 to 32767 mm							

byte 15	7	6	5	4	3	2	1	0
angle (low byte)	range -32768 to 32767 mm							

FIGURE 2-6: ROI sensor packet group 2, buttons and internal sensors

The remote control codes are undocumented in the ROI specification, but Table 2-4 shows the remote codes for the standard remote. There are 255 possible codes, allowing for many different possible commands. The more powerful Scheduler remote likely uses some of them. The RoombaComm code, introduced below and available for download at http://roombahacking.com/, contains the most comprehensive list of remote codes.

Table 2-4   Remote Control Codes

Value	Hexadecimal values	Description
255	0xff	No button pressed
138	0x8a	Power button pressed
137	0x89	Pause button pressed
136	0x88	Clean button pressed
133	0x85	Max button pressed
132	0x84	Spot button pressed
131	0x83	Spin left button pressed
130	0x82	Forward button pressed
129	0x81	Spin left button pressed

Distance and Angle

Roomba computes the distance and angle values by watching its wheels move. This doesn't always correspond to actual motion. The two-byte values are combined into a single 16-bit signed number through the standard method of:

```
value = (high_byte << 8) | low_byte
```

Or in plain English: Take the high byte, shift it up 8 bits and overlay (OR) it with the low byte.

Both the distance and angle are cumulative values and are cleared after they are read. This means that in order to get accurate readings, they must be read often enough that they don't overflow. In practice this isn't that much of a problem. A larger problem with not reading often enough is the loss of awareness of the true motion. Distance and angle give a single vector direction from point A to point B. In actuality, Roomba may have moved in a multi-point zigzag motion (or any other path) between points A and B. In Chapter 6 you'll delve into the details of dealing with these issues.

Power Sensors

Figure 2-7 shows the details of the 10 bytes that make up the power systems sensor packet group. Roomba spends a lot of time thinking about power, and that is reflected in this ROI packet group. Perhaps the two most useful values in the power sensors packet for hacking are the battery current and battery charge. By observing the current you can tell how much power the project is consuming when it does different things. By watching the charge level you can tell when Roomba is almost out of juice and needs to be recharged.

byte 16	7	6	5	4	3	2	1	0
charging state	values: 0:not charging, 1:charging recovery, 2:charging, 3:trickle charging, 4:waiting, 5:charging error							

byte 17	7	6	5	4	3	2	1	0
battery voltage (high byte)	range 0 to 65535 mV							

byte 18	7	6	5	4	3	2	1	0
battery voltage (low byte)	range 0 to 65535 mV							

byte 19	7	6	5	4	3	2	1	0
battery current (high byte)	range -32768 to 32767 mA							

byte 20	7	6	5	4	3	2	1	0
battery current (high byte)	range -32768 to 32767 mA							

byte 21	7	6	5	4	3	2	1	0
battery temp- erature	range -128 to 127 degrees Celsius							

byte 22	7	6	5	4	3	2	1	0
battery charge (high byte)	range 0 to 65535 mAh							

byte 23	7	6	5	4	3	2	1	0
battery charge (low byte)	range 0 to 65535 mAh							

byte 24	7	6	5	4	3	2	1	0
battery capacity (high byte)	range 0 to 65535 mAh							

byte 25	7	6	5	4	3	2	1	0
battery capacity (low byte)	range 0 to 65535 mAh							

FIGURE 2-7: ROI sensor packet group 3, power and charging sensors

Introducing the RoombaComm API

When first experimenting with the ROI, it's common to use some sort of a serial terminal program that can send binary sequences to try out various ROI commands. However, this quickly gets tiresome. The next thing to try is to write a small program to send these commands. This, too, proves problematic because typos can slip in and wrong commands can be sent. And with all these random serial bytes flying, it's not uncommon to "wedge" the Roomba, getting it into a state where it seems like it doesn't work any more. This can be remedied by rebooting Roomba by removing and re-inserting its battery, but that's a pain and destroys the hacking groove.

It would be a lot easier if there was a library to codify the exact recipe needed to make something work. The RoombaComm API is just such an encapsulation of the ROI binary commands into a more easy-to-use Java class. The following chapters show you how to build up the RoombaComm API. Specifically, Chapters 3 and 4 discuss starting communication with Roomba, Chapter 5 covers driving Roomba around, and Chapter 6 is all about reading Roomba sensors. The full version of the RoombaComm API and example programs using it can be downloaded from www.wiley.com/go/extremetech and http://roombahacking.com/.

Instead of coding something like:

```
// start up, drive straight at 400mm/s
serialport.send(0x80);
serialport.send(0x82);
serialport.send({0x89,01,0x90,80,00});
```

you would do something like:

```
// start up, drive straight at 400mm/s
roombacomm.startup();
roombacomm.driveStraight(400);
```

The obvious benefit is more human-readable code. This becomes even more evident when dealing with reading sensor packets, as you will see in Chapter 6.

The other main benefit is how the ROI commands are then used as "primitives" to create more complex behaviors. A good example of an added complex behavior would be implementing the task "at a speed of 100 mm/s, go forward 150 mm, then spin 90 degrees right." Using basic RoombaComm commands, that becomes:

```
roombacomm.goForwardAt(100); // go forward at 100mm/sec
roombacomm.pause(1500);      // wait 1.5 seconds,to get 150 mm
roombacomm.spinRightAt(100); // spin right at 100mm/sec
float ptime = 45.0 * roombacomm.millimetersPerDegree / 100.0;
roombacomm.pause(ptime);     // wait to spin thru the angle
```

However, when using more complex RoombaComm commands, that becomes simply:

```
roombacomm.setSpeed(100);   // set speed to 100 mm/sec
roombacomm.goForward(150);  // go forward 150 mm
roombacomm.spinRight(45);   // spin 45 degrees right
```

If this looks a little bit like the computer language Logo, this is on purpose. Logo was a great idea, a computer language targeted at kids. It was graphical and based around an on-screen turtle that could be told to go "forward 100" and "right 45." Look familiar? With Roomba and the RoombaComm API, one can re-create Logo in real life.

Java was chosen so that a single library could be used on Mac OS X, Windows, or Linux. The RoombaComm library and programs have been tested on all these platforms.

Summary

The Roomba ROI protocol is a pretty complete interface for controlling Roomba. You can see how to start implementing algorithms to make Roomba do new things and respond differently to its environment. If you've played with serial protocols before, now you know how to command the robot, you just need an adapter to make it happen. If you're still unsure how to start controlling Roomba, you now know that the RoombaComm API library will help conceal the low-level details and provide a scaffold upon which to build a richer environment for Roomba programming.

Building a Roomba Serial Interface Tether

in this chapter

☑ Explore how to design and build electronics

☑ Read and understand circuit schematics for voltage regulators, LEDs, and serial transceivers

☑ Build an RS-232 serial adapter

☑ Send commands to Roomba

When attempting to communicate with a new device for the first time, always go for the simplest, most direct communication path. Eventually the Roomba robotic vacuum will be on a Wi-Fi net, but you'll begin with a simple serial cable.

In the later Roomba projects, if nothing seems to work and it's unclear if the robot is even alive, revert back to the serial port tether. Verify the basics. The first troubleshooting question should always be, "Is it plugged in?" In the case of Roomba, this translates into, "Can I talk to the Roomba?" The serial tether is the simplest method of verifying computer-to-Roomba connectivity. From a solid foundation of known-good systems, work back up until the problem is found.

The serial tether circuit is simple, but this chapter goes into more detail describing and building it than you might expect. This extra detail is to give explanations as to why the parts of the circuit are designed the way they are and to show some of the rules-of-thumb that hackers use every day when designing circuits. This chapter is used as an introduction to designing and building electronics for those who don't do it regularly. If you're an experienced hacker, you can probably just look at the schematic, build it, and skip to the last section.

Alternatives

The serial tether presented here is just one possible design. There are several ways to solve this problem. The particular design is meant to be compatible with as many computers and embedded systems as possible. Also, it was designed to use such standard components that hackers are likely to have all the parts they need in their junk drawers. However, if you don't have a well-stocked junk drawer, you want to start playing robot with your Roomba *right now*, or you want a slightly more compact tether, there are alternatives. These alternatives cost more money, but they all function the same with the software in the rest of the book.

Alternative #1: RoombaDevTools RooStick

The industrial robot company RoboDynamics created the RoombaDevTools site (http://roombadevtools.com) to supply Roomba interface adapters and other Roomba hacking products. One of the most useful products is RooStick, a USB version of the serial tether. It functions exactly the same as the serial tether, appearing as a serial device to the OS (Windows, Mac OS X, or Linux). It is available for around $25, with an accompanying 7-pin Mini DIN for about $16. Be sure to use the Mini DIN cable from RoboDynamics unless you want to perform a pin-by-pin verification and rewiring of a non-approved cable. The current 8-pin Mini DIN cables from Jameco will *not* work without modification. To be safe, always use the RoboDynamics cable.

Demonstration code with source is provided in Visual Basic for Windows computers, available for download from the RoombaDevTools web site. Figure 3-1 shows what RooStick looks like.

FIGURE 3-1: RooStick from RoombaDevTools

Alternative #2: Cell Phone Sync Cable Hack

A potentially easier serial tether to build is to use a USB cell phone sync cable. Before pervasive Bluetooth and built-in USB ports on phones, phones had serial ports to allow data syncing. Of course now computers may not have a serial port on them, so the sync cable for these phones has evolved into a USB device with an embedded USB-to-serial converter in it.

These sync cables are currently available at Radio Shack for about $22 as a Future Dial Mobile Phone Data Cable. They are easily hackable to provide a 0–3.3V positive-logic serial port. This hack was originally discovered by the Linux router hackers, as they wanted access to the 0–3.3V serial console of these devices. For use with Roomba, some simple voltage converters are usually needed to convert the 0–3.3V used by the cell phone to the 0–5V used by Roomba.

Because these sync cables are becoming harder to find as fewer people need them and because this isn't as universal as a true RS-232 solution, it's not the focus of this chapter. However, instructions are presented in the "Building a USB Serial Tether from a Phone Sync Cable" sidebar in Chapter 15 if you want to go down that route.

Safety

This project and many others in this book entail building electronic circuits. Doing so exposes you to heat hot enough to burn your skin, electricity that may zap you or your projects, and lead that can poison you. It's easy to be safe, but if you feel unsure about what you're doing, stop and read Appendix A. It briefly covers how to solder and how to properly ground yourself.

Parts and Tools

Building electronic circuits is a lot like baking in the kitchen. The recipe is the circuit schematic and the ingredients are the various electrical components and parts used. Like the cooking utensils needed in a kitchen, you'll need a small collection of tools to make your circuit creation. The following list of tools will be used not just for this project, but for all projects in this book, and you can use them to build almost any electrical projects you'll find on the Internet.

If you're new to hacking, the following list may seem a bit overwhelming. But the component parts are simple (and cheap) and easy to get from a variety of suppliers. In the Introduction, I mention several good part suppliers. Jameco (http://jameco.com/) part numbers are used below simply because they carry both the parts and tools needed and have a friendly web site to order from.

The next section will show you how the entire project can be broken down into three easily digestible chunks. These chunks, or sub-circuits, will show up again in subsequent projects in this book and other circuits that you can discover on the Internet. No circuit is entirely new and unknown: It's composed of sub-circuits you will have seen before once you've built a few. Part of the fun of learning new circuits is to see how each one incorporates the bits and pieces you already know. And like baking, you'll find that variations to make a circuit your own are not only possible but recommended.

You will need the following parts for this project:

- Mini-DIN 8-pin cable, Jameco part number 10604
- 10 ft long serial cable with DB-9 female connector, Jameco part number 155521
- General-purpose circuit board, Radio Shack part number 276-150
- 78L05 +5 VDC voltage regulator IC, Jameco part number 51182
- MAX232 RS-232C transceiver IC, Jameco part number 24811
- 220 ohm resistor (red-red-brown color code), Jameco part number 107941
- Six 1µF polarized electrolytic capacitors, Jameco part number 94160PS

And you will need these tools:

- Soldering iron, stand, and solder, Jameco part numbers 170587CK, 192153CK, 141795
- Hot glue gun and hot glue
- Wire cutters and wire strippers
- IC Hook test leads, Jameco part number 135298
- Third-hand tool, Jameco part number 26690
- Digital multimeter
- DC power supply (wall wart) between +9V and +24V, Jameco part number 199566PS
- Mini DIN 8-pin socket, Jameco part number 207722
- Keyspan USA-19H or similar USB-to-serial adapter
- PC (Mac OS X, Windows, Linux) capable of running Java programs
- RoombaComm software package downloaded from www.roombahacking.com/
- Terminal emulation program (ZTerm for Mac OS X, RealTerm for Windows, minicom for Linux)

The Circuit

Figure 3-2 is the schematic of the entire circuit to be built. There are essentially three circuits in that schematic: a power supply, an RS-232 transceiver, and an LED lamp. The power supply converts the unregulated approximately +16 VDC Vpwr power line from the Roomba into the +5 VDC needed by the RS-232 transceiver. The RS-232 transceiver converts the 0-5 VDC signaling used by Roomba into the +/-12 VDC used in RS-232. And the LED circuit is there to let you know that power exists (and, besides, everything needs an LED).

Tip   If Figure 3-2 looks like hieroglyphics to you, see Appendix B for how to read schematics.

Understanding Voltage Regulators

The voltage regulator circuit, shown in Figure 3-3, is the same voltage regulator circuit seen in countless hobbyist projects. The 78L05 voltage regulator takes any voltage input between 7 and 35 VDC and converts it to 5 VDC. And it can supply up to 100 mA (0.1 Amp) of current. Its brother, the 7805, can supply up to 1 Amp of current. Why 100 mA of current? Why are the capacitors there? And why were those particular capacitor values chosen?

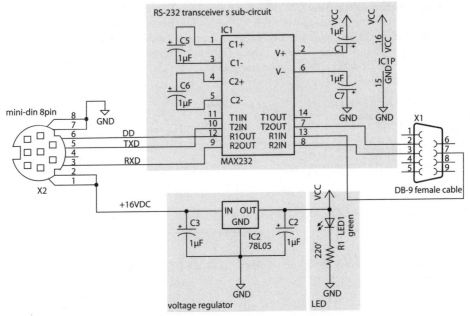

FIGURE 3-2: Schematic for serial tether, with sub-circuits highlighted

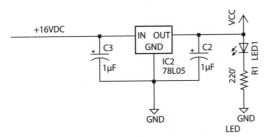

FIGURE 3-3: Voltage regulator circuit

Capacitor Values for Voltage Regulators

In circuit design, if you can make something not work as hard as it needs to, you do it, because your circuit will be more efficient and more reliable. In this case the input capacitor C3 is added to reduce any noise or dropouts on the input voltage coming from the Roomba. Figure 3-4 shows examples of noise and dropouts. A common source of noise is RF interference caused by other electronic devices or the motors. A common source of dropouts is some device like a motor quickly pulling too much power from the power supply. The power supply cannot keep up so its output voltage sags. The capacitor gives the voltage regulator a more steady power supply to work from, filtering out noise and smoothing out small dropouts. It smooths out dropouts by acting like a little charge reservoir for them, and it filters out noise by averaging out small variations in the voltage. Capacitor values are measured in Farads (symbol: F), which is a measure of how large their charge reservoir is.

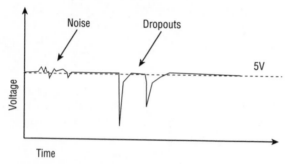

FIGURE 3-4: Noise and dropouts on an otherwise stable DC voltage

The output capacitor C2 performs a similar role for the users of the +5 VDC power it creates. Since this output voltage is used as the positive supply voltage to an IC (specifically the RS-232 transceiver IC), it's historically called Vcc, Vdd, or V+. Vcc will be used here.

Capacitor Voltage Ratings

Another parameter of capacitors is their voltage rating. This is often 16V, 50V, or 100V. There's no great trick to choosing this value: the general rule-of-thumb is to pick a voltage rating about twice the maximum voltage the capacitor will see. In this circuit, the maximum voltage is around 16V. Double that is 32V, so the 50V capacitors will work fine.

Understanding LEDs

The next sub-circuit is the status LED, shown in Figure 3-5. Its simple purpose is to shine when there is power present. This sub-circuit is not strictly necessary to make the serial tether function, but it does provide some visual indication as to whether there is a current running through the circuit. Also, always follow the general rule: if an LED can be added to a circuit without otherwise affecting its functionality, add it! It's fun, and it helps you troubleshoot whether there is any power in your circuit.

In order to light an LED, you must pass current through it. The amount of current determines the brightness of the LED, up to some maximum current. Beyond that maximum, the LED blows up. This is entertaining once or twice but doesn't really solve the problem of letting you know when your circuit is functioning.

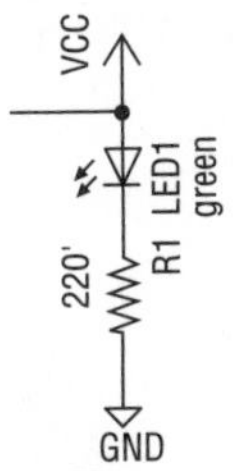

VCC

220' R1 LED1 green

GND
LED

FIGURE 3-5: LED circuit

If an LED is just connected directly to a power supply, it would draw as much current as possible, because it acts like a short-circuit. Standard LEDs have a maximum current of around 50 milliamps (mA). You want to be below that, say 25 mA. To control the amount of current so it doesn't go rushing around in a short-circuit, add a resistor.

Ohm's Law

Resistors, like all electrical components, obey Ohm's Law: $V = I \times R$, or flipping around to solve for the resistor value: $R = V/I$. R is the resistor's value, measured in ohms (symbol: Ω). V is the voltage applied across the resistor, measured in Volts (symbol: V), and I is the current in Amps (symbol: A). Ohm's Law always applies for any part of a circuit and it's a really useful tool to help analyze circuits.

You know I from above: 25 mA.

So what is V then? You may think +5V since that's the power supply, but that's not quite it. An LED (or any diode) drops some amount of voltage because of how it is made. This voltage drop is different for every diode, but is usually around 1.4V. You can measure it with a multimeter that has a diode setting, or you can measure it yourself by picking some resistor value you think may be correct, making the LED circuit and measuring the voltage drop across the LED. Because the LED drops 1.4V, that means that 5V - 1.4V = 3.6V, so 3.6V goes across the resistor.

This means the R=V/I equations for the resistor becomes: $R = 3.6/0.025$, or $R = 144$ ohms. Resistors come in certain fixed values and often the getting the exact correct value isn't important. In this case, since you want to err on the side of safety, you choose a value greater than 144 ohms. A common value is 220 ohms and is often the smallest value hobbyists have on hand. So it becomes a common value for LED circuits. That means the current through the LED is: $I = V/R = (5-1.4)/220 = 16$ mA.

LED Orientation

LEDs only conduct current in one direction. Therefore, the orientation of an LED is very important. In a schematic, an LED's "bar" is the negative side of the LED, and its "arrow" should always point toward ground. Refer back to the LED schematic symbol in Figure 3-5 for a clear representation of this. When physically laying out an LED, the flat part on side of the LED corresponds to the "bar" part of the schematic.

Understanding MAX232 RS-232 Transceivers

The MAX232 transceiver IC originally developed by Maxim (not the men's magazine, but the creator of some of the coolest interfacing ICs out there) performs the magic of converting the 0–5V positive logic signals from the microcontroller in the Roomba to the approximately +/- 12V negative logic signals that are part of the RS-232 standard. Instead of accomplishing this conversion with a tricky circuit using several transistors, resistors, and capacitors, you just plop down the MAX232 and a few capacitors and the problem is solved.

Virtually every microcontroller has a serial port on it, so many hackers are familiar with the MAX232. If you want your little gadget to talk to your computer, chances are you've used a MAX232. There are many circuit schematics on the Internet and in books with the MAX232, but they tend to vary regarding which value of capacitors to use. Some use 10 µF capacitors, some use 1 µF, and others use 0.1 µF. Which is the right value? Why do people use different values?

The pedantic but true answer is that the datasheet for the MAX232 tells what capacitor values to use. The trick is that there are slightly different versions of these transceivers that can take different capacitors. One variant, the MAX233, has internal capacitors, so no extra parts are needed. (It's expensive though.) Some parts are MAX232 clones and are also called MAX232 but are slightly redesigned. If you have the datasheet for the exact part being used, use the capacitors described in the datasheet. If unsure, use 1µF capacitors.

The MAX232 works by using the capacitors to create a charge pump that boosts the input voltage from 5V to either -12V or +12V. The capacitors store the charge needed to make this voltage. Since it takes more charge to drive long serial cable lines, generally the longer the cable, the larger the capacitors will need to be. And in RS-232, long means several hundred feet, not the 10-feet cable you'll be using here.

Maxim will help you use their parts by sending you free samples. Just go to the Maxim web site (www.maxim-ic.com), find the part you want, and click sample. This is really handy if you're a starving student and want to try out a few interesting parts. If you're in a hurry or need many Maxim parts, it's quicker and easier to buy them from a place like Digikey or Jameco. Most of their parts are only a few dollars.

Building the Serial Tether

Now that you have some understanding of the circuit, it's time to build it.

Caution   It's easy to burn yourself with a soldering iron. Be careful, always know where it's at, and always make sure to turn it off when done. Also be sure to be properly grounded so you don't zap anything. See Appendix A for some guidelines on soldering techniques.

Getting Ready

Figure 3-6 shows the parts needed all laid out. From here you can see the difference between the Mini DIN cable (round ends) and the DB9 cable (flat ends). From this particular set of parts, the 1 µF capacitors are the five lighter cylinders and the 10 µF one is the smaller black cylinder. The MAX232 chip is the black square with 16 pins, and the voltage regulator is the little 3-pin thing. The LED is to the right of the voltage regulator, and its little resistor barely visible next to it. The circuit board these parts will all be mounted on is in the middle. Many of these components, particularly the capacitors, may look different when you build this project, as there are many different styles of parts. As long as the values are correct, everything's fine.

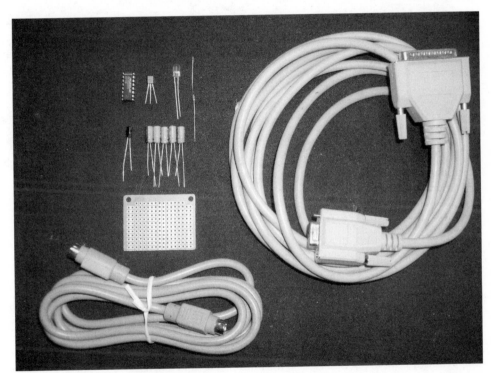

FIGURE 3-6: The parts needed for this project

Figure 3-7 shows the tools I used when building this project. The exact version of these tools isn't important. They're just the ones I've been using for a while. But it's nice to know that with only the tools shown here, you can build almost any circuit. At the top left of Figure 3-7 are the "helping hands" (with built-in magnifying lens). Next to them is the multimeter and soldering iron. Weller makes good irons. This particular iron is temperature controlled, which is why it has a base station, but it's a feature not needed for these projects. At the lower left are some cheap cutters, needle nose pliers, and glue gun I got at a swap meet. Finally there's the test leads and wall wart power supply rescued from a broken cordless phone.

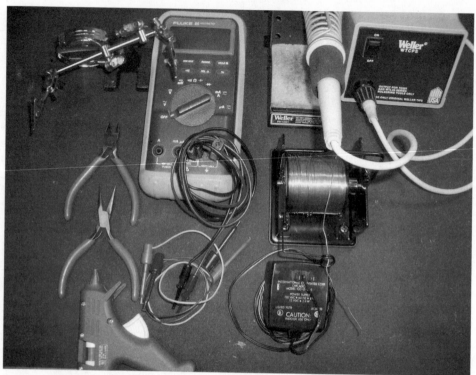

FIGURE 3-7: The tools needed for this project

Step 1: Preparing the Cables

The Mini DIN 8-pin cable and the DB-9 cable must first be prepared. Cut the Mini DIN cable six inches from the plug, and cut the DB-9 as far from the female DB-9 end as possible. To get at the wires, strip off about two inches of the big plastic sheath from each cable and then strip off about 1/4″ of the plastic insulation from all the wires inside.

It usually helps to put the cables in the third-hand clamp tool before continuing. Using the soldering iron, lightly tin each wire with solder. Perform a continuity test on each wire to figure out which colored wire goes to which pin on the jack. It seems every cable has had a different color-to-pin mapping, which is why this is necessary. The DB-9 cables seem to have a more standard color scheme, but you should always test to be sure. A bit of wire a few inches long used to poke into the DB-9 socket and using a Mini DIN 8-pin socket makes it easier to check continuity.

Figure 3-8 shows cables in the third-hand tool after being stripped and tinned. Notice how the Mini Din cable is only about 6 inches long and the DB-9 is about 15 feet long.

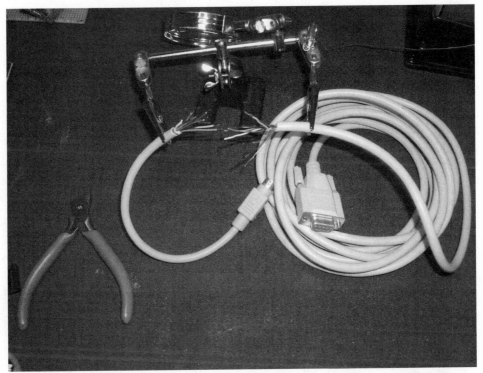

FIGURE 3-8: Mini DIN 8 and DB-9 cables in the third-hand tool, stripped and tinned for the circuit

Step 2: Laying Out the Parts

The prototyping boards often have holes that are joined together electrically. This is a great time saver since it means less soldering, but it also means a little more planning must be done to figure out how to use the board space efficiently. These particular boards from Radio Shack are great for IC-based projects because they have three holes (or pads) connected for every pin of an IC if the IC is inserted along the board's axis, and they have two bus lines down the middle, between the pads for the IC. The pad connectivity can be seen from the top thanks to the useful printing around the holes.

With that in mind, lay out the parts according to how they're connected in the schematic. To save physical space, cut the prototyping board in two, since only half of it is needed. (This means you have another board in case you want to build one for a friend.) Place the MAX232 chip so it straddles the two big vertical bus lines, then start placing parts around it, using the connected pads to minimize the amount of wiring needed. Of course, a few jumper wires are always needed. For these small jumpers, use snipped leads from parts.

Also, create test points using snipped leads to check voltages. Sometimes the jumpers can double as test points. Test points for Vpwr, Vcc, and GND should be created.

Figure 3-9 shows one possible layout that worked well.

FIGURE 3-9: Laying out the parts

Step 3: Soldering

With the parts placed, carefully bend the leads of the passive components (capacitors, resistors, and LEDs) to hold them in place and solder them down. When bending the leads, bend them toward where they need to connect. It's usually possible to use the leads as the connecting wire. Then insert and solder the MAX232 chip. Some people prefer to solder an IC socket instead and later insert the IC into the socket. Doing this is preferred, especially if you're unsure of your soldering skills, but the MAX232 is a pretty tough chip and can take around five seconds of direct soldering per pin. If you linger on a pin with the soldering iron, let the chip cool down a little before going to the next pin.

Figure 3-10 is the reverse side of the board, with the parts soldered down. Notice how the IC straddles the bus lines and component leads are bent to form connecting wires.

FIGURE 3-10: Soldering down the parts

Step 4: Checking Voltages

With the circuit built, it's time to hook it up to power and see if it blows up. Actually this is one of the most important steps and should never be skipped. It's easy to make soldering mistakes, and this is the step where those mistakes are caught. For this circuit the worst case would be to have the Vpwr line connected to the serial lines of the Roomba. This +16V applied to the Roomba's +5V-compatible lines would most certainly destroy them. But with a few quick checks, you can be assured everything is okay.

Here you use a standard DC wall wart of around +9V to +24V to emulate the +16V Vpwr line. The exact value isn't that important, because the whole point of the 78L05 voltage regulator is to turn whatever is on the input into +5VDC.

Using the test points created, hook up the multimeter to Vcc and GND. Connect the wall wart power supply to the Vpwr and GND test points on the circuit. The LED should light up. If it doesn't, disconnect power immediately and look to see why. Usually it's because the LED is wired backward. If the LED lights, the multimeter should read 5V. Figure 3-11 shows the circuit being tested. Once Vpwr is verified, check all the pins of the IC to be sure that only the input of the voltage regulator is getting Vpwr.

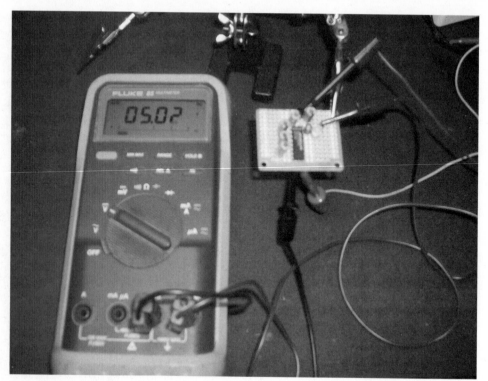

FIGURE 3-11: Testing the voltages of the circuit

Step 5: Soldering the Cables

Soldering the cables can be tricky because they're so big compared to the circuit. To make it more manageable and to give the cables some strain-relief in case they get pulled, hot glue them to the edge of the board. For an extra bit of added protection, before hot gluing, loop some stray insulated wire around the cables and into the circuit board holes and twist tight.

With the cables anchored securely, route them to the corresponding points in the circuit and solder them down. Figure 3-12 is an example of a way to anchor the cables and solder them. In that figure you can also see that I eventually opted for a socket for the IC.

Step 6: Testing Connections

The circuit is officially finished and ready to use, but to be extra paranoid, check the connections again. Connect the wall wart power supply from Step 4 to the Mini DIN connector using the Mini DIN socket. The LED should still light. Now test all the pins of both cables, using

FIGURE 3-12: Connecting the cables

the same techniques as in Step 1, but this time measure voltages. The main thing to watch for is that +16V is only on the two Vpwr pins of the Mini DIN cable that will plug into the Roomba.

Step 7: Putting It in an Enclosure

Although having a naked circuit looks pretty cool (in a nerdy way), it's usually a good idea to put the circuit in some sort of enclosure. A simple enclosure could just be some electrical tape wrapped around it or a cardboard box. If you're handy with a Dremel or similar hand tools, you can take everyday container objects and convert them into useful enclosures for your projects. Figure 3-13 is an example of putting the circuit in a floss container. The floss container has the benefit of being able to be opened and closed easily to inspect the circuit.

Once it's plugged into the Roomba, you may want to keep the circuit in place so the robot's wheels don't catch it. A small square of velcro fastener taped to the bottom of the circuit enclosure and to the Roomba encasing allows easy attachment and removal of the circuit to the robot's surface.

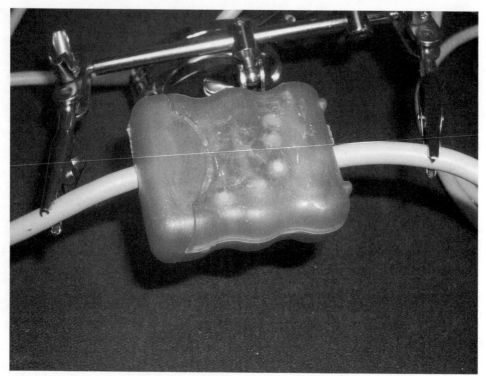

FIGURE 3-13: The circuit in a nifty blue enclosure made from a dental floss container

Connecting to a Computer

The serial tether has a standard RS-232C DB-9 serial port connector on it and many PCs (older ones anyway) have these serial ports on their motherboards. It's tempting to just plug in the tether and start going at it. However, because in this chapter you're going to try to avoid breaking things, it's instead prudent to use a USB serial adapter to guard against any bad voltages entering your computer. No-name USB serial adapters can be had for less than $10 online and so offer a cheap kind of protection for this and any future serial port–based projects you may build.

In this book, the Keyspan US-19HS USB serial adapter will be used (see Figure 3-14). Unlike some of the cheaper adapters, the Keyspan works on all operating systems and has never given me any trouble, unlike some of the more generic ones I've used.

All of these USB serial adapters require drivers to be installed. These are very minor drivers that won't mess up your system.

FIGURE 3-14: Keyspan US-19HS USB serial adapter

Simple Echo Test

Before plugging the serial tether into Roomba, do one final test. This test will be an end-to-end test from the computer, through the USB serial adapter, through the circuit, to the Mini DIN cable and back again. This end-to-end test enables you to check the entire communication path without worrying about the particulars of the Roomba protocol.

Using the Mini DIN jack connected to the Mini DIN cable, connect the DC wall wart power supply to the tether to emulate Vpwr like in Step 6 of the previous section. Using a test clip, connect pins 3 and 4 (RXD and TXD) of the Mini DIN jack together. That creates the loop that will echo back any data you send. This is called a *loopback* connection; it is one of the most powerful debugging techniques for serial data transmission.

With the USB serial adapter installed on your computer, start up a terminal program that can speak to serial ports. Not all terminal programs can do this, as it's not something that's normally needed. To communicate with Roomba, you need to set your serial programs to 57,600 bps with 8N1, no hardware or software handshaking, and no local echo.

Note Besides baud rate, turning off hardware and software handshaking is the most critical setting for serial-based projects. Handshaking (signals that indicate when it's okay to send data) is rarely used. Unless explicitly mentioned, assume no handshaking when configuring a serial port.

Mac OS X

On Mac OS X, the otherwise wonderful built-in Terminal.app cannot speak to serial ports, but ZTerm can, and it is the preferred terminal program. It's free and available from `http://homepage.mac.com/dalverson/zterm/`.

After you download ZTerm, launch it and hold down the Shift key to select the port (see Figure 3-15). Pick the serial port that corresponds to your USB serial adapter (KeySerial1 if using the Keyspan adapter) and click OK. A blank terminal window will open. Go to Settings and set the data rate and other parameters to what you'll use when talking to Roomba (see Figure 3-16). This is the 57600, 8N1 serial port setting mentioned in Chapter 2. Click OK and you're ready to echo.

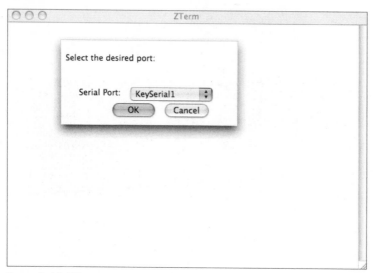

FIGURE 3-15: ZTerm startup

Windows 2000/XP

Microsoft Windows has the built-in HyperTerminal program — avoid it; it has some problems when dealing with non-modem devices. Instead the program RealTerm is great and available for free at `http://realterm.sourceforge.net/`.

After downloading, installing, and running the RealTerm program, you'll be presented with a window like the one shown in Figure 3-17. RealTerm automatically opened COM1, which is probably not the serial port you want opened. Click the Port tab and set the various settings like Figure 3-18 to match the 57600, 8N1 setting Roomba expects as described in Chapter 2. Click the Open button so that it pops up to close the current port. Click it again to open the correct port.

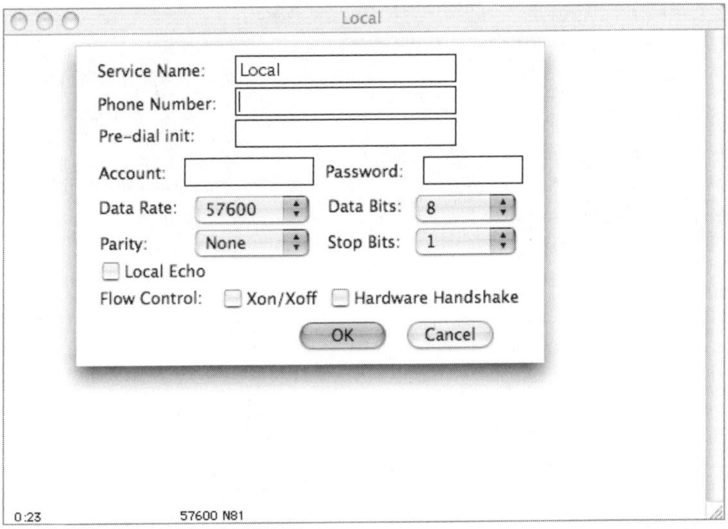

FIGURE 3-16: ZTerm configuration

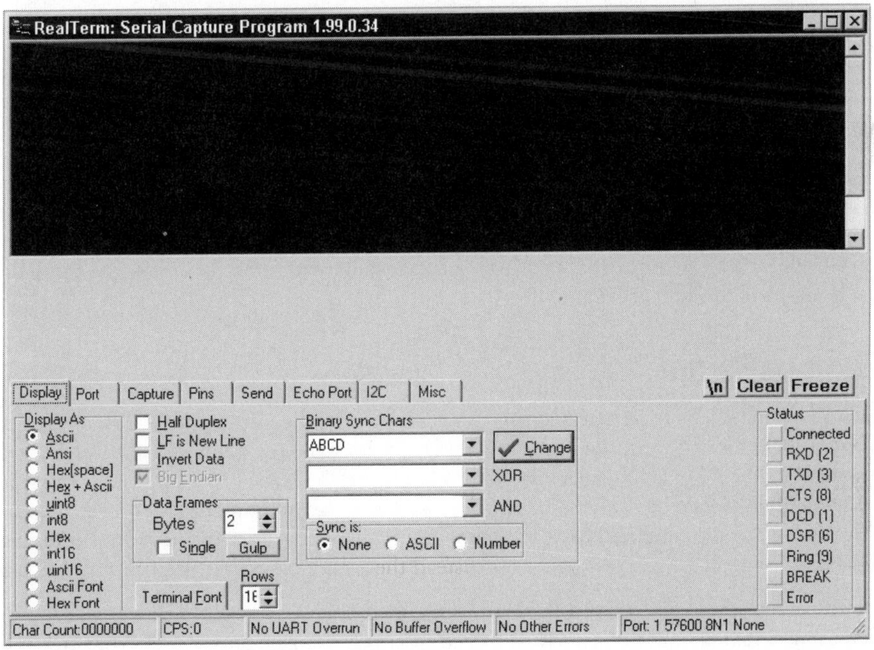

FIGURE 3-17: RealTerm startup

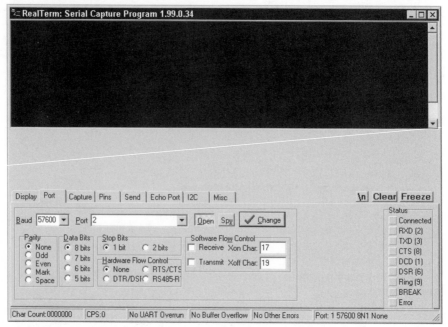

FIGURE 3-18: RealTerm configuration

Echo...Echo...Echo...

With either ZTerm or RealTerm set up correctly, click the terminal window and type a few characters. You should see what you type. Disconnect the test lead between pins 3 and 4 and type a few more characters. You should not see what you type. Reconnect the test lead and talk to yourself for a bit. Congratulations! You've successfully sent serial data from the computer to the Roomba plug and back again.

RoombaCommTest

Now that you've exhaustively tested this thing, you can finally plug it into the Roomba. Exit the terminal program from the previous test. Disconnect any test leads and the wall wart power supply. Plug the Mini DIN cable into Roomba. The LED on the serial tether should light up. Turn on Roomba and press the Clean button. The robot should function normally. Reset Roomba by turning it off and back on again. Now you're ready to actually control Roomba from the computer.

Download the RoombaComm software package from http://roombahacking.com/ and find the RoombaCommTest directory. This directory contains a program for both Mac OS X and Windows to test the tether by letting you control Roomba. Double-click the appropriate RoombaCommTest option for your OS. You should see a screen like Figure 3-19. It will look slightly different on each OS, but it will function the same. RoombaCommTest is a Java program written with the RoombaComm Java library that you'll be creating in the next few chapters and which was introduced at the end of Chapter 2. Normally Java programs don't act like double-clickable applications, but there exists wrapper software to bundle up a Java program into a format your OS will see as a real application. You'll get into how to make these later. If you'd like to get a head start on how to write Java programs, see the sidebar about it in Chapter 5.

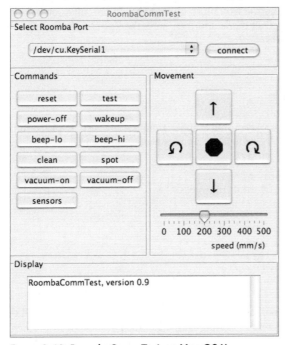

FIGURE 3-19: RoombaCommTest on Mac OS X

Choose the same serial port as you chose for the echo test and click connect. Messages will appear in the display area giving the results of the connect attempt. If all goes well, you'll see Roomba connected, and Roomba will beep.

Note On Mac OS X and Linux, the serial port may appear twice: once as a /dev/cu.something port and again as a /dev/tty.something. If it appears twice, choose the /dev/cu version of the port.

Note On Mac OS X, you may need to run the macosx_setup_command script contained in the RoombaCommTest directory. This makes a few minor changes that enable the Java serial library RXTX to work. A future version of the RXTX library used in RoombaCommTest will not need this setup command. You can still run the command if you want; it won't harm anything.

When connected, the rest of the buttons in RoombaCommTest are ready to use.

Click the Test button. RoombaCommTest will play a few notes on the robot's beeper and move Roomba in a little dance. Congratulations! You've just controlled Roomba with your computer. If you like, play around with Roomba using the other buttons. Whenever you want, quit the program.

Tip   When Roomba is being commanded through the ROI, the physical buttons on Roomba no longer work. This means the Power button won't turn off Roomba. To turn off Roomba, either lift the cleaner to trigger the safety fault mode and then press the Power button or click the power-off button in RoombaCommTest.

Commanding Roomba

The RoombaCommTest program is a Java GUI wrapper around some very simple commands. Take for example the first things it does when the connect button is clicked:

```
RoombaComm roombacomm = new RoombaCommSerial();
if( !roombacomm.connect(portname) ) {
  System.err.println("Could not connect");
  return;
}
roombacomm.startup();
roombacomm.control();
roombacomm.pause(30);
roombacomm.playNote(72,10);   // C
roombacomm.pause(200);
```

The preceding code creates a RoombaComm object. The RoombaComm object contains all the protocol-independent methods for communicating with a Roomba. It is the ROI protocol embodied in code. Subclasses of RoombaComm implement a few low-level methods to send and receive data. One example of such a subclass is RoombaCommSerial, for dealing with communicating with a Roomba over a serial port. Another example is RoombaCommTCPClient, for talking to a TCP-connected Roomba cleaner.

When an appropriate RoombaComm object is created, the next step is to try to connect to Roomba with the connect() method. In RoombaCommSerial, connect() hides all the Java serial messiness that is needed.

If the connection succeeds, then the START and CONTROL commands are sent to the robot through the startup() and control() methods. This puts Roomba in safe mode, ready to be controlled.

The playNote() method plays a single note of a given duration. It does this by sending first a SONG command with a one-note song defined and then a PLAY command to play that song. This is the first, albeit simple, example of the RoombaComm API building higher-level functions out of the basic ROI building blocks.

If the forward button is clicked, the following command is issued:

```
roombacomm.goForward();
```

This sends Roomba forward at the current speed setting.

Feel free to delve into the internals of the RoombaComm library. If you are familiar with Java JAR files, there is a `roombacomm.jar` that contains the full library so you can write your own programs. How to do this is discussed in Chapter 7.

Summary

Congratulations! You have hacked your Roomba! You have created a Roomba serial interface tether, the foundation for all future Roomba robotics hacking. Now you can begin experimenting with Roomba as a computer-controlled robot, using either the RoombaCommTest program or writing your own code. Even just driving the robot around from your computer is a lot of fun.

If this was one of your first circuits to build, congratulations are in order again. You now can build your own electronic circuits and know the fundamentals of three of the most important concepts in circuit design: power supplies, visual indicators, and signal-level translators. Virtually all circuits (certainly all the ones in consumer devices) employ at least two of those sub-circuits. When examining other circuits, you can look for these sub-circuits to learn how the whole works. The Roomba, for example, has a few variations of all three sub-circuits, and you could take apart other devices to find examples of them there too.

At a system level, with serial terminal programs and loopback cables, you can debug almost any kind of RS-232 or similar serial connections. Serial communication is the basis for most gadget communication. You'll run into it in many of the other projects. Serial is also the basis for more advanced computer communication like USB, Firewire, Ethernet, or Wi-Fi. The basic concepts (receive and transmit signals, communication speed, loopback, and others) you're now familiar with, and the diagnostic techniques used are translatable to these more advanced serial protocols.

Start thinking about designing your own circuits to work with other gadgets around your house. Maybe add a status light for something you wished you knew was on, or build a battery-replacement power supply, or even build PC interfaces for other devices you may have. Almost every gadget with an embedded microcontroller like Roomba has a serial port, although it may not be as accessible as the Roomba port.

Building a Roomba Bluetooth Interface

Although the serial tether is eminently useful and will constantly be used while debugging, it's not as magical as having a totally wireless connection to the Roomba robotic vacuum cleaner. With a small, discreet wireless adapter, Roomba appears to be functioning entirely on its own, while still being able to call upon the vast resources of a full-blown desktop PC.

Wireless communication is becoming more pervasive every day. In your house you probably have a dozen devices that have computers in them and some sort of wireless digital communication. You may be thinking of just the WiFi devices you have, but those are only the most obvious devices. Don't forget the wireless mouse or keyboard you might have. And what about your cell phone or mobile phone, or the infrared remote controls that talk to your TV and stereo? Or the RF remote controls for the garage door opener or ceiling fan? All of these devices contain tiny computers that talk to other tiny computers. Even Roomba has a wireless communications device in the form of its infrared remote control sensor.

As computational hardware becomes smaller and cheaper, portable objects that were previously dumb begin to get smarter. And when they're smart and portable, wirelessly connecting them inevitably follows. Increasingly, things we never considered as having a computer or needing to be networked are becoming that way. The dull, static objects of our lives are becoming smart and dynamic. In the past 20 years, the vacuum cleaner has gone from a simple bag with electric motor to a small industrial robot with more computing power than the original PC. Imagine what the next 20 years will bring. First objects become smart; then they start talking to each other.

This chapter demonstrates one path to that next stage by adding a Bluetooth interface to Roomba. Bluetooth is one wireless protocol out of many. You will use it here because the rise of built-in Bluetooth technology in new laptops has resulted in cheap (enough) Bluetooth-to-serial adapters.

in this chapter

- ☑ About Bluetooth
- ☑ Build a Bluetooth Roomba adapter
- ☑ Set up Bluetooth serial ports
- ☑ Design the RoombaComm API

Alternatives

As with the RS-232 serial adapter, RoombaDevTools.com has made a work-alike of the circuit presented here, called RooTooth (see Figure 4-1). If you're anxious to get your Roomba working wirelessly, RooTooth does the trick. As in most cases of build-vs.-buy choices, RooTooth is more expensive than building it yourself. Otherwise it's functionally identical to the circuit presented here and will work with all the example programs in this book.

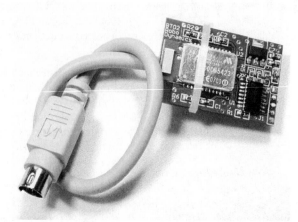

FIGURE 4-1: RoombaDevTools RooTooth Bluetooth Roomba adapter

Why Bluetooth?

Bluetooth is a technology for creating wireless *personal area networks* (*PANs*). Unlike Wi-Fi networks, which can have hundreds of users and extend over great distances, PANs were designed to be close to a single person. The original idea is that many single-purpose devices like headsets, displays, network adapters, PDAs, personal storage devices, and so on could all talk to one another wirelessly in an easy-to-use and secure fashion. When it was first designed in the late 1990s, Bluetooth was originally meant to be a low-power, low-bandwidth protocol. But when protocol issues arose that made various manufacturers incompatible with one another coupled with Bluetooth chipsets that weren't cheap enough, Bluetooth languished for several years, only becoming popular among a few cell phone makers as a wireless headset technology.

The Bluetooth designers have been working continuously to update the protocol, and the cost of implementing Bluetooth in a device has been plummeting. Today Bluetooth is more pervasive than ever. Bluetooth mice and keyboards are incredibly common. Many cars now have Bluetooth hands-free interfaces for cell phones. The next generation of game consoles uses wireless controllers, using Bluetooth or Bluetooth-like technology. All new computers and PDAs seem to have Bluetooth. And on the hacking front, the Lego Mindstorms NXT (a great companion kit to Roomba hacking) uses Bluetooth to download programs into its brain.

Recently SparkFun (http://sparkfun.com/) created a relatively inexpensive little board called the BlueSMiRF, which incorporates a Bluetooth serial interface. Now it's easy for hackers to add Bluetooth to the gadgets they create. The BlueSMiRF is what is used in this project.

How Bluetooth Works

Bluetooth is a radio protocol for transmitting digital information. It sits in the 2.4 GHz spectrum along with many other protocols and devices like 802.11b/g (Wi-Fi), cordless phones, X10 video cameras, wireless mice and keyboards, and microwave ovens. The 2.4 GHz spectrum is one of the few places on the airwaves where one can broadcast without needing an FCC license. The 2.4 GHz band is part of the ISM (industrial, scientific, medical) set of bands that are free to use by anybody.

And that makes it a crowded place to live for a wireless protocol. Having so many technologies existing in close physical proximity using the same band can cause interference. Bluetooth attempts to avoid interference through *frequency hopping spread spectrum*. This is a technique whereby instead of transmitting on a single channel, it transmits a little, hops to another frequency, transmits a little more, hops again, and so on. By hopping very fast (many times a second), the chances of hitting interference go down since not all frequencies are equally noisy.

One End of the Spectrum

This spread spectrum technique was first invented during World War II by movie star Hedy Lamarr. Her version used a piano roll to switch between 88 frequencies and was meant to foil enemy eavesdropping. She probably never imagined that in the following millennium people would use her invention to enhance listening in millions of households.

Bluetooth Power Classes

Bluetooth can transmit information at a variety of speeds, from 1.1 kbps to 2.1 Mbps (depending on the application) and at one of three power levels, or classes:

- **Class 1:** 100 mW power, about 100 meter range
- **Class 2:** 2.5 mW power, about 10 meter range
- **Class 3:** 1 mW power, about 1 meter range

Many early Bluetooth devices were Class 3 or badly designed and Bluetooth got a bad reputation for being too low power and not very useful. In actuality, Bluetooth devices can have a range that equals or exceeds Wi-Fi. And thanks to advances in low-power circuitry and improvements to the Bluetooth specification, Bluetooth has started to deliver on its promise of PANs.

Bluetooth Profiles

What is Bluetooth being used for? Like USB, Bluetooth defines a set of device classes to solve common tasks. Devices within the same class speak the same sub-protocol of Bluetooth and thus need no drivers to interoperate. Bluetooth calls these device classes *profiles*. Table 4-1 lists a few of the most commonly used profiles.

Table 4-1   Common Bluetooth Profiles

Profile Name	Description
Headset Profile (HSP)	Most common profile, used for those tiny headsets for cell phones
Hands Free Profile (HFP)	Used in cars to enable hands-free mode by integrating with the car stereo
Human Interface Device Profile (HID)	Supports keyboards and mice and other input devices
Object Push Profile (OPP)	Used mostly by cell phones to send vCards (virtual business cards) and vCalendar (virtual appointment book) entries
Basic Printing Profile (BPP)	Allows structured access to printers
Dialup Networking Profile (DUN)	Provides a standard way to access the Internet
Serial Port Profile (SPP)	Serial port cable emulation

Computers with Bluetooth implement some subset of all available Bluetooth profiles. This used to be a big problem, as the operating system drivers would only implement the HID and DUN profiles and nothing else, making the claim *Bluetooth compatible* a misnomer at best. Most operating systems now support all the profiles you would be likely to use.

The profiles provide a common task-centric set of languages that allow devices from disparate manufacturers to work together. In fact, Bluetooth is named after the 10th-century Danish King Bluetooth, who was famed for getting warring parties to negotiate with one another. It's a fitting sentiment for any wireless technology.

Parts and Tools

Many of the parts used in this project will be familiar from Chapter 3 since the circuit is very similar. You won't need a MAX232 transceiver chip to convert between 0–5V logic and the +/-12V logic of RS232, but you will need a header socket for the BlueSMiRF module (and the BlueSMiRF module itself). So for this project all the parts you will need are:

- Mini-DIN 8-pin cable, Jameco part number 10604
- General-purpose circuit board, Radio Shack part number 276-150

- 78L05 +5VDC voltage regulator IC, Jameco part number 51182

- 220 ohm resistor (red-red-brown color code), Jameco part number 107941

- Two 1µF polarized electrolytic capacitors, Jameco part number 94160PS

- 8-pin header receptacle, Jameco part number 70754

- BlueSMiRF Bluetooth modem, SparkFun part number RF-BlueSMiRF

You also need a Bluetooth-capable computer. If your computer doesn't have Bluetooth capability built in, then you can get a USB Bluetooth dongle. For Windows and Linux, SparkFun sells a generic dongle (SparkFun part number RF-BT-USB). For older Mac OS X machines, you can get the D-Link DBT-120 from the Apple Store. Figure 4-2 shows what these dongles look like. They're cheap and unobtrusive. The D-Link one is neat because it has an LED that blinks when Bluetooth activity occurs. The generic one from SparkFun is neat because it's a Class 1 device and so should be good for up to 100 meters.

FIGURE 4-2: USB Bluetooth dongles for computers without Bluetooth

All the tools from Chapter 3 will be used here too to build the project in this chapter.

Naturally, all of the usual safety precautions should also be followed as described in the previous chapter and in Appendix A.

The Circuit

Figure 4-3 shows the circuit for this project. Compared to the serial tether, it is much simpler. It mostly consists of two sub-circuits that you are already familiar with from Chapter 3: the voltage regulator and the LED status light. The only unknown part of this circuit is the BlueSMiRF header. Fortunately, due to how the BlueSMiRF is designed, the data lines from the Roomba ROI port can connect directly to the BlueSMiRF through the header. In fact, if the BlueSMiRF's built-in power supply could take up to 16V, this circuit would merely be an adapter cable. As it is, a voltage regulator must be supplied. And if you add a chip that makes a voltage, might as well add a light to let you know the voltage is there.

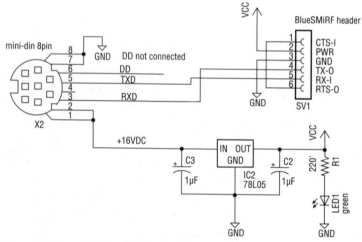

FIGURE 4-3: Roomba Bluetooth adapter schematic

The BlueSMiRF is a Bluetooth modem that implements the Bluetooth Serial Port Profile (SPP) and presents a normal 5V logic set of serial lines. When the BlueSMiRF is paired to a computer, the serial lines look like they are connected directly to the computer as a normal serial port. Since the output lines of the BlueSMiRF are at normal logic levels, they can be connected directly to the Roomba's ROI port.

The BlueSMiRF has a set of AT commands, like normal modems do, and various controls and configuration settings are accessible through these commands. For this project, only one setting needs to be changed: the speed of the serial lines connected to the Roomba. The BlueSMiRF can operate from 9,600 bps to 115,200 bps, and it needs to be configured to speak at the speed of Roomba: 57,600 bps.

There are many other Bluetooth serial adapters, some meant for hacker use like the BlueSMiRF and some with normal RS-232 ports for wirelessly connecting existing serial devices. In the latter category are devices like the IOGear GBS301, which looks eminently hackable and was going to be the core of this project if the BlueSMiRF hadn't come along. In the former category is the blu2i module available from Tek Gear. It looks a lot like the BlueSMiRF but is almost twice the price.

Note   The circuit presented here is just one of many ways of adding Bluetooth to the Roomba. The above devices could be used instead. If you have easy access to one of them, try playing around with it. If you have more than one, try adding Bluetooth to something else in addition to Roomba, like your coffee maker or stereo. These little wireless communication gadgets are fun to hack on to a variety of things.

Caution   The BlueSMiRF module is sensitive to static electricity. Make sure you're properly grounded before handling it. See Appendix A for techniques for staying grounded.

Building the Bluetooth Adapter

Construction of this circuit will be a good deal easier than the serial tether because of the lower part count. The BlueSMiRF does all the hard work.

Caution   It's easy to burn yourself with a soldering iron. Be careful, always know where it's at, and always make sure to turn it off when done. Also be sure to be properly grounded so you don't zap anything.

Getting the Parts Together

Figure 4-4 shows the parts needed for this project. The Mini DIN cable connector is shown at the top. Right below it is a horizontal rectangle that is the BlueSMiRF Bluetooth module. The circuit board below that is the one you've seen in Chapter 3. The two 1 µF capacitors are at the left bottom. For this particular version of the Bluetooth adapter, two matching capacitors couldn't be found in the junk box, so a mismatched set was used. Next to that is the three-legged 78L05 voltage regulator. And on the bottom right is the resistor and LED for the power on light. The little black square to the right of the circuit board is the header receptacle that the BlueSMiRF will plug in to. It's safer to solder a receptacle down than to solder directly to the somewhat sensitive BlueSMiRF.

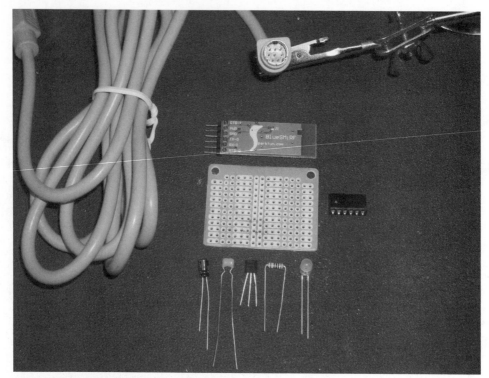

FIGURE 4-4: Parts needed for this project

Step 1: Preparing the Cable

The Mini DIN 8-pin cable preparation is exactly the same as for the serial tether. Cut the cable about six inches from the Mini DIN connector. To get at the wires, strip off about two inches of the cable's big plastic sheath and then strip off about 1/4″ of the plastic insulation from all the wires inside. It usually helps to put the cables in the third-hand clamp tool before continuing. Using the soldering iron, lightly tin each wire with solder. Perform a continuity test on each wire to figure out which colored wire goes to which pin on the jack. It seems every Mini DIN cable has had a different color-to-pin mapping. Even if you've built the serial tether, don't assume the color-to-pin mapping you discovered when building it applies to another Mini DIN cable. Even different ends of the same cable can have different pinouts. Refer to the left side of Figure 3-8 for details.

Step 2: Laying Out the Parts

The BlueSMiRF constrains the part layout, since it is so long and flat. Place the black header receptacle at one end of the circuit board facing in, so the circuit board provides some support to the BlueSMiRF. You may want to cut down the 8-pin receptacle to a 6-pin one to save space. With the header placed, arrange the voltage regulator and the other parts so you can minimize the number of wires needed to solder. Figure 4-5 shows one possible layout, with the BlueSMiRF plugged into the receptacle to gauge the layout. Notice that the header receptacle has been cut down to be 6-pin, and the pins have been bent at a right angle so the receptacle lays flat against the board.

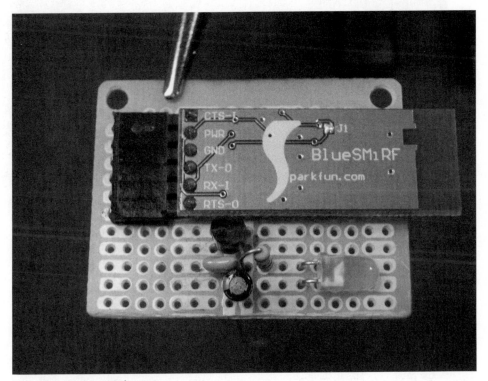

FIGURE 4-5: Laying out the parts

Step 3: Soldering

When you have a layout you think you like, turn the board over and start soldering things down. When soldering the header receptacle, remove the BlueSMiRF so you don't damage it. The whole point of the receptacle is to keep the BlueSMiRF away from the harshness of circuit building. Figure 4-6 shows the previous layout mostly soldered down. Notice the small jumpers made from cut part leads (the horizontal silver-colored wires soldered between pads). The figure also shows the cable beginning to be attached to the board.

FIGURE 4-6: Soldering down the parts

Step 4: Checking Voltages

As in the serial tether, when things are soldered down, you should hook up the wall wart power supply and test the voltages. See the section "Step 4: Checking Voltages" in Chapter 3 for more details, as the process here is almost exactly the same as there.

Your standard DC wall wart of around +9V to +24V is used to emulate the Vpwr +16V power coming from the Roomba. The 78L05 voltage regulator will turn that unregulated voltage into the +5VDC needed by the BlueSMiRF and the LED.

Using the test points created, hook up the multimeter to Vcc and GND. Connect the wall wart power supply to the Vpwr and GND test points on the circuit. The LED should light up. If it doesn't, disconnect power immediately and check to find out why. Usually it's because the LED is wired backward. If the LED lights, the multimeter should read 5V. When Vpwr is verified, check all the pins of the BlueSMiRF header receptacle. Pay particular attention when testing the voltages in the header receptacle. It should only have +5V going to it, and only on the one header socket meant for +5V.

Step 5: Soldering the Cable

This project is also easier than the last one because you only have to prepare and solder down one cable, the Mini DIN 8-pin that will plug into the Roomba. The same technique in Step 5 of the previous chapter applies here, too. Figure 4-7 shows the Bluetooth adapter almost complete. To make it more manageable and to give the cable some strain-relief in case it is pulled, hot glue it to the edge of the board. For an extra bit of added protection, before hot gluing, loop some stray insulated wire around the cable and into the circuit board holes and twist tight. Also, add a bit of hot glue to the header receptacle to secure it to the board so plugging and unplugging the BlueSMiRF doesn't stress the solder joints.

With the cables and header receptacle secured down with hot glue, solder the wires and receptacle pins down. In general you want to secure connectors and wires before soldering them so any flex of those parts doesn't affect the solder joints.

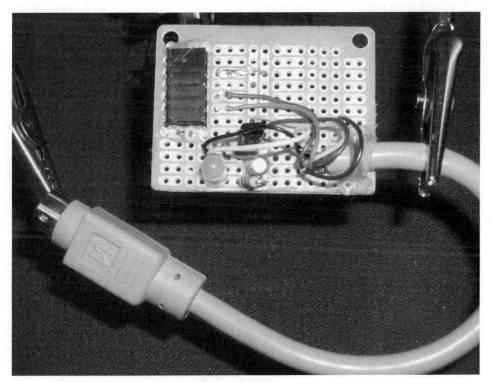

FIGURE 4-7: The Mini DIN 8-pin cable soldered down

Step 6: Testing Connections

The adapter is now complete. However, before plugging in the BlueSMiRF and connecting it to the Roomba, perform one last set of continuity and voltage checks. When they check out, plug in the BlueSMiRF and power the circuit with the wall wart power supply in lieu of the Roomba. The LED should still light and the BlueSMiRF shouldn't get warm or smoke or do anything else bad. If you're measuring the current consumption of the entire circuit, it shouldn't be more than 50 mA. If you're measuring the voltage output of the regulator, it should still be at 5V. If it drops, disconnect power immediately. All of this testing and retesting may seem overkill, but it would be a shame to fry a BlueSMiRF. In addition to being cute, they are a little expensive to destroy. After getting used to performing these kinds of tests, doing them for other projects becomes faster and basically second nature.

Figure 4-8 shows a finished Roomba Bluetooth adapter, tested and ready to be plugged into the Roomba. You may notice that this adapter is slightly different than the one in Figure 4-7. I have created many of these, each with slightly different layouts, but they all implement the same schematic and work the same.

Tip   At this point the Bluetooth adapter and the serial tether should act exactly the same when plugged into the Roomba. That is, they'll both light their LEDs and the Roomba is otherwise still usable.

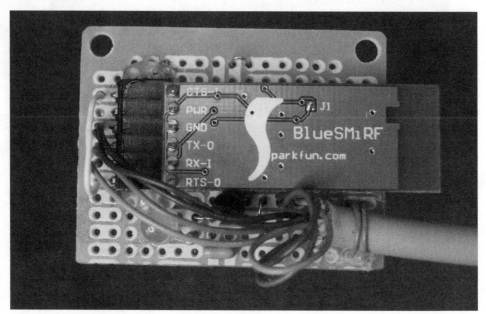

FIGURE 4-8: The finished circuit

Step 7: Putting It in an Enclosure

Every project needs an enclosure. Following the dental floss idea from the previous project, this project uses one too. Figure 4-9 shows the adapter in its new home, plugged into the Roomba and ready to go. Of course, a blue dental floss box seems perfect for a "blue-tooth" adapter. One side effect of using floss containers for all your projects is you accumulate more floss than you could ever use. So an alternative enclosure is shown in Figure 4-10. This enclosure is made from a blue container for a sugary grape-flavored gum that comes rolled up in a meter-long length. So it's still a "blue-tooth" adapter.

FIGURE 4-9: Bluetooth adapter in a "blue-tooth" enclosure

FIGURE 4-10: Another "blue-tooth" enclosure, only this time sweeter

Setting Up Bluetooth

In order to talk to the Roomba Bluetooth adapter, you need to configure Bluetooth properly on your computer. If your computer doesn't have Bluetooth support built in, you'll need to get a small USB Bluetooth interface fob (which costs less than $20). The steps are:

1. Install the USB Bluetooth interface on your computer (if it isn't built-in already)

2. Install the drivers for the interface (may be already installed if built in).

3. Search for and pair your computer with the Roomba Bluetooth adapter.

4. Create the virtual serial port on the paired device.

5. Configure the Roomba Bluetooth adapter.

In Windows, all those steps look incredibly different depending on which Bluetooth interface and which variant of Windows is used. In Mac OS X, it's all standardized, even when using an older Mac without built-in Bluetooth support. The screenshots throughout this section show the steps in OS X, but be assured they are the same in Windows, even if they look different. In Windows, Steps 1 and 2 are the standard hardware install process, complete with reboot after installation. For OS X the drivers are already installed even for non–built-in Bluetooth interfaces. So I'll skip Steps 1 and 2 and go directly to Step 3.

Pairing with Roomba Bluetooth Adapter

Pairing is the mechanism Bluetooth devices use to securely recognize each other. The first time this is done it requires a human to help out. Plug the Bluetooth adapter you just made into the Roomba (or into the wall-wart power supply; it doesn't matter for these tests). Then open the Bluetooth Preferences dialog box, turn on Bluetooth, and click Set Up a New Device. You'll see a screen similar to the one shown in Figure 4-11, asking to choose the device type. These are the most supported Bluetooth profiles mentioned earlier. Since the Serial Port Profile isn't listed, select Any Device and click Continue.

The computer starts scanning for Bluetooth devices and the Roomba Bluetooth adapter should be in the list as shown in Figure 4-12. It is listed as BlueRadios or BlueSMiRF. Select it and click Continue.

At this point the computer attempts to pair with the BlueSMiRF in the Roomba Bluetooth adapter. The BlueSMiRF responds that to pair it needs a passkey. That brings up the next screen, asking for a passkey. The default passkey for all BlueSMiRF devices is the string *default*. Type that in as shown in Figure 4-13 and click Continue. If you've used Bluetooth much, you've seen that passkeys are usually numeric.

Tip If you are on Windows and you're given the option to save the pairing passkey, be sure to select Yes so you don't have to type in the passkey default each time.

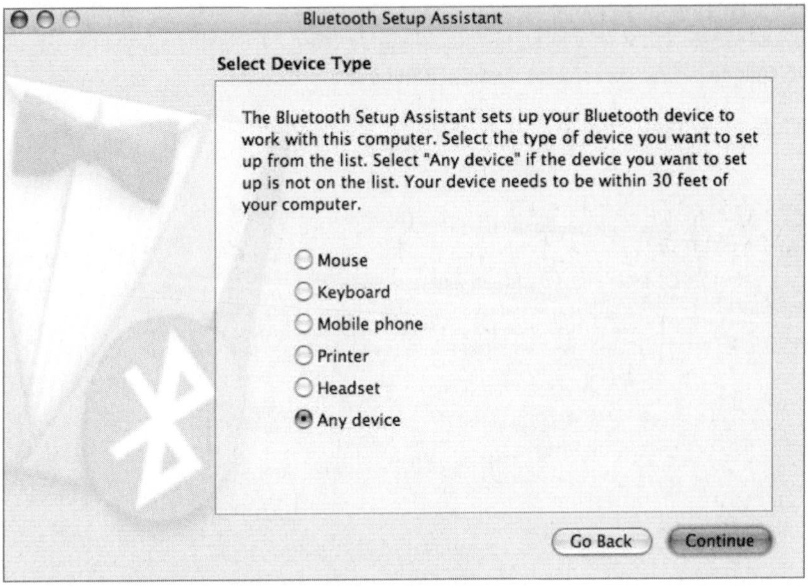

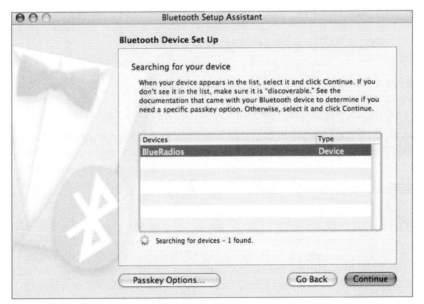

FIGURE 4-11: Starting up Bluetooth Setup Assistant

FIGURE 4-12: Bluetooth devices found after scanning

Note On Mac OS X 10.3 (Panther) and earlier, you can only input numeric passkeys. The best solution is the upgrade to Mac OS X 10.4 or higher. If you cannot upgrade, see the BlueSMiRF datasheet for how to change the passkey it uses by default to be a numeric one.

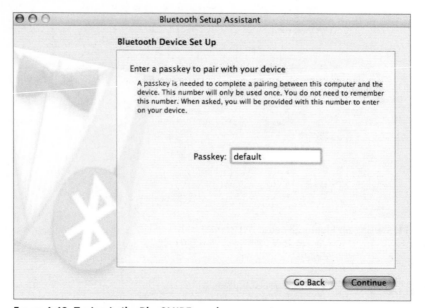

FIGURE 4-13: Typing in the BlueSMiRF passkey

The Mac OS X Bluetooth Setup Assistant then gets a little confused because it does not find a GUI setup agent for serial ports. This is fine; just ignore it and click Continue. In contrast, the Windows pairing just does its job without the somewhat confusing message.

At this point, setup is done and you can go back to the known Bluetooth device list and see the BlueSMiRF (or BlueRadios). The screen should look similar to the one in Figure 4-14.

Creating Bluetooth Virtual Serial Port

Select the newly paired BlueSMiRF from the device list and click Edit Serial Ports. You should see a window similar to Figure 4-15. If you do not see the same info, change it so it matches. In Mac OS X, all serial port devices have unique names. For virtual serial ports, the name is constructed from the device type. So in the example BlueRadios-COM0-1, the device type is BlueRadios-COM0 and it's the first one of that type. (COM0 is the first serial channel for BlueRadios; apparently there can be others.) Click OK and the virtual serial port is ready to use.

Note In Windows, you'll have to go through the extra step of binding the virtual serial port to a COM port (like COM7 or another unused port). The Bluetooth software for your adapter should help with this. Whichever COM port it is, remember it, because you'll need to know it later.

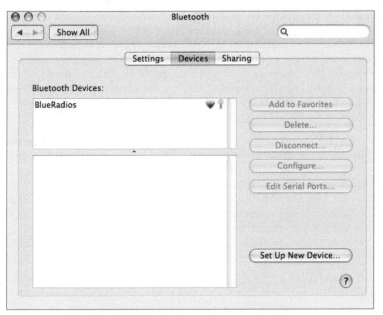

FIGURE 4-14: The BlueSMiRF device is now known and paired.

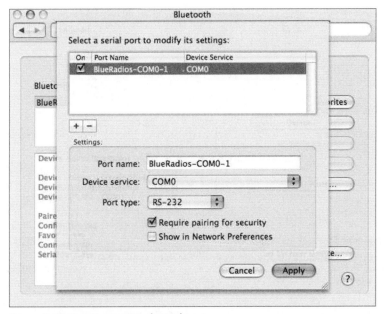

FIGURE 4-15: Creating a virtual serial port

Configuring the Bluetooth Adapter

Now that the BlueSMiRF in your Roomba Bluetooth adapter is paired up and a virtual serial port is created, the virtual serial port acts just like a normal serial port. You can connect to it like any other serial port on your computer. Unlike a normal serial port, the speed at which you connect to it has nothing to do with the speed at which it sends data over the physical wires connected to the Roomba. For that you need to configure the BlueSMiRF to the speed Roomba expects.

As in Chapter 3, open your serial terminal program (for example, ZTerm or RealTerm). Make sure you connect to the virtual serial port that represents the BlueSMiRF module. When connected, type +++ and press Return (that's three plusses and one Return). The BlueSMiRF should respond with OK. Then type the magic incantation **ATSW20,236,0,0,1** and press Return again. The BlueSMiRF should respond with OK again. Type **ATMD** and press Return, and if you've set up the adapter in a loopback test mode (as in Chapter 3), you can type characters and see them echoed back to you. (See the following section for more information on echo tests with BlueSMiRF.) Figure 4-16 shows the results when using ZTerm.

That magic incantation changes the baud rate of the BlueSMiRF to 57,600 bps 8N1, which is what Roomba expects.

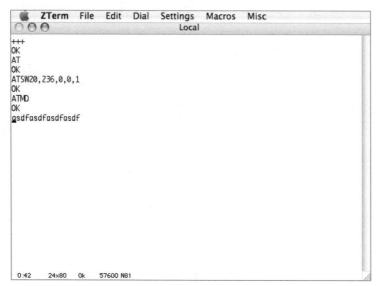

FIGURE 4-16: Changing BlueSMiRF baud rate to Roomba-compatible 57,600 8N1

Note The BlueSMiRF has many other AT commands besides the two shown here. It is a fairly capable device, suitable for many tasks. To see some of the other commands, download the BlueSMiRF datasheet at `www.sparkfun.com/datasheets/RF/BlueSMiRF_v1.pdf` or send an e-mail to SparkFun asking for the complete command set.

Testing Bluetooth

All the tests performed for the RS-232 serial adapter should be done with the Bluetooth adapter, too. Here's why: It enables you to get a feel for how the Bluetooth adapter responds compared to the serial cable. Because Bluetooth is a wireless protocol, it can suffer from interference, dropouts, and signal strength issues like any other wireless protocol. This is reflected in various pauses in data transfer between the computer and the Bluetooth device. Normally it's not really an issue, but it is something to be aware of.

The most noticeable difference between using a cable and using Bluetooth to control the Roomba is the initial connect time can be longer if the computer has to re-connect and re-pair with the BlueSMiRF. It can be several seconds if the Bluetooth adapter has been idle or turned off.

While you're doing echo tests, it is a good idea to walk around while typing, get a feel for the range you can achieve. Since the BlueSMiRF is a Class 1 Bluetooth device, it theoretically has a range of about 100 meters (about 300 feet). Imagine controlling your Roomba from down the street.

Tip If your connection tests repeatedly fail, go back to the serial tether and verify that it still works. If everything seems like it should work but you still cannot connect, try a different computer.

Using the Adapter

With the adapter fully tested with your computer, you can run RoombaCommTest or any of the other tests from the previous chapter, and they should work exactly the same. The beauty of the Bluetooth serial adapter is that it appears just like a serial port to your operating system. All software that can use RS-232 serial ports can also use these Bluetooth serial ports.

There is one caveat for Windows. In the RoombaComm software you'll find an option called hwhandshake that should be used when connecting over Bluetooth. This only applies to Windows and is used to get around an odd interaction between the serial library RoombaComm uses and the operating system. This will probably eventually be fixed. After you download RoombaComm, read the release notes to see if the hwhandshake situation has changed. The hwhandshake option is discussed further in the section "RXTX Serial Port Library."

Making RoombaComm

RoombaComm is a Java API library you'll be creating to talk to Roombas. It will contain a large number of functions to make talking to the Roomba easy. It will also have a bunch of higher-level commands to let you do hard things simply. It wraps the ROI protocol and allows you to program the Roomba without needing to know hexadecimal, bit manipulation, or all the details of the ROI. These details will be discussed in the next few chapters as you build up RoombaComm. To get started using it right away, you can download the latest version of RoombaComm from http://roombahacking.com/.

RoombaComm is open source, a continual work in progress by several people, and community support is appreciated. So far RoombaComm has been run successfully on Mac OS X, Windows 2000/XP, and Linux on desktop computers and embedded systems. If you like, you can add new functionality to RoombaComm to become part of the official release.

The basic RoombaComm idea was introduced in Chapter 2. In this chapter you'll begin actually writing RoombaComm by creating the foundation methods to communicate with any Roomba over a serial port (RS232 or Bluetooth).

Code Structure

Java requires all code to be in a package. The package is a namespace for all code with a similar concept. The Java package name for all RoombaComm code can be just roombacomm. The library then can consist of a base class called roombacomm.RoombaComm. In this base class you can put all the communication protocol-independent functions. You can then make subclasses of it, like roombacomm.RoombaCommSerial and roombacomm.RoombaCommTCPClient, to implement the additional functionality needed to talk to a Roomba over a serial cable or a network cable, respectively.

In the roombacomm.RoombaComm base class, you can translate the entire ROI specification into Java code. Such methods as startup(), control(), and drive() can map directly to the START, CONTROL, and DRIVE ROI commands. Then you write Java code as if they were ROI commands like this:

```
RoombaComm roombacomm = new RoombaComm();
roombacomm.startup();
roombacomm.control();
roombacomm.drive(0x8000,0x0200);
```

With even just a few ROI commands implemented and a communication-specific subclass, an infinite number of programs can be created to interact with Roomba in all sorts of fun ways. The RoombaCommTest GUI program from the previous chapter is one simple example. You can create many others. But you first must get Java to talk to serial ports.

RXTX Serial Port Library

Most of the projects here that use RoombaComm will use the RoombaCommSerial class to connect to the Roomba over a serial port. Java doesn't have a way to talk to serial ports so an external library is needed.

RXTX is open-source implementation of the Sun Microsystems Java Communications API (CommAPI). The CommAPI is used to talk to serial and parallel port devices. RXTX works on all the operating systems that the CommAPI does, as well as many others. The only difference in use between RXTX and the official Java CommAPI is the `import` statement at the top of the code. Otherwise the two function identically.

Note For more information about the Java Communication API, see `http://java.sun.com/products/javacomm/`. And to learn more about RXTX, visit `http://rxtx.org/`.

With the above code structure, you'll end up using the same boilerplate setup code at the beginning of any RoombaComm program:

```
String portname = "/dev/cu.KeySerial1";
RoombaCommSerial roombacomm = new RoombaCommSerial();
if( !roombacomm.connect(portname) ) {
  System.out.println("Couldn't connect to "+portname);
  System.exit(1);
}
roombacomm.startup();
roombacomm.control();
roombacomm.pause(100);
```

This chunk of code does several important things:

- Specifies a serial port to use by name
- Creates a new RoombaCommSerial object
- Attempts to connect to that serial port
- Sends the START and CONTROL ROI commands to the Roomba

Of these, the `connect()` method is the most important and most difficult. To create it you use the various aspects of the RXTX library to find the serial port by name, open it, and set the port parameters to match what the Roomba expects. It's shown along with the other methods used above in Listing 4-1.

The internals of `connect()` are based on what is a standard way of using RXTX to open serial ports. You might be inclined to pass in the serial port name as a string (such as `/dev/cu.KeySerial1`) and expect `connect()` to work, but RXTX uses more complex `CommPortIdentifier` objects and can't support a simple string-based lookup. So the first part of `connect()` deals with searching through the list of all ports to find one that is both a serial port and has the desired name. When the right port is found, the port is opened, and its communication parameters are set to what Roomba expects.

The main deviation from the standard way of opening serial ports with RXTX is the `if(waitForDSR) {}` clause after the port is opened. That clause loops for up to six seconds waiting for the DSR (Data Set Ready) line to be true. For some reason when using

Bluetooth adapters on Windows, opening a port will succeed, but trying to write to it will fail unless enough time has elapsed for the virtual serial port to be initialized. Waiting for DSR to be true seems to fix this problem. When writing your own code, be sure to include `roombacomm.waitForDSR=true;` before connecting because this will trigger the `waitForDSR` clause in `connect()`. In all example programs in this book, the command-line ones take a -hwhandshake flag, and the GUI ones have a h/w handshake check box to set `waitForDSR` to true. It's called h/w handshake for hardware handshake, which is what the DSR is part of. Some slower serial port systems or those with very fast data transfer use extra hardware signals like DSR to indicate when it's okay to send or receive data. On modern computers this handshaking is rarely needed. For now you can hack around the virtual serial port initialization problem by waiting for the DSR line. This will probably eventually be updated, and when it is you'll be able to simplify your code by removing the `waitForDSR` clause and related code.

Tip   When using Windows and Bluetooth adapters, use the -hwhandshake flag or select the h/w handshake check box to make RoombaComm deal with Windows virtual serial ports correctly.

Listing 4-1: Important Methods in RoombaComm and RoombaCommSerial

```
private boolean connect(String portid) {
  portname = portid;
  boolean success = false;
  try {
  Enumeration portList =
CommPortIdentifier.getPortIdentifiers();
    while(portList.hasMoreElements()) {
      CommPortIdentifier portId =
                  (CommPortIdentifier) portList.nextElement();
      if(portId.getPortType() == CommPortIdentifier.PORT_SERIAL)
{
        if (portId.getName().equals(portname)) {
          port = (SerialPort)portId.open("roomba serial",2000);
          input  = port.getInputStream();
          output = port.getOutputStream();

port.setSerialPortParams(rate,databits,stopbits,parity);
          port.addEventListener(this);
          port.notifyOnDataAvailable(true);
          logmsg("port "+portname+" opened successfully");
          if(waitForDSR) {
            int i=40;
            while( !port.isDSR() && i-- != 0) {
              logmsg("DSR not ready yet");
              pause(150); // 150*40 = 6 seconds
            }
            success = port.isDSR();
          } else {
```

Listing 4-1 *Continued*

```
                success = true;
              }
            }
          }
        }
      } catch(Exception e) {
        logmsg("connect failed: "+e);
        port = null;
        input = null;
        output = null;
      }
      return success;
    }
    public boolean send(int b) {   // will also cover char or byte
      try {
        output.write(b & 0xff);   // for good measure do the &
      } catch (Exception e) { // null pointer or serial port dead
        errorMessage("send", e);
        return false;
      }
      return true;
    }
    public void startup() {
      send(START);
    }
    public void control() {
      send(CONTROL);
    }
    public void pause(int millis) {
      try { Thread.sleep(millis); } catch(Exception e) { }
    }
    // Roomba ROI opcodes
    // these should all be bytes, but Java bytes are signed
    public static final int START   =  128;  // 0 bytes of data
    public static final int BAUD    =  129;  // 1 byte of data
    public static final int CONTROL =  130;  // 0
    // ...
```

The send() command takes the output stream obtained from connect and tries to send a byte out it. The startup() and control() commands use send() to send the appropriate ROI byte, using an easy-to-remember alias instead of the raw byte values. Finally, the pause() command implements a simple pause function via the standard Java mechanism of Thread .sleep().

The code in the listing puts the Roomba in safe mode ready for more ROI commands, but doesn't yet make the Roomba move. In Chapter 5, you'll add to RoombaComm enough to allow you to drive around the Roomba.

Summary

The BlueSMiRF adapter is an awesome hacker toy that you no doubt have ideas for using elsewhere in your house. Rumor is that the cost of Bluetooth serial adapters like BlueSMiRF will become extremely cheap soon. Having a Bluetooth device that acts like a simple serial port makes it a breeze to hack together gadgets that talk wirelessly to your computer. Using RXTX you've seen how serial port access from Java works and how it's really not that bad, especially if you've done serial port programming in C. Plus, coding in RXTX enables your serial port code to work on just about any operating system out there. Although the current state of RoombaComm doesn't enable you to do much with Roomba, you can no doubt see how to expand the code to incorporate more ROI commands.

With both a serial tether for debugging purposes and a wireless adapter for showing off, you can now develop custom Roomba software and use your computer to control your Roomba from all over the house.

Driving Roomba

A mobile robot doesn't get very far without some means of locomotion. The Roomba has its two drive wheels. These two wheels form a differential steering system. Differential steering is similar to the differential gearing used in a car, but instead of one source of power divided between two wheels with gearing, both wheels are powered independently.

By having both wheels powered, the Roomba can do things no differentially geared car could ever hope to do, like pivot in place and move in perfect circles. A differential steering system also poses some problems. The biggest one is driving in exactly a straight line. To drive straight, exactly the same power must be applied to both wheel motors, if not, the robot moves in a subtle arc.

The ROI command used to drive the motors hides much of these issues by presenting a simple DRIVE command that takes a velocity and radius, and the radius can be set to zero to drive in a straight line. Behind the scenes Roomba is actually reading its distance and angle sensors (odometry sensors) and compensating the power to each motor. If it did not do this, simply driving the Roomba from the computer would be much more of a hassle because one would have to be continually reading the odometers, calculating errors, and updating the motor outputs to compensate.

Since drive systems are so important, this chapter shows a taken-apart Roomba motor unit and discusses the gearing and other mechanisms. Because the following figures show a taken-apart Roomba, there is no need for you to take a screwdriver to your own Roomba. The whole point of the ROI (and this book) is to allow Roomba hacking without voiding your warranty. If, however, you don't mind having a potentially non-working Roomba at the end, I highly recommend taking one apart to see how everything all goes together.

in this chapter

☑ Roomba drive system internals

☑ ROI DRIVE command

☑ Grow RoombaComm

☑ Move Roomba with RoombaComm

The Roomba Motors and Drive Train

Figure 5-1 shows what the left Roomba drive motor unit looks like; the right unit is the same but mirror imaged. Immediately noticeable is the rather large cable bundle emanating from the motor unit. Only two wires of the cable bundle directly control the motor. These two wires plug into a high-current motor driver. The motor driver is controlled by the Roomba's microcontroller through a digital-to-analog (D/A) converter with approximately 10-bit resolution.

FIGURE 5-1: Roomba drive motor unit

The other four wires are used to sense the wheel's rotation. I discuss those wires and the sensors they go to in Chapter 6. Also visible in Figure 5-1 is the spring and pivot point (the X-shaped bit toward the bottom) that enables the spring-loaded aspect of the Roomba wheels. By making the wheels spring-loaded, the Roomba is always pressing down on the ground and getting positive traction, even if it's not entirely level (for example, when it drives over cables or transitions from a rug to a hard floor).

With the main cover unscrewed (see Figure 5-2), it's easy to see that the unit consists of an electric motor and a gearbox, joined by a rubber drive belt. The belt was chosen instead of gearing probably because a belt offers more resistance to dirt and offers a fail-safe ability to slip in case the wheel can't turn but the motor continues to run. The slotted wheel on top of the gearbox is part of the drive sensor. That it's after the belt means that it won't measure any of the slipping, which makes for more accurate odometry.

Inside the triple-sealed gearbox is the planetary (also called epicyclic) gearing system (see Figure 5-3). The shaft driven directly by the motor is the "sun" gear and the two sets of three "planet" gears connect to the wheel. The gearing on the inside of the gearbox is called the *annulus* and keeps the planet gears in place.

FIGURE 5-2: The Roomba drive motor unit with cover removed

FIGURE 5-3: Planetary gear system inside drive unit

A planetary gear system has several benefits. For a robotic vacuum cleaner, the two most important are that it is a sealed gear system (making it dirt resistant) and that it is very high torque (making it able to climb steep inclines or thick carpet). The particular gearing used here creates a 25:1 reduction. It takes 25 rotations of the main shaft driven by the motor for the wheels to make one rotation. Electric motors are efficient only at high speeds, and even then they do not have very much torque. A planetary gear system trades off high speed for high torque. Such gearing is found in many electric drive systems, perhaps the most famous recently is in the Toyota Prius.

The ROI DRIVE Command

In Chapter 2 and in the ROI specification, you might have noticed that the DRIVE command takes a speed value in millimeters per second (mm/s) instead of something more direct like an arbitrary motor speed value that may be familiar to you if you have programmed robots before. This is because the DRIVE command triggers a pretty complex algorithm in the Roomba firmware that takes into account the input from the drive motor sensors, performs math on the sensor data, and feeds that data back into the routine that adjusts the voltage going to the motors to get a consistent speed. All this work is done for you and all you have to do is specify a speed in units you're familiar with (in this case millimeters).

Recall from Chapter 2 that the DRIVE command is five bytes long: the command byte of 137 (0x89) followed by two bytes for velocity and two bytes for radius.

Velocity

The velocity value is specified in millimeters per second and describes the averaged velocity of the two drive wheels: ((Vleft + Vright)/2). From that equation you can see that a positive velocity makes the Roomba go forward and a negative velocity makes it go backward. But since this is an average and Roomba, like all real machines, takes time to come up to a speed and slow down from a speed, any command will result in positional error, and series of commands with rapid starts and stops will accumulate position errors to an even greater degree. This is the nature of using a single value to specify what is at least a three-value problem: acceleration time, velocity, and deceleration time. In practice the error is minimized if the speed is kept below 200 mm/s (as always, boring is safe). You can also programmatically make Roomba speed up and slow down to and from cruising speed.

Tip When moving straight, to make Roomba go a particular distance, move it at a particular velocity for a particular amount of time. For example, going 200 mm/s for two seconds should result in Roomba going 400 mm. In practice it's not that exact, but you get the idea.

Note Although the minimum velocity is +/- 1 mm/s, in actuality it appears Roomba only responds to values of at least +/- 10 mm/s.

Radius

The second parameter is the turn radius value, which is specified in millimeters. This radius describes a circle upon whose circumference the Roomba will drive. Driving in arcs enables Roomba to make easy and efficient turns around many things like wall corners, table legs, and so on.

Positive radius values turn the Roomba toward the left, and negative radius values turn it toward the right. Postive velocity values turn Roomba in a forward direction, and negative velocity values have it back up as it turns. Depending on the sign of the velocity and radius values, there are four different ways the Roomba can move in a curve (see Figure 5-4).

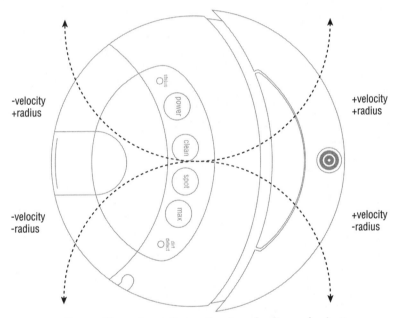

FIGURE 5-4: Types of turns depending on the sign of radius and velocity

The turn radius always makes the Roomba travel along a circle. The only exceptions to this are the three special case radius values of: straight (32768, which is also known in hexadecimal as 0x8000), spin left (1, which is also known as 0x0001), spin right (-1, which is also known as 0xffff). If you've done a little homebrew robotics where you directly control each wheel, this method of dealing with driving two motors in a robot seems a little strange. A common first task in homebrew robotics is to have direct control of each wheel, allowing you to drive only one wheel at a time. How is it possible to drive the Roomba in this way using the velocity and radius values of the DRIVE command?

Converting Radius/Velocity to Left/Right Speeds

The Roomba's two-wheel differential steering is common for robots. If you want to make it move in nice curves instead of the simple "drive straight, pivot, drive straight" method, you have to do some math to convert the curves to speeds for the left and right motors. Luckily the ROI does this for you, but it's useful to know how it does it in case you want to do some special-case movements.

Roomba has a wheelbase diameter (the distance between the centerline of its two drive wheels) of 258 mm. The diameter of the Roomba itself is around 340 mm (see Figure 5-5). All turning motion is based off of the 258 mm wheelbase. (The front caster wheel is not used for turning; it's a free spinning wheel that skids during non-straight turns.)

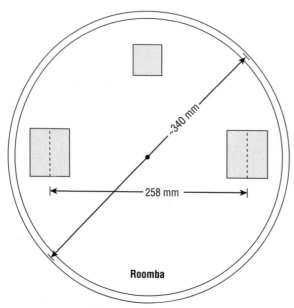

FIGURE 5-5: Schematic of important drive distances

The special-case move of turning with only one drive wheel moving can be accomplished by setting the DRIVE radius to one-half the wheelbase diameter, or 129 mm. This seems intuitively obvious, but the description needs a diagram and a little math. Figure 5-6 shows a differential steering system with wheelbase b turning through an arc with angle θ and radius r.

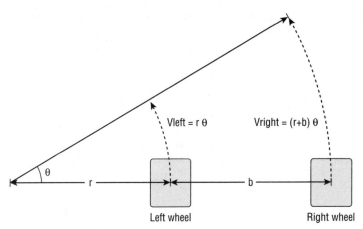

FIGURE 5-6: Deriving left/right speeds

From basic geometry, the speeds of the left wheel, the right wheel, and the average speed are:

```
Vleft = rθ
Vright = (r+b)θ
V = (Vleft + Vright) / 2
```

From this, it's possible to derive a relation to the radius and wheelbase. This derivation is a bit more complex, but the result is:

```
r = b(Vleft+Vright)/(2(Vright-Vleft))
```

And with two equations and two unknowns, you can solve for Vleft and Vright to get:

```
Vleft = V(1-b/2r)
Vright = V(1+b/2r)
```

The original question is really, "when does Vleft=0 or Vright=0?" That happens when r=±1/2b. Since you know b=258, if r=129 then Vleft=0, and the Roomba spins on its left wheel. If r=-129 then Vright=0, and the Roomba spins on its right wheel. See? The math you learned in high school is actually useful for something.

Note If your geometry skills are a little rusty, the θ character is called the Theta and is a measurement of the angle between two lines. A good resource for brushing up on your math skills (not just geometry but all math) can be found at `http://mathforum.org/`.

If you have previous code that expects to directly control the motors, you can solve the equations for V and r and use them in the DRIVE command.

Tip

You will notice some differences in actual Roomba movements compared to the commands issued. Mostly this is due to variable friction with the environment. If you want more accurate movements, remove the bin and brushes and operate the Roomba on a low profile but high-traction surface like office carpeting. Roomba is designed to vacuum home carpets, but the variability of home carpets can mess up trying to accurately move the robot.

Table 5-1 summarizes some common DRIVE commands. The byte sequence column is the exact set of hex bytes that make up a particular command.

Table 5-1   Some Common and Special-Case DRIVE Commands

Description	Velocity	Radius	Byte sequence
Straight forward, medium speed	250	32768	0x89,0x00,0xfa,0x80,0x00
Straight backward, slowly	-50	32768	0x89,0xff,0xce,0x80,0x00
Spin left in place	500	1	0x89,0x01,0xf4,0x00,0x01
Spin right in place	500	-1	0x89,0x01,0xf4,0xff,0xff
Turn left forward (left wheel stopped)	200	129	0x89,0x00,0xc8,0x00,0x81
Turn right forward (right wheel stopped)	200	-129	0x89,0x00,0xc8,0xff,0x7f

A Word about Writing Java Programs

Writing, compiling, and running Java programs is pretty easy, but it can be a little daunting if you've never done it before. Java is a good language for learning the basics of object-oriented programming. If you're totally unfamiliar with Java, visit the main Java website and check out some of the tutorials at http://java.sun.com/learning/tutorial/.

There are many ways of going about working with Java, from simple command-line commands to large integrated development environments (IDEs) to professional Unix-build script systems. This book assumes that you'll be building and running code from the command line (Terminal.app in Mac OS X, CMD.EXE in Windows, and xterm in Linux). When you have Java installed and have downloaded and unzipped the RoombaComm code, the steps are:

1. Write code. Create or edit the source code files. Go into the roombacomm directory and edit roombacomm/Drive.java. Make any changes you want or just poke around.

2. Compile code. For simple Java programs, just typing **javac Drive.java** would work. RoombaComm uses an additional Java library called RXTX to allow serial port access, so the compile line becomes a little more complicated:

```
% javac -classpath .:rxtxlib/RXTXcomm.jar
roombacomm/*.java
```

A Word about Writing Java Programs *Continued*

3. Run the code. For simple Java programs, typing **java Drive** would work. But again, because of the additional library, the command becomes:

```
% java -Djava.library.path=rxtxlib
        -classpath .:rxtxlib/RXTXcomm.jar roombacomm.Drive
```

Everyone has his or her own preferred style for writing software. For this project, I made simple scripts that wrap up the preceding commands to save typing and reduce typos. There is `makeit.sh`, which encapsulates Step 2, and `runit.sh`, which encapsulates Step 3. My workflow then looks something like:

```
% unzip roombacomm-latest.zip   // downloaded from
roombahacking.com
% cd roombacomm                 // get into the directory
% emacs roombacomm/Drive.java   // edit the file
% ./makeit.sh                   // build it
% ./runit.sh roombacomm.Drive   // run it
```

The scripts will work on any Unix-like system (Mac OS X, Linux, Windows with Cygwin). For vanilla Windows, you can use `makeit.bat` and `runit.bat`, which work the same way.

Command-Line Roomba Driving

The example program `roombacomm.Drive` enables you to play with the DRIVE command by itself. Like all the example programs, run it with no arguments to get usage:

```
% ./runit.sh roombacomm.Drive
Usage:
  roombacomm.Drive <serialportname> <velocity> <radius> <waittime>
```

You should know the serial port name from the previous chapters. The `velocity` and `radius` values are direct arguments to DRIVE, as signed integers, and `waittime` is the number of milliseconds to run before stopping. To go straight forward at 250 mm/s for two seconds, run the program:

```
% ./runit.sh roombacomm.Drive /dev/cu.KeySerial1 250 32768 3000
```

There are many other simple example programs that are all run in the same manner (but with different arguments after the serial port name).

The following sections show the evolution of the RoombaComm API and how the Drive program is created.

Simple Tank-Like Motion

Roomba is capable of some fairly complex moves. The most useful (and simplest) are those like that of a toy tank: forward, back, spin left, and spin right. These moves aren't the most elegant that Roomba is capable of, but they certainly do the job.

Moving with the send() Command

All movement commands are implemented by sending byte sequences. RoombaComm provides a direct way of sending DRIVE commands to the Roomba via the low-level send() method. This method takes either a single byte or an array of bytes. All more advanced commands ultimately use send(). The more advanced commands are what will normally be used, but sometimes they cannot do what you want. So instead you can pop down to the lowest level and do it by hand.

The most basic command sends a single byte, say for instance the CLEAN command:

```
roombacomm.send((byte)RoombaComm.CLEAN);
```

The byte is sent immediately; no queuing is performed in the RoombaComm code (although there may be OS-level queuing). The casting to byte is required because Java has no way to do unsigned bytes. That is perhaps the most frustrating thing about Java when using it with an embedded computer system like a robot.

 Note In the code samples, remember from the end of Chapters 3 and 4 that roombacomm is a roombacomm.RoombaComm object that's been initialized and connected to a serial port and Roomba. All the functions described in the next few chapters are in RoombaComm (or the subclass RoombaCommSerial for serial port specific functions).

The more commonly used variant of send() takes an array of bytes. For example, to send the byte sequence that starts the Roomba spinning in place to the right:

```
byte cmd[] = {(byte)RoombaComm.DRIVE,(byte)0x01,(byte)0xf4
                              (byte)0xff,(byte)0xff};
roombacomm.send(cmd);
```

Repeating the above to do a large sequence of movements would get really tiresome, so you will create a better way.

 Note The constants CLEAN, DRIVE, and many others are in the RoombaComm class to assist with these kinds of tasks. Look in roombacomm/RoombaComm.java or read the Javadocs to see them all.

Moving with the drive() and ...At() Commands

A first step would be to encapsulate the above DRIVE command into a method like in Listing 5-1.

Listing 5-1: RoombaComm.drive()

```
public void drive(int velocity, int radius) {
  byte cmd[] = {(byte)DRIVE,
                (byte)(velocity>>>8),(byte)(velocity&0xff),
                (byte)(radius  >>>8),(byte)(radius  &0xff)};
  send(cmd);
}
```

Using it to replace the DRIVE command created before results in the following:

```
roombacomm.drive(500,-1);
```

This command is much easier to read and understand.

The internals of the drive() method may look a little strange if you are not familiar with bit-manipulation. The situation is this: There's a 16-bit number (say, velocity) and you need to break it up into two 8-bit bytes. In Java, as in most computer languages, the high byte (the most significant byte) of a two-byte value is thought of as being on the left and the low byte (the least significant byte) is thought of as being on the right. The statement velocity >>> 8 says "Shift velocity's high byte to the right 8 bits" which results in the low byte being discarded, moves the high byte into the low byte's place, and fills the high byte with zero. Any 16-bit number's value without a high byte is the value of the low byte.

Knowing how to do bit manipulations is key to programming with robotics and microcontrollers. It's easy to forget exactly how to form proper bit-manipulation code, so the first thing you do is create a function to remember it for you, which is what drive() does.

The drive() command is what the Drive example program is based around. In fact, the entirety of Drive, except the parsing of the command-line arguments, is in Listing 5-2. Controlling Roomba is getting easier.

Listing 5-2: A Condensed Drive.java

```
String portname = "/dev/cu.KeySerial1";
int velocity = 500, radius = -1, waittime = 2000;
RoombaCommSerial roombacomm = new RoombaCommSerial();
if( !roombacomm.connect(portname) )
  System.exit(1);
roombacomm.startup();
roombacomm.control();
roombacomm.pause(100);
roombacomm.drive(velocity, radius);
roombacomm.pause(waittime);
roombacomm.stop();
roombacomm.disconnect();
```

Once you have `drive()`, building new commands from it comes naturally:

```
public void spinLeftAt(int aspeed) {
  drive(aspeed, 1);
}
public void spinRightAt(int aspeed) {
  drive(aspeed,-1);
}
public void goStraightAt(int velocity) {
  drive(velocity,0x8000);
}
```

These three methods abstract away the three magic values of `radius`. In `RoombaComm` the convection of ending a method with `At` indicates its first argument is a speed or velocity. And the example `DRIVE` command now simply becomes:

```
roombacomm.spinRightAt( 500 );
```

Tip To make Roomba stop at any time, issue the `roombacomm.stop()` command or physically lift up the Roomba to trigger its safety feature.

Moving Specific Distances

Due to how the ROI works, all of these commands start the action going but do not say when it stops. After issuing the `spinRightAt()` command in the preceding section, Roomba begins to spin around and around and nothing stops it, even if you exit the program. If you want to go straight for a certain distance or spin left a given angle, you issue the command at a particular speed and then wait for a known time. For example, to go forward 300 mm at 100 mm/s, wait three seconds and stop. The following three methods generalize that concept, for generally moving in straight lines and going forward or backward:

```
public void goStraight( int distance ) {
  float pausetime = Math.abs(distance/speed); // mm/(mm/sec) = sec
  goStraightAt( speed );
  pause( (int)(pausetime*1000) );
  stop();
}
public void goForward( int distance ) {
  if( distance<0 ) return;
  goStraight( distance );
}
public. void goBackward( int distance ) {
  if( distance<0 ) return;
  goStraight( -distance );
}
```

The `goStraight()` method computes how long to wait based on how fast and how far the Roomba is supposed to go. It starts the Roomba up at a speed, waits for the right amount of time, and then stops. What is this speed value though? That's easy. Define a few more methods.

```
public void setSpeed( int s ) { speed = Math.abs(s); }
public int  getSpeed() { return speed; }
```

Now each Roomba has an internal speed that the distance-based At commands will work with. You don't often change the speed of the Roomba, so specifying every function is wasteful.

In use, the preceding new methods would be used like this:

```
roombacomm.setSpeed(500);    // 500 mm/sec, fastest possible
roombacomm.goForward(1000);  // forward 1 meter = ~39 inches
roombacomm.goBackward(254);  // backward 254 mm = ~10 inches
```

Rotating Specific Angles

So now you can move Roomba forward and backward specific distances. What about turns? In order to determine how much the Roomba has rotated through a spinLeftAt() and spinRightAt() command, you need to remember that when spinning in place, the velocity parameter to the DRIVE command becomes the velocity of the wheels along the circumference of a circle. That velocity is in mm/s and you need degrees/s. To convert it, you need to figure out how many millimeters per degree in the circle the Roomba is moving around:

```
mm/degree = (258 mm x pi) / 360 = 2.2515 mm/degree
```

Then to get angular speed from Roomba speed in mm/s, you use this formula:

```
angular_speed in degrees/s = speed in mm/s / (mm/degree)
```

To figure out how long to wait to spin a specific number of degrees, you flip the equation around to solve for seconds for a given Roomba speed and angle you want:

```
wait time = (mm/degree) x (angle/speed)
```

For example, to do a 180 degree spin at a speed of 300 mm/s, you spin for (2.2515) × (180/300) = 1.35 seconds.

In Java, these equations are implemented for spinRight() and spinLeft() as in Listing 5-3.

Listing 5-3: RoombaComm.spinRight() and spinLeft()

```
int wheelbase = 258;
float millimetersPerDegree = (float)(wheelbase * Math.PI /
360.0);

public void spinRight( int angle ) {
   if( angle < 0 ) return;
   float pausetime = Math.abs(millimetersPerDegree *
angle/speed);
   spinRightAt( Math.abs(speed) );
```

Continued

Listing 5-3 *Continued*

```
    pause( (int)(pausetime*1000) );
    stop();
}
public void spinLeft( int angle ) {
    if( angle < 0 ) return;
    float pausetime = Math.abs(millimetersPerDegree *
angle/speed);
    spinLeftAt( Math.abs(speed) );
    pause( (int)(pausetime*1000) );
    stop();
}
```

Pivoting around a particular angle looks like this:

```
roombacomm.spinLeft(90);      // turn anti-clockwise 90 degrees
roombacomm.spinRight(360);    // spin totally around
```

Moving in Curves

The straight motion and pivot turns are sufficient to move around, but arc turns are more efficient when dealing with obstacles. Watch the Roomba while it operates and you'll notice it performs three basic types of non-pivot turns (see Figure 5-7):

- **Small circle:** The Roomba moves in a circle with a radius a little bigger than itself. This is performed when the Roomba bumps against something (usually a table leg) and tries to drive around it.

- **Waggle:** The Roomba moves forward by moving in alternating arcs with radii larger than the Roomba. This is performed when the Roomba is searching for something (dirt, home base, and so on).

- **Expanding spiral:** The Roomba moves in a circle that gets bigger and bigger, accomplished by gradually increasing the radius over time. This is performed when the Roomba thinks it is in a large open area of floor.

The small circle is pretty evident now. You can implement it in RoombaComm with something like:

```
roombacomm.drive(200,300); // 250 mm/s, turn left on 300 mm radius
```

You can test that out with the `roombacomm.Drive` program.

The waggle is just the same, but with a larger radius, approximately 600, and it switches every quarter second or so. To emulate it, you could do something in a loop for however long you want, like Listing 5-4.

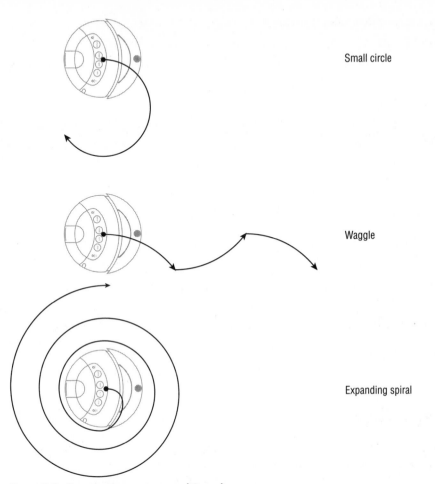

Small circle

Waggle

Expanding spiral

FIGURE 5-7: Common preprogrammed Roomba moves

Listing 5-4: The Main Loop of Waggle.java

```java
while( !done ) {
  roombacomm.drive(200,600);
  roombacomm.pause(250);
  roombacomm.drive(200,-600);
  roombacomm.pause(250);
}
```

You can test this out with the `roombacomm.Waggle` test program.

By far the most interesting curve the Roomba does is the expanding spiral, usually when it first starts up. You can implement a spiral by keeping a fixed velocity and incrementing the radius value over time. In the example program Spiral.java, you can see one way to go about this. Listing 5-5 shows the important parts of Spiral.java.

Listing 5-5: Important Parts of Spiral.java

```
int radius = 10;
int radius_inc = 20;
int pausetime = 500;
int speed = 250;
while( !done ) {
  roombacomm.drive(speed,radius);
  radius += radius_inc;
  roombacomm.pause(pausetime);
}
```

An initial `radius` is defined that is then incremented by `radius_inc` amount every `pausetime` milliseconds. Try modifying Spiral.java to choose different values for those three numbers. With the numbers in the listing, Roomba spirals out, but if you choose a large initial `radius` and a negative `radius_inc`, the Roomba would spiral inward.

Real-Time Driving

The problem with the distance-based methods is the `pause()` method in them that causes your program to wait (or block). Although it is paused, it can't do anything else, such as check for user input or read the Roomba's sensors. A refinement you could add to RoombaComm would be to implement a way to drive a certain distance without blocking the code. One way to do this might be to put all Roomba control code in one thread and sensor reading or user-interface checking in another thread.

However, if you want to drive your Roomba in real time, you don't necessarily need to drive specific distances or angles. By using the simplest methods like `goForward()`, `spinLeft()`, and `spinRight()`, coupled with `stop()`, you can create a program to drive the Roomba from your computer.

In Java the easiest way to respond to single key presses is to write a GUI program. Java makes it quite easy to build GUI programs with buttons and windows and such. The toolkit one generally uses to make GUI Java programs is called Swing and is part of Java. Normally these Swing GUI elements handle mouse clicks and key presses for you, but you can override that functionality and have your own program listen to key presses. This is done by making your code's class implement a `KeyListener` and having a method called `keyPressed()`.

Listing 5-6 shows the DriveRealTime program. It is launched from the command line as usual:

```
% ./runit.sh roombacomm.DriveRealTime /dev/cu.KeySerial1
```

Figure 5-8 shows what it looks like after a few seconds of use.

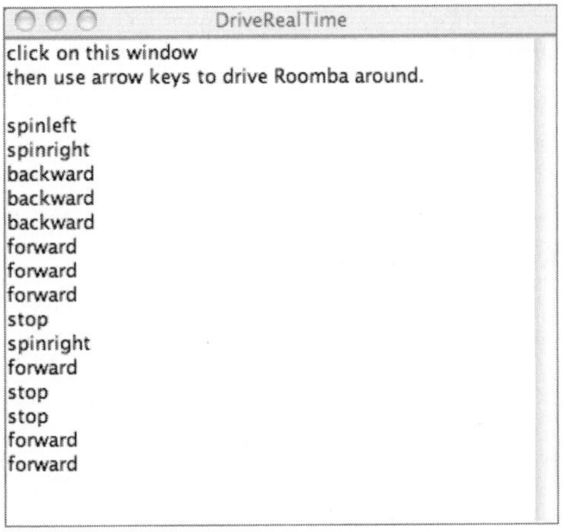

```
○ ○ ○          DriveRealTime
click on this window
then use arrow keys to drive Roomba around.

spinleft
spinright
backward
backward
backward
forward
forward
forward
stop
spinright
forward
stop
stop
forward
forward
```

FIGURE 5-8: DriveRealTime program in use

DriveRealTime subclasses a `JFrame` (an application window in Java Swing). The `setupWindow()` method holds all the things necessary to make the window work properly. Most important is `addKeyListener(this)`, which will make the `keyPressed()` method of `DriveRealTime` get called whenever you press a key. The last thing `setupWindow()` does is show the window with `setVisible(true)`. To make the program a little more visually interesting, a JTextArea is created and whenever you press a key, the text area is updated with the command you pressed through `updateDisplay()`.

Listing 5-6: DriveRealTime.java

```java
public class DriveRealTime extends JFrame implements KeyListener {
  RoombaCommSerial roombacomm;
  JTextArea displayText;
  public static void main(String[] args) {
    new DriveRealTime(args);
  }
  public DriveRealTime(String[] args) {
    super("DriveRealTime");
    String portname = args[0];
    roombacomm = new RoombaCommSerial();
```

Continued

Listing 5-6 *Continued*

```java
    if( !roombacomm.connect(portname) ) {
      System.out.println("Couldn't connect to "+portname);
      System.exit(1);
    }
    roombacomm.startup();
    roombacomm.control();
    roombacomm.pause(50);
    setupWindow();
  }
  public void keyPressed(KeyEvent e) {
    int keyCode = e.getKeyCode();
    if( keyCode == KeyEvent.VK_SPACE ) {
      updateDisplay("stop");
      roombacomm.stop();
    }
    else if( keyCode == KeyEvent.VK_UP ) {
      updateDisplay("forward");
      roombacomm.goForward();
    }
    else if( keyCode == KeyEvent.VK_DOWN ) {
      updateDisplay("backward");
      roombacomm.goBackward();
    }
    else if( keyCode == KeyEvent.VK_LEFT ) {
      updateDisplay("spin left");
      roombacomm.spinLeft();
    }
    else if( keyCode == KeyEvent.VK_RIGHT ) {
      updateDisplay("spin right");
      roombacomm.spinRight();
    }
  }
  public void updateDisplay( String s ) {
    displayText.append( s+"\n" );
  }
  public void setupWindow() {
    displayText = new JTextArea(20,30);
    displayText.setLineWrap(true);
    displayText.setEditable(false);
    displayText.addKeyListener(this);
    JScrollPane scrollPane = new JScrollPane(displayText,
                    JScrollPane.VERTICAL_SCROLLBAR_ALWAYS,
                    JScrollPane.HORIZONTAL_SCROLLBAR_NEVER);
    Container content = getContentPane();
    content.add(scrollPane, BorderLayout.CENTER);
    addKeyListener(this);
```

Listing 5-6 Continued

```
    pack();
    setResizable(false);
    setVisible(true);
  }
 }
```

Writing Logo-Like Programs

In the 1980s a computer programming language called Logo became popular, targeted toward kids or others with little computer experience. It was famous for *turtle graphics*, a method of drawing images on the screen by moving a virtual *turtle* around that had a pen strapped to it. Although anyone who knew Logo knew about this scribbling turtle, Logo originally didn't have it when it was invented in the late 1960s. The turtle idea came about afterward from a real radio-controlled floor roaming robot named Irving that had touch sensors and could move backward and forward and rotate left and right. Sound familiar?

Logo programs could be very simple, like this one to draw a square:

```
FORWARD 100
LEFT 90
FORWARD 100
LEFT 90
FORWARD 100
LEFT 90
FORWARD 100
LEFT 90
```

When the program was run, it would output an image like Figure 5-9.

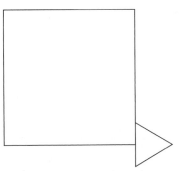

FIGURE 5-9: Logo output for the square program

One could even define functions to encapsulate behavior. The square above could be written as:

```
TO SQUARE
REPEAT 4[FORWARD 100 LEFT 90]
END
```

These RoombaComm commands to go specific distances and rotate specific angles mirror Logo's commands. You can create a Java version of the Logo square using RoombaComm like so:

```
public void square() {
  for( int i=0; i<4; i++ ) {
    roombacomm.goForward(100);
    roombacomm.spinLeft(90);
  }
}
```

Not as simple as Logo perhaps but still compact and understandable. And of course, like the original turtle Irving, the Java code controls a real robot roaming around on the floor.

Summary

Making the Roomba move around is the most important and most fun part of controlling it as a robot. From sending raw command bytes to high-level drive commands, you can now make Roomba move along any conceivable path. The ROI protocol's abstraction away from commanding each motor means you can easily move the Roomba in graceful circular curves, but with the relevant equations you can subvert that and control each drive wheel independently. You can even control the Roomba in real time with the cursor keys on a keyboard and can now make any even your computer control the Roomba.

The idea of encapsulating paths as functions in Logo is a powerful one, and you can start building functions to draw all sorts of shapes with Roomba. Understanding how to implement some of the existing paths performed by Roomba is the basis for creating alternative moves for it that will work better in your environment.

Reading the Roomba Sensors

in this chapter

☑ Send and receive sensor details

☑ Master the ROI SENSORS command

☑ Parse sensor data

☑ Make Roomba autonomous

☑ Measure distance and angles

☑ Spy on your Roomba

A robot without sensors is merely a fancy machine. Some robots use humans as their sensors (and brain). If you've ever watched Battle Bots or similar programs, you've seen what are really telepresence robots and not robots in the purer sense of the term.

Human senses have a huge range of fidelity. The ear and eye can detect information through a wide range of the electromagnetic spectrum. The sense of touch can detect heat and cold, variations in pressure, and damage. The vestibular system provides a high-resolution sense of balance and orientation. By comparison, a robot's sensors are typically very simple. They detect a single, very specific kind of information: a touch sensor that triggers a switch, an "eye" that detects only a single color of light and only one pixel, or an "ear" that can only hear on frequency of sound.

Currently, creating sensors that mimic the dynamic range of human senses is expensive and computationally complex. As we have learned from the simpler life forms, sensors don't need to be complex or high-fidelity to be useful. And as the subsumption architecture style of AI has shown, any interaction with the environment is better than none.

The Roomba has a wide variety of sensors for a robot that is so inexpensive and so single purpose. Roomba has touch sensors, rudimentary vision, and an internal sense of orientation, just like the people it works for, but you'll find its version of those senses are both unique to it and much simplified. And unlike people, you'll see it has a sense for dirt that could be best described as licking the carpet.

Roomba Sensors

Figure 6-1 shows the location of all the physical sensors present in Roomba. The entire front bumper is bristling with sensors. It is the Roomba's "head" in more than one way. In normal operation, the Roomba always moves forward, so it can observe the world.

Turning over the Roomba, as in Figure 6-2, shows what some of these sensors look like. Actually, there's not much to see. There are no probes or whiskers or obvious sensors besides the fact that the bumper moves. The two silver squares near the front caster are the charging plates for the recharge dock. With the brushes removed, you can see the two silver discs that compose the dirt sensor. By just looking at the Roomba, it's hard to tell exactly how it manages to detect cliffs or walls.

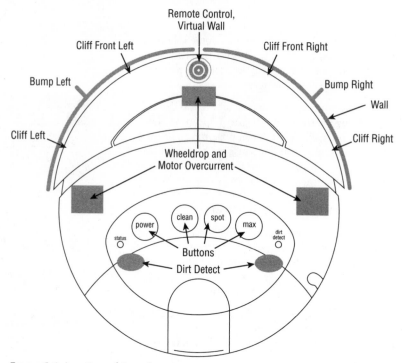

FIGURE 6-1: Location of Roomba sensors

Almost all of the physical sensors in the Roomba are implemented as an optical emitter/detector pair. Even sensors that could be switches like the bumpers are implemented optically. Optical sensors have the advantage of being high-wear due to less physical interaction. They are usually more complex (and thus expensive) to implement.

Figure 6-3 shows three of the most common examples of emitter/detector configurations.

FIGURE 6-2: Underside of Roomba, showing sensors

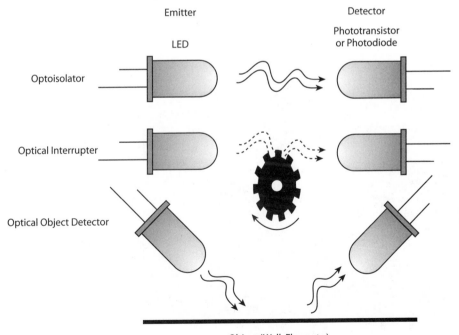

FIGURE 6-3: Common optical emitter/detector configurations

Optoisolator: Remote Controls, Virtual Wall, and Dock

The emitter in an optoisolator is usually an infrared LED and the detector is a matching infrared phototransistor or photodiode. The first type shown in Figure 6-3 is an optoisolator. The light from the LED travels unobstructed to the detector. Optoisolators are used to separate circuits that would be too dangerous or too physically far apart to connect electrically. Common optoisolator uses are motor driver circuits (to separate high-power circuits from low-power circuits) and remote controls for TVs and stereos. You could also consider the high-speed fiber optic links that drive the Internet as a type of optoisolator. In the case of Roomba, the clear circular bump on the front top is the detector half of an optoisolator. It receives remote control commands and "sees" the virtual wall and docking base.

Optical Interrupter: Wheel Speed and Bump Detection

The optical interrupter configuration is the next type shown in Figure 6-3. It adds a physical barrier that is inserted between the emitter and detector to indicate some event. The Roomba's motors have a toothed disc (shown in Figure 5-2) that rotates between an emitter/detector pair. Each pulse of light received by the detector indicates some number of degrees of rotation. For the Roomba's bump sensors, a plastic tab is attached to the bumper and when the bumper is pushed in, the tab interrupts the beam of light between the LED and the detector. Figure 6-4 shows what the bump sensor actually looks like. It's the black plastic box attached directly to the main green circuit board (oriented vertically with wires attaching to it in Figure 6-4). The emitter and detector are oriented vertically, with a horizontal black plastic tab as the interrupter. The black tab is part of the whole black bumper and moves when it does. You can just see the spring on the left (the shiny coil of steel) that pushes the bumper back out.

FIGURE 6-4: Bump sensor, an optical interrupter

Optical Object Detector: Cliff and Wall Detection

The third configuration rotates the emitter and detector so they don't directly see each other and instead point toward something that may be there. If it is there, the LED's light is reflected by it and the detector sees it. This is how the Roomba's cliff and wall sensors are implemented. Figure 6-5 shows the wall sensor on the right side of the Roomba. It is two wells in the bumper: the long deep one on the right houses the LED and the small one on the left houses the photodetector.

FIGURE 6-5: Wall sensors, an optical object detector

Tip Although the optical sensors all use infrared light, some digital cameras can see that light. Turn your Roomba over when it's on and point your camera at it. Through the electronic viewfinder you should see the glow of LEDs.

Micro-Switches: Wheeldrop and Buttons

The wheeldrop and button sensors are implemented with standard micro-switches. They get much less use, so a mechanical switch was deemed sufficient by iRobot engineers. Figure 6-6 shows what one of these micro-switches looks like. This particular one is the wheeldrop sensor for the left wheel. The switch is the small rectanglar box glued to a screwed-down mount. It has a black tip with two light-colored wires coming from it. The tip is what moves to trigger the switch. The wheel unit pushes against it when the wheel is up, and stops pushing it when the wheel drops. The black disc behind the wheeldrop switch with dark wires coming from it is the Roomba's speaker.

Other Sensor Implementations

Figure 6-7 shows the internals of the Roomba. The two large silver boxes located right below the vacuum brush motor are the dirt sensors. Inside each of those magic boxes is a dense analog circuit that implements something similar to those capacitive touch sensors on table lamps. The motor over-current sensor is presumably determined by a current sense somewhere in the motor driver circuit.

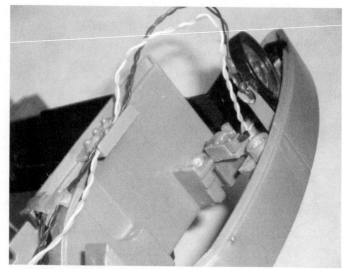

FIGURE 6-6: Wheeldrop sensor, a microswitch

FIGURE 6-7: Roomba internals

ROI SENSORS Command

The ROI presents the sensor data in a lightly processed form available through the SENSORS command. This command can retrieve subsets of the entire sensor payload, but for all the code presented here, you will fetch the entire set.

Sending the SENSORS Command

Sending the SENSORS command is easy; it's the receiving of the data and parsing it that can be problematic. To send the SENSORS command that retrieves all of the sensor data using RoombaComm, any of the following are valid:

```
// using send() and literal bytes in an array
byte cmd = {(byte)0x8e,(byte)0x00};
roombacomm.send(cmd);

// using two sends and RoombaComm defines
roombacomm.send((byte)RoombaComm.SENSORS);
roombacomm.send((byte)RoombaComm.SENSORS_ALL);

// using high-level command
roombacomm.sensors(RoombaComm.SENSORS_ALL);

// using high-level command default
roombacomm.sensors();
```

The last one is what's normally used. When this command is sent, Roomba responds by sending the sensor packet in usually less than 100 milliseconds. The actual time it takes depends on the serial interface used (Bluetooth adapters have a higher latency at times) and other factors that aren't entirely clear.

Receiving the Sensor Data

The SENSORS command is the only one that returns data. All the other commands are one-way. Receiving data can be tricky because the only way to know if you've read all the data is to wait for all the bytes to come back. An initial implementation of getting sensor data might be (in pseudo-code):

```
send(SENSORS, SENSORS_ALL);
byte data[] = receive(26); // waits for 26 bytes of data
```

Unfortunately if this hypothetical receive() function never gets all 26 bytes, then it either waits forever or times out. Even if it works every time, every time you get data, you block the rest of your program until all the data is obtained.

A common way to deal with this problem in Java is to spawn another thread and tell it to go deal with the waiting around. It performs its task and then notifies the main code, returning the result. This callback or event pattern is used a lot when dealing with the real world. In Java, you'll use the RXTX library's ability to call back to you when the Roomba sends data.

Receiving Serial Events

To get serial events, you first have to tell the RXTX SerialPort object that you want to listen to events. The open_port() method in RoombaCommSerial shows how this is done. The important parts are:

```
port = (SerialPort)portId.open("roomba serial", 2000);
port.addEventListener(this);
port.notifyOnDataAvailable(true);
```

That code snippet gets the serial port object, adds RoombaCommSerial as an event listener, and then specifies the kind of event you care about. You'll find several types of events. The most useful for RoombaComm (and the most reliably implemented in RXTX) is the event that tells you when serial input data is available.

The serialEvent() method that will be called when data is available looks like Listing 6-1.

Listing 6-1: RoombaCommSerial.serialEvent()

```
byte[] sensor_bytes = new byte[26];
byte buffer[] = new byte[26];
boolean sensorsValid = false;
int bufferLast;

synchronized public void serialEvent(SerialPortEvent
serialEvent) {
  try {
    if( serialEvent.getEventType() ==
SerialPortEvent.DATA_AVAILABLE ) {
      while( input.available()>0 ) {
        buffer[bufferLast++] = (byte)input.read();
        if( bufferLast==26 ) {
          bufferLast = 0;
          System.arraycopy(buffer,0,sensor_bytes,0,26);
          sensorsValid = true;
          computeSensors();
        }
      } // while
    }
  } catch (IOException e) {
    errorMessage("serialEvent", e);
  }
}
```

This is one of the most complex bits of code in the entire book, and it really isn't that bad. The first two lines are variables inside of the RoombaCommSerial object that create two byte buffers. The sensor_bytes buffer holds known good sensor data from the last sensor read, and sensorsValid tells you if sensor_bytes is good or not. The other buffer, just called buffer, is the staging area for incoming data. There is no guarantee that you'll receive all 26 bytes of sensor data at once, so what is received is buffered up. To keep track of progress, the bufferLast variable is a count of how many bytes received so far.

The data received does not contain start- or end-markers; so you have to assume that every 26 bytes is a whole sensor packet. This presents some interesting synchronization problems if the SENSORS command is used a lot and you need a guaranteed response time. Just trust that things are okay. If you were building a robot for industrial use, this would be entirely unacceptable, but for hacking it's okay and rarely messes up.

In the serialEvent method itself, first make sure the event is the right type (DATA_AVAILABLE) and then loop while data is available() and read data from the input into buffer. If it gets 26 bytes, reset the counter and copy the finished data into sensor_bytes. Finally, report that there's data with sensorsValid and update the internal state of the RoombaComm object with computeSensors(). The computeSensors() method encapsulates any processing of a newly received sensor packet. Right now, computeSensors() does nothing. If you wanted to add functionality to RoombaComm that acts on sensor data, the place to add it would be computeSensors().

So as far as the code goes, it's straightforward. What is difficult to deal with is the multi-threaded nature of this setup. The serialEvent() method may be called at any time, like when your code may be in the middle of driving the Roomba or updating a user interface. You mitigate this problem with the buffer[] holding area and communicating through the sensorsValid flag. When you want to update your sensors info, set sensorsValid=false, send the SENSORS command to the Roomba, and then watch for sensorsValid to become true again.

Making Serial Events Invisible

By decoupling the sending of the SENSORS command and when sensor_bytes gets updated, you can have your code go about doing other things like dealing with a user interface. This makes the code more responsive to the user. But, if you don't mind sitting and waiting, you can do what updateSensors() does in RoombaCommSerial, as in Listing 6-2:

Listing 6-2: RoombaCommSerial.updateSensors()

```
public boolean updateSensors() {
  sensorsValid = false;
  sensors();
  for(int i=0; i < 20; i++) {
    if( sensorsValid ) {
      break;
    }
    pause(50);
  }
  return sensorsValid;
}
```

If you saw this code without knowing about serialEvent(), it would appear nonsensical. First sensorsValid = false and then it's repeatedly checking to see if it's true. But knowing about serialEvent(), it's easy to see what's going on: The sensors() method sends the SENSORS command to the Roomba and then it waits for up to 20 × 50 = 1000 milliseconds for the Roomba to send back data. If sensorsValid results in true, the loop is exited early. In either case, the value of sensorsValid() is returned so that the caller of updateSensors() can quickly know if it worked or not.

With updateSensors() your Roomba control code (which doesn't have a user interface or anything else time-sensitive) can look similar to what the hypothetical pseudo-code looked like. In use, you do this:

```
boolean valid = roombacomm.updateSensors();
if( !valid ) {
  System.out.println("Couldn't get sensors from Roomba.");
}
```

Further Improvements

When not using updateSensors(), you must poll the sensorsValid flag regularly to make sure the data you read is reliable. A way to improve on this is to present an event-based interface to users of RoombaComm. Just like RXTX's SerialPort and its listeners, there could be RoombaComm sensor listeners that get notified whenever sensor data is valid. This hasn't been implemented, but feel free to experiment and make it work. You already have a start on it with the callback to the computeSensors() method after receiving a full sensor packet. Changing this to be a general event-based system would simply mean you add a method like the RXTX addEventListener() method. Then, in addition to computeSensors(), call a method on each listener (maybe called sensorDataAvailable()?) to notify them of new sensor data.

Parsing Sensor Data

Now that you have the sensor data in a little 26-byte array, the task is to figure out how to parse it. Recall from Figures 2-5, 2-6, and 2-7 the organization of the sensor data packet. For SENSORS_ALL, Roomba sends the data as a single 26-byte chunk. The first byte is the bumps and wheeldrops bit field, the second byte is the wall value, and so on. The data has different formats:

- **Word values:** Two bytes joined together form a 16-bit value.
- **Bit fields:** A single sensor byte contains several independent binary values.
- **Enumerations:** A non-numeric byte with arbitrary values.
- **Byte values:** A byte contains a byte. This is the easy case, mostly.

And for the non-bit values, there are the unsigned and signed (positive and negative) variants. Dealing with just the raw byte array with all these different types of data would be a big pain in the neck, so it's time to design some wrappers.

The ROI has a variety of ways of representing numbers because it has a variety of different types of data to represent. Some sensors produce simple on/off data that can be represented as a single bit. Other types of sensors (like the current sensor) have a high resolution and thus need many bits to represent its value. The variety of data sizes allows Roomba to store its data compactly, since it has limited space. The data you receive over the ROI is broken up into byte-sized chunks that then must be teased apart or combined together to reconstruct the original sensor data.

Word Values

Word (16-bit) values are sometimes thought of as two bytes concatenated together, but it's truly a single 16-bit entity. The 16-bits of a word can be used to represent positive numbers from 0 to 65535, in which case it's called an *unsigned word*. Alternatively, those bits can represent a positive and negative number range from -32768 to 32767. That's called a *signed word*. You use words when you can't fit the values you want to represent in a single byte (range 0 to 255 or -128 to 127). You choose signed or unsigned depending on whether you need to represent negative numbers.

The general method of converting two bytes to a 16-bit (word) values is:

```
int word = (high_byte << 8) | low_byte
```

In Java, things aren't so simple because technically bytes are signed (-127 to 127) and doing the preceding sort of combination of two bytes yields very strange and incorrect results. Also, the preceding pattern doesn't quite work for the signed values of distance and angle. Instead of having one mechanism for dealing with word values, you need two. Listing 6-3 shows the Java version of these two. A Java `short` is a 16-bit signed integer, thus capable of holding a 16-bit signed word. To hold an unsigned 16-bit word, you have to use a Java `int`.

Listing 6-3: RoombaComm.toShort() and .toUnsignedShort()

```java
static public final short toShort(byte hi, byte lo) {
  return (short)((hi << 8) | (lo & 0xff));
}
static public final int toUnsignedShort(byte hi, byte lo) {
  return (int)(hi & 0xff) << 8 | lo & 0xff;
}
```

They are defined as `static public final` because they are just utility methods usable by anyone.

Word values were first mentioned in Chapter 2, and in Chapter 5 you saw how to decompose a 16-bit word value into two bytes for use with the `DRIVE` command.

Bit Fields

A bit holds a single on/off (Boolean or binary) value. In many computer languages, Java included, the concept of a Boolean value is distinct from a numeric value with the value of 0 or 1. To convert a bit in a bit field to a Boolean requires two steps. First, you get out the bit you want, and then you convert the bit to a Boolean value.

The pattern for getting out a bit in a bit field is:

```
boolean bit = somebyte & (bit_position << 1)
```

By using the logical AND operator (&), you mask off all the other bits to leave only the one you want. In practice you don't usually do the left-shift of a bit position but just use the numeric value it represents. For example, a bit in position 2 has a value of 0x04, because (2<<1) == 0x04. You can remember these bit masks by making them usefully named constants, for example BUMPLEFT_MASK to get the bump left bit.

To help in remembering which byte of the array to work on, also define constants for the positions in the array, like BUMPWHEELDROPS for the first byte in the array. These constant definitions aren't necessary, but they do cut down on the typos.

The preceding equation is actually a little wrong, especially for Java. The masking returns the byte value of the bit set. So if it were the left-most bit, the value after masking would be either zero or 127 (0x80). This can lead to all sorts of nasty, hard-to-find bugs. Java tries to help you out with this by forcing you to convert a number to a Boolean value. In Java, the preceding would really be written as:

```
boolean bit = (somebyte & (bit_position << 1)) != 0;
```

The test against zero results in a Boolean value, effectively doing a conversion. With this correction and using defines, the test for left and right bumps become Listing 6-4.

Listing 6-4: RoombaComm.bumpLeft() and .bumpRight()

```
public boolean bumpLeft() {
  return (sensor_bytes[BUMPSWHEELDROPS] & BUMPLEFT_MASK) !=0;
}
public boolean bumpRight() {
  return (sensor_bytes[BUMPSWHEELDROPS] & BUMPRIGHT_MASK) !=0;
}
```

All binary sensor values in RoombaComm are implemented the same way. They are too numerous to list here; check the code or the Javadocs. As you might expect, they are cliffFrontLeft() and wheelDropCenter().

Chapter 2 and the ROI specification show the various bit values available in the sensor data. Chapter 2 also mentions briefly in a Tip how to test single bits. The converse of reading bits, writing them, is discussed in the sidebar "Setting and Clearing Bits" in the same chapter. Many of the same techniques are used to either read or write bits.

Enumerations

Enumerations are a way of using a number to encode a list of things. The remote codes in Table 2-2 are an example of such an enumeration. In RoombaComm they are dealt with by creating a list of constants that can be checked against. For example, if you want to check to see if the spin left button on the remote was being pressed, you can do this:

```
if( roombacomm.remoteOpcode() == RoombaComm.REMOTE_SPINLEFT) {
  // do something
}
```

If you're not inspecting the values of the enumeration, you can treat them as ordinary numbers, which is useful when you need to store them (like you're recording all remote control button presses over a period of time).

Byte Values

Byte values present a special problem in Java because it doesn't have the concept of an unsigned byte. For example, if you want to implement a wrapper method that returns the left dirt sensor value (a byte value from 0–255), your first inclination may be to implement it like this:

```
public byte dirtLeft() {
  return sensor_bytes[DIRTLEFT];
}
```

This works well until the value goes over 127; then it becomes negative. This is because of the decision to make all bytes signed in Java. Instead the following must be done:

```
public int dirtLeft() {
  return sensor_bytes[DIRTLEFT] & 0xff;
}
```

From an efficiency standpoint, such a change is horrible because it adds an extra computation and doubles the storage used. Java is rarely about computing or storage efficiency, however, because it chooses to make a language that's easier to use for most people. For the normal uses of RoombaComm, the extra overhead doesn't matter, but you should be aware of it when porting RoombaComm to a more constrained platform (like a microcontroller).

Using Sensor Data

Any program you write to control the Roomba for more than a few seconds runs in a loop of the following form:

```
while( !done ) {
  lookAround()
  doSomething()
}
```

In this loop, the hypothetical function lookAround() contains sensor reading functionality and doSomething() moves the Roomba or otherwise controls its actuators. Advanced programs may have multiple loops of this type and ways to jump among them.

With the `updateSensors()` manner of observing the world, there's always going to be big chunks of time where your program just sits and waits. For example, in the loop:

```
while( !done ) {
   roombacomm.pause(1000);
   roombacomm.updateSensors();
}
```

the actual time through the loop varies between 1050 milliseconds and 2000 milliseconds, depending on whether or not the Roomba is plugged in. At the expense of having little complexity, you can have an optimally responsive loop that is able to do both things not related to Roomba and Roomba sensor-related things as soon as the sensor data is available. Listing 6-5 shows an example of such a loop.

Listing 6-5: Non-blocking Periodic Sensor Reading

```
int ts = 1000;
int t1 = System.currentTimeMillis();
roombacomm.sensors();
while( !done ) {
  t2 = System.currentTimeMillis();
  if( (t2 - t1) > ts && roombacomm.sensorsValid() ) {
      actOnSensors();
      roombacomm.sensors();
      t1 = t2;
  }
  doNonSensorThings();
}
```

The variable `ts` defines the time interval that sensors should be checked. Variables `t1` and `t2` are used to measure when `ts` has occurred. Every time through the loop, it will execute `doNonSensorThings()`. But if `ts` amount of time has happened, it will execute `actOnSensors()`, resend the sensors command, and reset the time pointers. Just to be sure, before acting on the sensors, it makes sure that sensors are valid.

This non-blocking will be used in Chapter 7 as part of the RoombaView application.

BumpTurn: Making an Autonomous Roomba

Armed with the techniques from above, you can now start creating some programs that make an autonomous Roomba. By reading sensors, you can make Roomba know about its environment, and by commanding its motors, you can make it move around. Listing 6-6 shows the entirety of the `main()` method for `BumpTurn.java`, a program that turns Roomba into a simple object-avoidance robot. (The `keyPressed()` method isn't shown in the listing but implements a Java-standard way of detecting a keypress.)

Compile and run this program and watch the Roomba wander around the room. It displays an amazing amount of agility and is able to get out of tight corners with ease. The code has only three actions:

- Go forward.

- If bumped on the left, turn 90 degrees to the right.

- If bumped on the right, turn 90 degrees to the left.

It's amazing that a robot with only three directives can be so agile. The Roomba bumper design assists in making code simpler. This sounds strange, but observe how Roomba reacts when it bumps things and see how even bumps along the side trigger the bumper. The bumper is sensitive on approximately 160 degrees of its circumference. That's over 44 percent of its body that is responsive to pressure.

Try modifying BumpTurn to do different things when bumped, or try using different sensors. Note that because the Roomba is in safe mode it will automatically go into a safety fault (turning off all motors) if a cliff or wheel drop is detected. If you want to get around that, you need to switch Roomba to full mode by adding the line roombacomm.full() after the pause. Of course, if you do this be sure to keep the Roomba away from stairs or other possible drops!

Listing 6-6: The main of BumpTurn.java

```
String portname = "/dev/cu.KeySerial1";
RoombaCommSerial roombacomm = new RoombaCommSerial();
if( ! roombacomm.connect(portname) ) {
  System.out.println("Couldn't connect to "+portname);
  System.exit(1);
}
roombacomm.startup();
roombacomm.control();
roombacomm.pause(100);
roombacomm.updateSensors();
boolean done = false;
while( !done ) {
  if( roombacomm.bumpLeft() ) {
    roombacomm.spinRight(90);
  }
  else if( roombacomm.bumpRight() ) {
    roombacomm.spinLeft(90);
  }
  else if( roombacomm.wall() ) {
    roombacomm.playNote(72,10);  // beep!
  }
  roombacomm.goForward();
  roombacomm.updateSensors();
  done = keyIsPressed();
}
roombacomm.stop();
roombacomm.disconnect();
```

Measuring Distance and Angle

The distance and angle values in the sensor data are odometry data, just like your car's odometer. Unlike a car odometer, the values can be either positive or negative. For distance, that's forward and backward; for angle, that's counter-clockwise and clockwise.

Distance

The distance is obtained from the optical interrupter sensor on the wheels. The value comes from counting the number of beam interruptions caused from the toothed interrupter disc. In Roomba the disc has 32 slots (shown in Figure 5-2) and you know from Chapter 5 that the gearing provides a 25:1 reduction between wheel and disc. Thus the sensor's angular and distance resolution of the turning wheel is:

```
360 degrees / (32 teeth × 25 gearing reduction) = 0.45 degrees
(68 mm diameter × pi) / (32 teeth × 25 gearing reduction) = 0.27 mm
```

The ROI specification gives a distance resolution of 1 mm, and that's probably because 1 mm is the standard distance unit inside the Roomba and the distance value has been averaged to give a more stable (but lower resolution) value. In any case, getting actually 1 mm of precision from any real-world electromechanical device is tough, let alone a consumer-grade vacuuming robot moving on variable surfaces. In practice, don't expect anything below 100 mm resolution. Even with that you can have a lot of fun.

Angle

Although the distance value is a straightforward measurement like a normal odometer, the angle value is an odometrical difference. Roomba has a distance sensor on each wheel, and the angle value in the sensor data is the difference in the distance traveled by each wheel. Specifically, it is the right wheel difference minus the left wheel difference.

This *difference* describes a rotation around the center point between the two wheels. It's as if the wheels are moving partially along a circumference determined by the distance between the wheels. If this sounds a lot like that millimetersPerDegree stuff from Chapter 5, you're right. It's the same thing. The angle sensor value is a fraction of the entire number of millimeters per degree:

```
Angle in degrees = angle value / millimetersPerDegree, or
Angle in degrees = angle value × (360 degrees / (258 mm × pi))
```

Measuring Velocity

Both the distance and angle values returned in the sensor data are based on the last time the sensors were read. Each time they are read, they are reset to zero. This makes the values relative measurements (as opposed to absolute measurements). It also makes velocity calculations easier. By reading the sensors at a time interval, you can determine the velocity (both translational and angular) by simply dividing the distance and angle by the time interval.

Accurate Readings

Roomba only has the two numbers, distance and angle, to describe the path it travels. Reading them periodically enables you to build up that path by connecting the little path sections that each reading represents. Figure 6-8 shows an example of reading odometry values at two different time intervals.

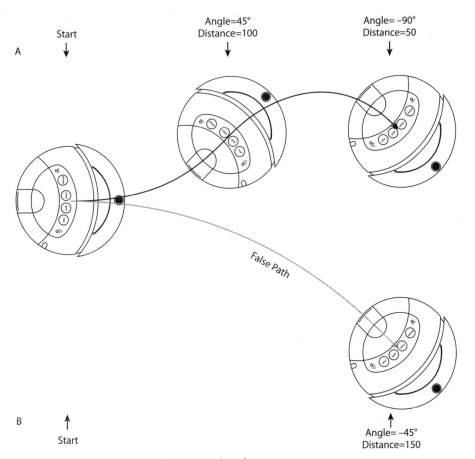

FIGURE 6-8: Sensor reading for distance and angle

In case A, odometry data is read at an interval that captures an accurate view of the Roomba's motion. In case B the data is read at a much slower interval, and the result is a distorted view of where the Roomba really is.

Spying on Roomba

With all these new sensor tools at your disposal, you can now create a program to spy on
Roomba as it goes about its business. Listing 6-7 shows the main() method of a simple
Spy.java program.

Listing 6-7: The main() Method of Spy.java

```java
static int pausetime = 500;
public static void main(String[] args) {
  // standard RoombaComm setup
  boolean done = false;
    while( !done ) {
       roombacomm.updateSensors();
       printSensors(roombacomm);
       roombacomm.pause(pausetime);
       done = keyIsPressed();
    }
}
public static void printSensors(RoombaCommSerial rc) {
   System.out.println( System.currentTimeMillis() + ":"+
        "bump:" +
        (rc.bumpLeft()?"l":"_") +
        (rc.bumpRight()?"r":"_") +
        " wheel:" +
        (rc.wheelDropLeft()  ?"l":"_") +
        (rc.wheelDropCenter()?"c":"_") +
        (rc.wheelDropLeft()  ?"r":"_") );
}
```

By now you know what all the bits do, and the result is pretty simple, with all the hard parts
hidden away. The following is what you might see if you ran the Spy program for a few seconds, it bumped into you, and then you picked it up:

```
1148355917865:bump:_r wheel:___
1148355918417:bump:lr wheel:___
1148355918969:bump:__ wheel:___
1148355919529:bump:__ wheel:___
1148355920089:bump:l_ wheel:_c_
1148355920649:bump:__ wheel:___
1148355921202:bump:_r wheel:_c_
1148355921754:bump:_r wheel:_c_
1148355922306:bump:_r wheel:lcr
1148355922907:bump:_r wheel:lcr
1148355923459:bump:_r wheel:lcr
1148355924011:bump:_r wheel:lcr
```

The `printSensors()` function is a simplified version of the `RoombaComm.sensorsAsString()` method, which dumps out *all* the sensor data in a human-readable form.

When running Spy.java, you'll notice that transitory events (like quickly tapping the bump sensor) will get missed. This is due to the fact that if an action happens between calls to `updateSensors()`, it will be missed. To minimize this problem, you can decrease the `pausetime` setting between sensor readings, at the expense of the output scrolling by very quickly. If you're recording the output of Spy.java to a file for later analysis, set `pausetime` to whatever you feel is appropriate for your application.

You may also notice the time between sensor readings varies a little. If you want a more accurate fixed time between readings, compute how long `updateSensors()` takes to execute by wrapping it in `System.currentTimeMillis()` calls, and subtract the execution time from `pausetime`.

Summary

The RoombaComm library is built up enough now to be truly useful. You can read all of the Roomba sensors and control all of its functions. The hardest part of dealing with any device is having conversations with it. It's one thing to only send or only receive, but to do both can be tricky. You now have two ways of dealing with this trickiness: an easy but slightly inaccurate way and a more complex approach that allows for greater flexibility.

Roomba contains some interesting sensors, and you probably have ideas on how to implement similar techniques in other projects. And it's becoming obvious where the deficiencies in the sensors lie. For a vacuum cleaner, it has just enough ability to observe its environment to function, but for a non-vacuuming robot it should have more. I'll discuss adding new sensors later in this book.

Fun Things to Do

part

in this part

Chapter 7
Making RoombaView

Chapter 8
Making Roomba Sing

Chapter 9
Creating Art with Roomba

Chapter 10
Using Roomba as an
Input Device

Making RoombaView

in this chapter

☑ Code sketching with Processing

☑ Package RoombaComm

☑ Design a Roomba instrument panel

☑ Build a complete cross-platform application

With a complete Roomba API like RoombaComm, it's now possible to build an application on your computer that is both a control panel for Roomba with many more features than the standard remote and an instrument gauge for all the robot's sensors. You've already seen how to do parts of this, but this chapter shows how to tie together everything you've learned so far into one application.

If you've used the code from the previous chapters, you can see how the iterative process of write-compile-run is a bit clunky. It assumes a level of comfort with the command line that many people don't have. Java IDEs are an alternative, but they are so powerful they are often overkill for the hobbyist. Those tools are made for professional software engineers and the staggering array of options and controls is daunting.

When you do have a working Java program, creating an actual double-clickable application can be problematic. Java programs run in a Java Virtual Machine (VM) that must be installed on the computer on which you want to run your program. If the computer doesn't have a Java VM, or has the wrong version, your program won't work. If you're on Windows, you had to sit through the enormous download of the Java software development kit (SDK) to compile the programs in the previous chapters.

Java made the promise of "write once, run anywhere." Code written on one type of computer could be run on any other. That sentiment is mostly true except for the above rather large caveats. It turns out there is a solution to many of these issues, and it came from an unlikely source.

About Processing

Processing (http://processing.org/) is a free open source programming language and environment for people who want to write graphical programs quickly. It was started by alums of the MIT Media Lab, and includes contributions from hundreds of others. Processing arose out of frustration with the state of software for generating and manipulating art. The existing software was either costly and easy to use or free but difficult to use.

Processing is an application for Windows, Mac OS X, or Linux. When first run, it presents a simple text-editing window, with a few controls on top. Figure 7-1 shows what it looks like on Mac OS X; Figure 7-2 shows it on Windows. The interface is identical, save for the platform-specific menu items. The Linux version looks almost identical to the Windows version.

In the text-editing window, you write your *sketch*. A sketch consists of a list of Java-like commands like `line()` to draw lines, `ellipse()` to draw ellipses, and `text()` to draw text. A sketch can create either static graphics or dynamic animations that respond to user input.

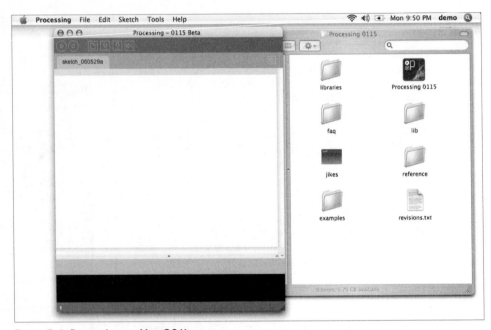

FIGURE 7-1: Processing on Mac OS X

The sketch can then be run by clicking the Run button (right-pointing arrow button) and stopped with Stop (square box icon button). If there are syntax problems with the sketch, they will be described in the black status window at the bottom and highlighted in the sketch window. A sketch can be saved, named, and reloaded, just like with any text editor. When you're happy with the sketch, you can export it as a web page with embedded applet. You can also export it as an application, and Processing will create true executables for all three operating systems.

In many ways, Processing is a descendant of the Logo programming language. Both are visually focused and provide a number of functions to make drawing graphics easier. However, Processing can do much more. It can operate in 3D, work with video and sound, talk over the Internet and MIDI, perform physics simulations, and do many other things. It's continuously being expanded and improved through libraries created by anyone with a good idea.

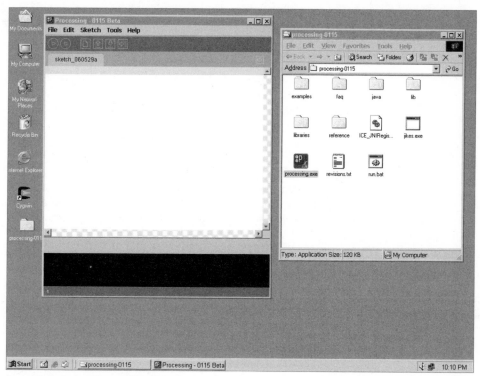

FIGURE 7-2: Processing on Windows

How Processing Works

Processing is implemented in Java. The simple commands that are part of the sketch are actually method names and variables of the Java class within which sketches run. The Processing language is really no different from Java at all; it just removes the visible overhead and complexity from Java.

So Processing is a Java IDE of sorts, albeit a simplified and specialized one. It enables you to write quick graphical code sketches that respond to user input. For example, Listing 7-1 is the entire code for a Processing sketch that creates a window, paints it gray, and then draws a blue box that follows the user's mouse.

Listing 7-1: Example Processing Sketch

```
void setup() {
  size(300,300);
  framerate(15);
}
```

Continued

Listing 7-1 *Continued*

```
void draw() {
  background(51);
  fill(0,0,255);
  rectMode(CENTER);
  rect(mouseX,mouseY, width/5,height/5);
}
```

When the Run button is clicked, the sketch content is embedded in a Java class, compiled, and run. These are the exact steps that you do when writing normal Java programs, but the process has been streamlined and made transparent to the user. If you've ever tried to write a Java program that just draws on the screen, you'll appreciate how much Processing does for you.

The Processing installation includes many sample Processing sketches and you can download even more samples from the Processing website (http://processing.org/learning/). Load a few of them and play with them.

Note   You can access the Processing samples by selecting File ⇨ Sketchbook ⇨ Examples.

Why Use Processing?

Processing enables you to create, run, and share dynamic graphical programs. If you are a professional software engineer, Processing may at first seem unnecessary. But many software professionals use Processing just because its sketching metaphor and helper functions enable them to try out ideas fast. Artists and animators find its direct approach to painting on the screen attractive. And it's a good tool for beginning programmers due to its simplified environment and language.

There are many tools somewhat similar to Processing. In the Java development space you have Java IDEs like the free and awesome Eclipse (www.eclipse.org/). For animation there's Adobe Flash (www.adobe.com/products/flash/), which has been getting more powerful as a general-purpose tool as its ActionScript language has become more standards compliant. For audio/video processing, there's Max/MSP/Jitter (www.cycling74.com/products/maxmsp). This is just a small representative sampling of the tools out there, and all of these tools are better than Processing in some ways and worse in others.

Cost

Processing is not only free; it's open source. Anyone can see exactly what Processing is doing and modify the code to suit his or her needs. Or borrow it. Some of the original RoombaCommSerial code was taken from the Serial library in Processing. Reading other people's code is a great way to learn. Flash, Max, and most Java IDEs cost several hundred dollars and are entirely proprietary. One of the goals of this book is to enable you to write Roomba code for free.

Ease of Use

The fact that Processing calls its programs sketches is telling. Even if you are an expert programmer, sometimes you just want to whip up something quickly. Processing is a great environment for getting coding ideas up and running fast, especially if they have an animated graphics component.

Full Access to Java

Although Processing presents a simple command set for doing common things, it doesn't prevent you from including and invoking other Java classes. In fact, it's pretty easy to use the full Java class library or wrap up any other Java class into a Processing library. This is what is done to allow RoombaComm to be used.

Cross-Platform Compatibility

When Sun introduced Java, one of the company's slogans was "write once, run anywhere." Java promised a world of easily shared code, compiled on a single machine and sent through the network to run on any machine. This is mostly true, assuming the Java VM is the same version on every platform. In practice, there are several different versions of Java VMs out there, all with different capabilities and bugs. Much effort is being expended in the Java developer community to solve this problem, but it still persists. Processing outdoes Java in a few ways. First, it obviates the Java VM versioning issue by ensuring that whenever you use Processing, you get a consistent Java VM environment. Second, it enables the sketch author to make dynamic web pages out of his or her sketch with a single button press, making sharing of running code a piece of cake. Third and most importantly, it goes the extra distance that Java never did and produces true applications for Windows, Mac OS X, and Linux. To the user of Processing, the difference between making a web page applet and a full application is the difference of deciding which Export button to press.

When Not to Use Processing

In general, Processing is meant as a code-sketching environment. You wouldn't want to build a large application with it. Processing is simply a tool that helps you try out ideas fast.

Processing supports an animation metaphor but it is not a full replacement for an animation package like Flash. Processing's vector art support is relatively weak and drawing complex shapes with it can be exhausting. It can work with video, but not with the same alacrity as Max/MSP/Jitter. It can produce applets like Flash, but Java applets seem to always load slower and tax the system harder than the equivalent Flash applet.

Although almost every aspect of Java can be used within Processing, using the Swing or AWT GUI components of Java can be challenging, because Processing assumes a certain level of control over the graphics environment.

If you find that doing something in Processing is becoming too difficult, don't hesitate to drop it and adopt a different methodology.

Using RoombaComm in Processing

Processing can use libraries created by third parties. There are no restrictions on who can create a Processing library, and there are few restrictions on what a Processing library can do. Most importantly, Processing libraries can contain native code. Although RoombaComm is 100 percent Java, it uses the RXTX serial library, which does contain native code.

Packaging RoombaComm

RoombaComm the API library is just a collection of Java class files in a directory. This is fine for development and testing, but when deploying code for use elsewhere, it helps to have some sort of packaging system to treat the collection as a unit. With Java, that packaging system is a JAR file. And with Processing, the packaging system is a zipped directory containing JAR files.

Creating a Java Archive

Before RoombaComm can be used as a Processing library, it must first be made into a Java Archive (JAR) file. A JAR file is created with the `jar` command and its use is similar to the `zip` or `tar` command-line programs. To create a JAR file for the RoombaComm code, get a command-line prompt and in the directory above the `roombacomm` directory type:

```
% jar -cfm roombacomm.jar packaging/Manifest.txt roombacomm
```

That command creates the JAR file `roombacomm.jar` using the manifest file `Manifest.txt` and packages up the `roombacomm` directory. A manifest is an optional part of a JAR file that specifies any configuration or meta-information about the JAR file. In this case, it specifies that it needs the `RXTXComm.jar` file.

Note The preceding command is encapsulated in a script called `build-jar.sh` located in the root directory of the RoombaComm software.

A JAR file is just a ZIP/PKZIP-formatted file. It can be inspected with `zip/unzip` or any program that can read ZIP files. The `jar` program is essentially a cross-platform version of `zip` that also includes some extra options to process manifest files. Some developers like to include the dependent JAR files into the main JAR file. This makes for easy distribution, but it also makes updates of the dependent JAR files difficult. Since Processing allows multiple JAR files in its library, `roombacomm.jar` contains only the RoombaComm code, and the RXTX library is kept as its own JAR file.

Making a Processing Library

With RoombaComm packaged as a JAR file, a Processing library can be created. In the Processing root directory, there is a sub-directory called `libraries`. All libraries go here, both ones that come with Processing and any third-party libraries you install. Processing expects libraries to conform to a directory standard and have certain files present. When RoombaComm is installed in Processing, the directory structure looks like Listing 7-2. There are several libraries already in Processing, like for OpenGL or network access.

Listing 7-2: Processing Directory Structure for Libraries

```
Processing/
        faq/
        reference/
        lib/
        libraries/
                serial/
                net/
                opengl/
                roombacomm/
                        library/
                                export.txt
                                roombacomm.jar
                                RXTXComm.jar
                                librxtxSerial.jnilib
                                librxtxSerial.so
                                rxtxSerial.dll
```

The export.txt file contains a list of which files in the library are needed. For RoombaComm it looks like:

```
application.macosx = roombacomm.jar, RXTXcomm.jar,
librxtxSerial.jnilib
application.linux = roombacomm.jar, RXTXcomm.jar, librxtxSerial.so
application.windows = roombacomm.jar, RXTXcomm.jar, rxtxSerial.dll
```

When Processing processes export.txt it recognizes which files are JAR files and which are shared libraries and loads them appropriately. No more setting classpaths and Java environment variables. Notice that export.txt contains separate entries for each OS. This is needed only because RXTX needs platform-specific code to access the serial ports. If RoombaComm didn't need platform-specific code, then a single application line would be sufficient.

 Note   The RoombaComm distribution contains a script called `build-processing-library.sh` that builds a `roombacomm-processing.zip` file containing everything needed to replicate the directory structure of Listing 7-2 including all the necessary files.

Installing a Processing Library

When `roombacomm-processing.zip` is created (or downloaded off the Net), it needs to be installed in Processing. Assuming both `roombacomm-processing.zip` and Processing are in the same directory, installing the library through the command line is done with these commands:

```
% cp roombacomm-processing.zip "Processing 0118/libraries"
% cd "Processing 0115/libraries"
% unzip roombacomm-processing.zip
```

Or you can use a file explorer to drop the ZIP file into the libraries directory and double-click it to unzip the contents.

When Processing is restarted, you should see roombacomm as a library choice when selecting Sketch ➪ Import Library (see Figure 7-3).

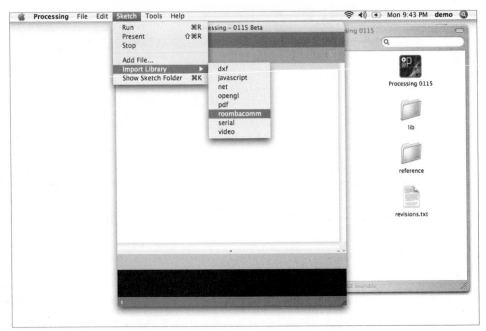

FIGURE 7-3: RoombaComm installed successfully as a Processing library

When you choose a library (like roombacomm) from Sketch ➪ Import Library, Processing inserts the relevant Java import lines into the sketch. You don't have to import libraries this way, but it makes things easier if you don't know the exact name of the classes in the library you're importing.

A First RoombaComm Processing Sketch

A good start might be the SimpleTest.java program seen before. Listing 7-3 shows a complete Processing program to test Roomba. It's a slightly abbreviated version of the regular SimpleTest program, but it accomplishes the same purpose. With Listing 7-3 input into Processing, click the Run button. Processing will compile and launch your sketch. If all's well, the robot will beep and move around. Your sketch will print out its messages to the black status panel below your sketch. If there are errors, the errors will be in the status panel.

Listing 7-3: SimpleTest in Processing

```
String roombacommPort = "/dev/cu.KeySerial1";  // or COM3, etc.
boolean hwhandshake = false;      // true for Windows bluetooth
RoombaCommSerial roombacomm = new RoombaCommSerial();
roombacomm.waitForDSR = hwhandshake;

if( !roombacomm.connect(roombacommPort) ) {
  println("Couldn't connect to "+roombacommPort);
  System.exit(1);
}
println("Roomba startup on port"+roombacommPort);
roombacomm.startup();
roombacomm.control();
roombacomm.pause(30);

println("Playing a note");
roombacomm.playNote(72,10);  // C
roombacomm.pause(200);

println("Spinning left, then right");
roombacomm.spinLeft();
roombacomm.pause(1000);
roombacomm.spinRight();
roombacomm.pause(1000);
roombacomm.stop();
roombacomm.disconnect();
```

The main difference from the regular Java SimpleTest is the lack of command-line argument parsing. For now, just enter in the serial port name directly into the sketch by setting the `roombacommPort` variable. Also different from regular Java is all the normal Java class infrastructure code. With Processing you just start writing the commands you want.

Dealing with setup() and draw()

You probably noticed when comparing Listing 7-1 and Listing 7-3 that the latter doesn't have the `setup()` and `draw()` functions defined. In Processing you can either operate in a single-pass mode or in animation mode. Listing 7-3 is an example of the single-pass mode. The code runs from top to bottom once and then exits. Listing 7-1 is an example of animation mode: `setup()` is called once, then `draw()` is called several times a second, determined by the `framerate()`.

Since `draw()` is called periodically, it is expected to complete before the next occasion of `draw()` is to be called. For example, when `framerate(15)` is specified in `setup()`, `draw()` will be called 15 times per second, which is every 67 milliseconds. Thus waiting around doing nothing for 1000 milliseconds with `roombacomm.pause(1000)` is entirely inappropriate when using animation mode.

When programming with RoombaComm, this means you should not use any of the API that contains `pause()` commands. This is actually fine because the only time you'll find yourself working in animation mode is when you are dealing with Roomba in real-time and don't want your code to just hang out for a few seconds when it should be doing something. You'll find in Chapters 8 and 10 that when using Roomba's sound capability, you normally will use `pause()` to wait for the sound to play. In this case you can either use `pause()` but set the framerate of your sketch to be slow enough that small pauses do not affect it, or you can check the time periodically with the `millis()` Processing function to see when enough time has elapsed.

Creating Applications with Export

When you have a Processing sketch you like, you may want to share it or even just make it runnable outside the context of Processing. Normally turning a Java program into an executable program for a particular operating system is difficult and time consuming. Generating an executable for three different operating systems is even more difficult. Processing has figured out all the issues with turning Java programs into applications on Windows, Mac OS X, and Linux. All that is required of the user is to select File ➪ Export Application, and the current sketch is converted to three applications, one for each OS. These applications are completely stand-alone and do not require Processing to run.

Note When developing your sketch, it's easy to forget to re-export the applications. When you are finished and attempt to run the application you just built, you will find that you are running an older version if you don't export the application you just built first. Be sure to select Export Application when done with your sketch.

Note Normally Processing will also create an applet, but since RoombaComm uses native code via RXTX, Java applets will not work.

Designing RoombaView

Processing makes graphical interactive programs easy to create, and RoombaComm allows easy access to controlling Roomba. The two can be combined in a number of ways to create interesting gadgets. The rest of this chapter will focus on creating RoombaView, a program to let one view graphically the robot's sensors while it goes about its business or while you control it.

What Features to Include

For RoombaView to be a usable Roomba instrument panel, it should show the main sensors the robot uses to do its job:

- Distance and angle odometry
- Bump and cliff sensors state

- Wheeldrop and drive motor over-current state
- Dirt sensor state
- Battery charge, voltage, and current consumption
- ROI mode

To remotely command Roomba, it should have:

- Buttons and/or keyboard control to drive the Roomba and various speeds
- Buttons to reset the ROI and Roomba to a known state
- Buttons to toggle between passive, safe, and full ROI modes

Additionally, RoombaView should know if it's connected and talking to Roomba or not and show that state.

Driving Roomba in Real-Time

Processing will call the `keyPressed()` method in your sketch whenever a user presses a key. The variables `key` and `keyCode` hold the key character pressed and the raw keycode, respectively. If you wanted the arrow keys to move the robot, the spacebar to stop it, and the Return key to reset it, the `keyPressed()` method might look like this:

```
void keyPressed() {
  if( key == CODED ) {
    if( keyCode == UP )
      roombacomm.goForward();
    else if( keyCode == DOWN )
      roombacomm.goBackward();
    else if( keyCode == LEFT )
      roombacomm.spinLeft();
    else if( keyCode == RIGHT )
      roombacomm.spinRight();
  }
  else if( keyCode == RETURN || keyCode == ENTER ) {
    println("resetting");
    roombacomm.reset();
  }
  else if( key == ' ' ) {
    roombacomm.stop();
  }
```

Add any other keys to command the Roomba to do whatever you want. Any RoombaComm command or combination of commands can be added, but be aware of the `pause()`-related issues mentioned earlier.

Note To test for key presses from keys without a character representation like the arrow keys, shift, enter, and so on, first check for key==CODED and then examine the `keyCode`.

MyGUI Processing Library

Besides keyboard control, it would be nice to also have GUI buttons that the user could click with the mouse. Using the standard Java GUI buttons is not easy in Processing. It can be done but you have to figure out how to step around the built-in Processing event loop and AWT-based graphics. There have been a few simplified GUI toolkits created for use in Processing. The one used in RoombaView is called MyGUI, written by Markavian and available from the Processing library page at http://processing.org/reference/libraries/. MyGUI uses Processing for all of its drawing and event handling.

Note   MyGUI registers with Processing at a level not available to sketches. Because of this, MyGUI controls always draw on top of any sketch graphics.

Listing 7-4 shows MyGUI in use and Figure 7-4 shows the result for the movement buttons.

Listing 7-4: Making Graphical Buttons with MyGUI

```
void makeMoveButtons(int posx, int posy) {
  MyGUIGroup buttonGroup = new MyGUIGroup(this, posx, posy);
  buttonGroup.setStyle(new MyGUIStyle(this, cBut));
  PImage forward    = loadImage("but_forward.tif");
  PImage backward   = loadImage("but_backward.tif");
  PImage spinleft   = loadImage("but_spinleft.tif");
  PImage spinright  = loadImage("but_spinright.tif");
  PImage stopit     = loadImage("but_stop.tif");
  PImage turnFL     = loadImage("but_turnleft.tif");
  PImage turnFR     = loadImage("but_turnright.tif");
  butForward   = new MyGUIButton(this,  0,-45, forward);
  butBackward  = new MyGUIButton(this,  0, 45, backward);
  butSpinLeft  = new MyGUIButton(this,-45,  0, spinleft);
  butSpinRight = new MyGUIButton(this, 45,  0, spinright);
  butStop      = new MyGUIButton(this,  0,  0, stopit);
  butTurnLeft  = new MyGUIButton(this,-45,-45, turnFL);
  butTurnRight = new MyGUIButton(this, 45,-45, turnFR);
  sliderSpeed = new MyGUIPinSlider(this, 0,80, 100,20, 0,500);
  sliderSpeed.setValue(200);
  sliderSpeed.setActionCommand("speed-update");

  gui.add(buttonGroup);
  buttonGroup.add(butForward );
  buttonGroup.add(butBackward);
  buttonGroup.add(butSpinLeft);
  buttonGroup.add(butSpinRight);
  buttonGroup.add(butStop);
  buttonGroup.add(butTurnLeft);
  buttonGroup.add(butTurnRight);
```

Listing 7-4 *Continued*

```
    buttonGroup.add(sliderSpeed);
  }

  void actionPerformed(ActionEvent e) {
    String cmd = e.getActionCommand();
    Object src = e.getSource();

    if( cmd.equals("speed-update") ) {
      roombacomm.setSpeed(sliderSpeed.getValue());
      return;
    }
    else if( src == butStop ) {
      roombacomm.stop();
    } else if( src == butForward ) {
      roombacomm.goForward();
    } else if( src == butBackward ) {
      roombacomm.goBackward();
    } else if( src == butSpinLeft ) {
      roombacomm.spinLeft();
    } else if( src == butSpinRight ) {
      roombacomm.spinRight();
    } else if( src == butTurnLeft ) {
      roombacomm.turnLeft();
    } else if( src == butTurnRight ) {
      roombacomm.turnRight();
    }
  }
```

MyGUI enables you to group buttons together in a group and style all the buttons in that group identically, which is what the first two lines of makeMoveButtons() do. The movement buttons are graphical icons created beforehand and put in the data directory of your sketch. MyGUIButtons can load a Processing PImage image object, so makeMoveButtons() loads the button images and creates buttons using them. Another GUI element MyGUI provides is a value slider it calls a PinSlider. This will be the speed control, similar to what RoombaCommTest has.

When MyGUI is used, it registers with Processing a new method called actionPerformed(). If you implement this method in your sketch, MyGUI will call it to tell you when GUI events happen. The actionPerformed() method for the movement buttons simply checks which button has been pressed and activates the appropriate Roomba action, very much like the keyPressed() method did in the previous section. However, one difference is dealing with the slider. If the event is from a slider move, the code updates the default speed used by the various RoombaComm movement commands.

The code for the text control buttons is not shown but is virtually identical to the movement buttons. Their function and implementation should all be self-explanatory except for the SPY button. This button sets Roomba into passive mode and performs a virtual button press of the CLEAN button. It enables you to quickly set the robot going about its business, while you get to watch what it's doing.

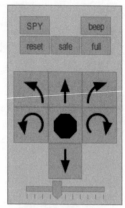

FIGURE 7-4: Buttons for RoombaView

Displaying Sensors

There are many ways to display sensor data. The simplest would be to just get the data out and print it as text, like in Spy.java from Chapter 6. This isn't very usable, however. Some of the most interesting bits of data are the distance and angle data. When you can draw graphics easily, the idea immediately comes to mind to draw a virtual Roomba on the screen and move it in concert with the real Roomba based on its odometry data. With a virtual Roomba on-screen, it also makes sense to display the sensor data on the virtual robot in the approximate location they appear on the real one.

Drawing a Virtual Roomba

The on-screen Roomba icon can be drawn at any rotation and position, in order to match the motion of the real robot. The icon version will also display sensor information. By using the Processing `line()`, `ellipse()`, and `rect()` drawing commands, it's possible to produce a little iconic Roomba outline as shown in Figure 7-5. The outer circle represents the entire Roomba shape, while the inner thick line segments each represent a different sensor, described by the nearest text. When the sensor isn't detecting anything, the corresponding line segment stays gray, but if it is detecting, it will change to a bright color, usually red. The code to draw the virtual robot is shown in Listing 7-5. It uses many Processing tricks and gives an example of how you can go about drawing a detailed object in Processing. The color change on sensor detection is accomplished by changing the line color with the `stroke()` command.

FIGURE 7-5: On-screen
virtual Roomba

Listing 7-5: Drawing Virtual Roomba

```
void drawRoombaStatus(float posx, float posy, float angle) {
  pushMatrix();
  translate((int)posx,(int)posy);
  rotate(angle);
  smooth();
  textFont(fontA, 10);
  textAlign(CENTER);

  fill(cBack); strokeWeight(2); stroke(100);
  ellipseMode(CENTER_RADIUS);
  ellipse(0,0, 50,50);

  fill(0); stroke(cTxt); strokeWeight(4);
  text("cliff", 0, -30);
  if( roombacomm.cliffLeft()        ) stroke(cOn);
  else stroke(cOff);
  line(-35,-25, -40,-15);
  if( roombacomm.cliffRight()       ) stroke(cOn);
  else stroke(cOff);
  line( 35,-25,  40,-15);

  if( roombacomm.cliffFrontLeft()  ) stroke(cOn);
  else stroke(cOff);
  line(-30,-30, -20,-30);
  if( roombacomm.cliffFrontRight() ) stroke(cOn);
  else stroke(cOff);
  line( 30,-30,  20,-30);

  stroke(cTxt);  strokeWeight(7);
  text("bump", 0,-17);
  if( roombacomm.bumpLeft()  ) stroke(cOn);
  else stroke(cOff);
  line(-25,-20, -20,-20);
  if( roombacomm.bumpRight() ) stroke(cOn);
  else stroke(cOff);
```

Continued

Listing 7-5 *Continued*

```
   line( 25,-20,   20,-20);

   stroke(cTxt);
   text("wheeldrop", 0,3);
   if( roombacomm.wheelDropLeft()   ) stroke(cOn);
   else stroke(cWhl);
   line(-35,8, -20, 8);
   if( roombacomm.wheelDropRight() ) stroke(cOn);
   else stroke(cWhl);
   line( 35,8,   20, 8);
   if( roombacomm.wheelDropCenter()) stroke(cOn);
   else stroke(cWhl);
   line(-2,-8,    2,-8);

   stroke(cTxt); strokeWeight(5);
   text("over",0,18);
   if( roombacomm.motorOvercurrentDriveLeft()   ) stroke(cOn);
   else stroke(cOff);
   line(-30,16, -20, 16);
   if( roombacomm.motorOvercurrentDriveRight() ) stroke(cOn);
   else stroke(cOff);
   line( 30,16,   20, 16);

   stroke(cTxt);  strokeWeight(7);
   text("dirt",0,42);
   if( roombacomm.dirtLeft()>0 )
 stroke(40,40,roombacomm.dirtLeft());
   else stroke(cOff);
   line(-22,40,-18,40);
   if( roombacomm.dirtRight()>0 )
 stroke(40,40,roombacomm.dirtRight());
   else stroke(cOff);
   line( 22,40, 18,40);

   popMatrix();
}
```

Drawing with Rotation and Translation

Normally drawing things at any arbitrary position and rotation would be very painful. Imagine trying to draw the virtual Roomba at any arbitrary location: every line() command would need to be modified. Processing includes a few tools to make it easier. One of these is the *transformation matrix*, which can be used to contain a set of drawing commands that are to undergo translation and rotation. This transformation matrix is saved and restored with the pushMatrix() and popMatrix() commands. You can see them in Listing 7-5.

The Processing commands `translate()` and `rotate()` act to move and rotate all subsequent drawing that is done. For simple sketches, using those commands by themselves is fine. When drawing multiple things that use `translate()` and `rotate()`, you need a way to save and restore the current drawing position. The `pushMatrix()` and `popMatrix()` commands save and store, respectively, the current drawing position and orientation.

For example, the `draw()` method in Listing 7-1 could be only slightly re-written so that it's a blue diamond that follows the mouse instead:

```
void draw() {
  background(51);
  fill(0,0,255);
  rectMode(CENTER);
  pushMatrix();
  translate( mouseX, mousey );
  rotate(PI/4);
  rect(0,0, width/5, height/5);
  popMatrix();
}
```

Notice that a square is still being drawn, but what turns it into a diamond is the `rotate(PI/4)` method, which rotates all subsequent drawing by 45 degrees. (In Processing rotation is measured in radians. 2 × PI radians == 360 degrees, thus PI/4 radians = 45 degrees. PI is π = ~3.14.) This is the exact technique used to rotate and move the Roomba icon without needing to parameterize all the Roomba drawing code.

Computing Position

The three values needed to specify the location of the virtual Roomba icon are `rx`, `ry`, and `rangle`. These are the three arguments to the `drawRoombaStatus()` method in Listing 7-5. From the current location and new distance and angle data from the Roomba, the task is to compute the next on-screen location. Listing 7-6 is the algorithm used every time the sensors are read to determine the new location.

The first thing it does once it gets the new sensor data is to determine the new `rangle` by subtracting it from the current angle. It's a subtraction because Roomba measures angles counter-clockwise, while Processing measures angles clockwise. Given the new angle, the x and y components (`dx`,`dy`) of the updated position vector can be computed. With a new `dx` and `dy`, the values of `rx` and `ry` can be calculated by adding `dx`,`dy` to `rx`,`ry`. The sign of `dy` is flipped because on-screen graphics are flipped top-to-bottom from normal coordinate systems. Both `dx` and `dy` are scaled by experimentally determined `scalex` and `scaley` values. This scaling effectively converts the sensor data in units of millimeters to units of pixels. Finally, if `rx` or `ry` extend beyond the boundaries of the screen, they are wrapped around, just like in Asteroids.

Note   Experiment with changing `scalex` and `scaley` to see how it affects the speed of the on-screen version of Roomba.

Listing 7-6: Computing Roomba On-Screen Position

```
float rx,ry,rangle;   // roomba position and angle
float scalex = 0.4;   // pixels per mm, essentially
float scaley = 0.4;

void computeRoombaLocation() {
  int distance = roombacomm.distance();
  float angle   = roombacomm.angleInRadians();
  rangle = rangle - angle ;
  rangle %= TWO_PI;
  float dx = distance * sin(rangle);
  float dy = distance * cos(rangle);
  rx = rx + (dx * scaley);
  ry = ry - (dy * scalex);
  // torroidial mapping
  if( rx > width  ) rx = 0;
  if( rx < 0 ) rx = width;
  if( ry > height ) ry = 0;
  if( ry < 0 ) ry = height;
}
```

Displaying Status

Some of the important Roomba sensor data won't fit on the Roomba icon. And displaying connection status would be difficult as an icon. So on the top of the screen, put a few lines of text that show the connection status, the current ROI mode, and the battery condition. Listing 7-7 shows that status bar. It computes battery charge as an easy-to-read percentage as well as showing you the instantaneous current draw on the battery so you can see what Roomba actions draw more current than others.

Listing 7-7: Displaying Roomba Status

```
void drawStatus() {
  String status = "unknown";
  int batt_percent=0, batt_charge=0, batt_volts=0, batt_mA=0,
batt_cap=0;
  if( roombacomm == null )
    status = "not connected. no roomba. please restart.";
  else if( ! roombacomm.connected() )
    status = "not connected. no roomba. please restart.";
  else if( ! roombacomm.sensorsValid() )
    status = "connected. sensors invalid. unplugged?";
  else if( roombacomm.safetyFault() )
```

Listing 7-7 *Continued*

```
      status = "connected. safety fault. reposition roomba and
reset.";
    else
      status = "connected. roomba detected. ok.";
    if( roombacomm != null ) {
      batt_mA     = roombacomm.current();
      batt_volts  = roombacomm.voltage();
      batt_charge = roombacomm.charge();
      batt_cap    = roombacomm.capacity();
      batt_percent = (batt_cap == 0) ? 0 :
(batt_charge*100)/batt_cap;
    }
    text(" status: "+status, 8,12);
    text("   mode: "+((roombacomm==null)?"no roomba" :

roombacomm.modeAsString()), 8,24);
    text("battery: "+batt_percent+"%
("+batt_charge+"/"+batt_cap+" mAh) "+
                   " voltage: "+batt_volts+" mV @ "+batt_mA+"
mA", 8,36);
  }
```

Wrapping It All Up

With all the important functionality wrapped up in methods, the draw() method becomes
very simple. Listing 7-8 shows what it looks like. The only methods not covered were
drawGridlines(), which draws a simple grid on the background, and updateRoombaState(),
which only calls roombacomm.sensors(). The finished and running application is shown in
Figure 7-6. If you'd like to save the trouble of typing in all the code, the full sketch with already
exported applications can be found in the examples directory of the RoombaComm software
distribution. This is a fairly complex Processing application, but even so is only approximately
400 lines of code.

Listing 7-8: The draw() Method of RoombaView

```
void draw() {
  background(cFlor);
  drawGridlines();
  drawStatus();
  computeRoombaLocation();
  drawRoombaStatus(rx,ry,rangle);
  updateRoombaState();
}
```

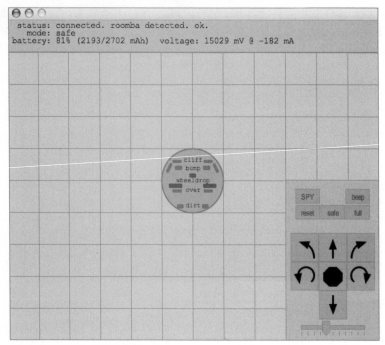

FIGURE 7-6: RoombaView running

Summary

With Processing and RoombaComm together, you can quickly create all sorts of Roomba programs to try out various ideas you might have. Maybe you can improve RoombaView to have a calibrated grid and try to map out your living room. Or you can push Roomba around and use its bump sensors and motion tracking as a big stylus to draw on the screen. Or use the mouse to control the robot somehow. Make a video game with Roomba. Even simple little 20-line sketches that experiment with different movement modes are easy and fun. Anything Roomba can sense, you can display graphically, and it's easy to take user input and cause Roomba to act on it.

You can also extend Processing to have more features. You have a choice of many existing Java libraries that could be useful when programming Roomba. You now know how to pack a Java library for use with Processing and invoke the library from within Processing. Search around on http:/freshmeat.net/ for interesting bits of Java code you could incorporate into your Roomba sketches.

Create real applications for yourself or your friends using the Processing Export ability. When you create an interesting application, upload it to http://roombahacking.com/ so others can experiment and play with it.

Making Roomba Sing

in this chapter

☑ Sound capabilities of Roomba

☑ ROI SONG and PLAY commands

☑ Making Roomba play ringtones

☑ Roomba as a MIDI instrument

Roomba is a very personable little robot. It plays happy tunes when it starts and finishes its tasks, and it plays doleful melodies when it runs into trouble. It even humorously plays the standard back-up alarm beep when reversing out of its charging dock. With the inexpensive addition of a beeper and some software to control it, Roomba gains a personality and elicits empathy from its owners. The original design spec surely only had some sort of auditory annunciator to let one find a hiding or stuck Roomba, but implementing that to include the ability to play little songs definitely increases Roomba's appeal.

When creating the ROI protocol, iRobot could have easily left out a mechanism for creating your own tunes. Instead, the same capability that the Roomba makers themselves have to make tunes is provided. This has some benefits and some drawbacks.

This chapter describes not only how to create little melodies similar to what is already present in Roomba; it also discusses how to "play" the Roomba interactively and use some of the other noise-generating parts of Roomba for musical purposes. You will probably want to use the serial tether from Chapter 3 instead of the Bluetooth adapter because the serial tether has slightly better timing. It's also easier to debug in case of problems.

And while you may not be able to make Roomba bwoop and bweep exactly like your own personal R2D2, it can be expressive musically.

Sonic Capabilities of Roomba

By its very nature of being a robot that vacuums, Roomba is a noisy beast. It has its beeper, but it can make other noises, too. If timed right, these other noises can be used as percussive accompaniment to the melodic lead of the beeper.

Piezo Beeper

The sound-generating device that plays the Roomba tunes is called a piezo beeper. The beeper is similar in function and sound to most other beepers you've experienced in consumer electronics such as watches, cell phones, computers, and toys.

Figure 8-1 shows where it is located on Roomba (in case you want to attach a microphone), and Figure 8-2 shows what the piezo beeper actually looks like. The beeper sits behind the left drive wheel. It is a silver-colored disc with a copper colored center disc, held in place by a black plastic ring.

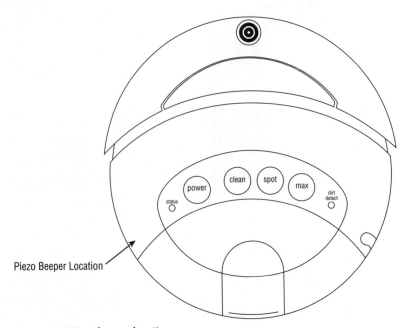

Piezo Beeper Location

FIGURE 8-1: Piezo beeper location

How Piezo Beepers Work

The beeper consists of a metal disc glued to a piece of piezoelectric crystal. Piezo crystals are amazing things. They are naturally found crystals that flex when voltage is applied to them. Turn off the voltage, and the crystal flexes back to its original shape. Pulse the crystal with voltage fast enough, and it generates sound. Pulse it 440 times a second, and it will generate a 440 Hz tone (A440, the musical note A above middle C). The sound is generated just like with any speaker, by moving air back and forth. Since the mass that a piezo can move is small, piezos are better at reproducing higher frequency sounds than lower frequency ones. The metal disc on the piezo beeper in Roomba is the mass used to push air and make noise. Piezo speakers are also used as the tweeters in some audio speakers.

The piezo crystals can also be used in the opposite way. If you flex or squeeze a piezo (from the greek *piezein*, "to squeeze or press"), it will generate electricity, up to thousands of volts in fact. But don't worry; the amount of current it can generate is tiny, so it can't harm you. It may give you a little shock similar to static electricity, however. This is how piezo crystals were first discovered: Squeezing certain rocks generated sparks. In a less drastic application, a flexed piezo can be used as a sensor to detect vibrations or even as a microphone to detect vibrations in the air.

FIGURE 8-2: Piezo beeper inside Roomba

Piezo crystals are used in many consumer items besides beepers and microphones. Ceramic record player cartridges use piezos to detect the grooves, piezo transducers vibrating at 2 MHz are the basis of medical ultrasound imaging, and the igniters on fireplace lighters are piezos. There's research on turning piezo crystals into portable power sources. In the future, your sneakers may contain little piezos that generate electricity to power your portable devices.

Other Sound Sources

Besides the beeper, Roomba has two other sources of noise: the vacuum motors and the drive motors. These sound sources are less precise than the beeper and not melodic, but they do make noise and so are fair game in Roomba musical compositions.

Vacuum Motors

The main vacuum motor produces the deepest bass sound due to the hollow dirt collection chamber. The vacuum can be the bass drum or metronome, but the timing of it is tricky since it takes a while to spin up.

Drive Motors

When driving the motors at normal speeds, a subtle *digital mewling* (for lack of a better term; it's a very indescribable sound) can be heard from them. When the motors are driven at very slow speeds, this mewling is much more pronounced and creepier. It's a great sound to have as another percussive noise for any compositions.

ROI SONG and PLAY Commands

To play musical tones on the beeper, you first define a song and then play it. Roomba has no mechanism to play a single note. Several songs can be defined and stored in the Roomba's memory. Once a song is stored, it can be played on demand.

SONG Command

The SONG command defines a song of up to 16 notes. It is the most complex command in the ROI. The number of notes in the song determines the length of the command. After the command header, notes are specified as a list of (note number, duration) pairs. The duration is in units of 1/64th of a second. Sixteen songs may be stored in Roomba.

The note number is the same as MIDI note numbers. In MIDI, a 12-note octave starts at C and continues up to B. There are ten octaves, and notes are given names to designate which note and octave they are. Thus, the note name A4 means the A musical note in octave four, and C7 means the C note in octave seven. Figure 8-3 shows this arrangement for three common octaves. To convert a note name to a note number requires a little math. The note letter becomes a number offset with an octave, numbered between 0 and 11 (C = 0, C# = 1, D = 2, and so on). The octave number indicates how many octaves' worth of notes to add to the note offset. In MIDI, notes start at C-1, an octave below the lowest in a piano (-1 is a negative 1). In general, to convert a note name to MIDI note number, the equation is:

```
note number = note offset + ((octave+1) x 12)
```

To convert the standard note name A4 to a MIDI note number, you use 9 + ((4+1) × 12) = 69. The lowest note that the Roomba can produce is G1 with a note number of 31 (7 + ((1+1) × 12). The highest note Roomba can produce is G9 with a note number of 127 (7 + (9+1) × 12). To Roomba, numbers below 31 are treated as rests (pauses) where no sound is made for the duration of the note.

Note MIDI will be covered more in the "RoombaMidi: Roomba as Roomba MIDI" section later in this chapter. MIDI is a standard that's been around for over 20 years. It was created by a bunch of nerdy musicians looking for a way to hook their synthesizers together. You can find a lot information about MIDI online, including the Harmony Central tutorial at www .harmony-central.com/MIDI/Doc/tutorial.html. Ricci Adams' Musictheory.net (www.musictheory.net/) has a wonderful set of step-by-step interactive notation and music theory lessons that are fun and quick.

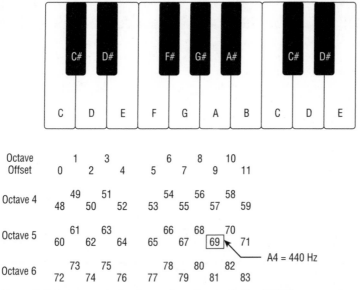

FIGURE 8-3: Some musical note numbers for use in the SONG command

PLAY Command

Once a song is defined, it can be played with the PLAY command. If you try to play a song that hasn't been defined, Roomba does nothing. In contrast to the SONG command, PLAY is a simpler command, taking a single argument, the song number 1-16.

Songs play immediately upon reception of the PLAY command. Sending a PLAY command while a song is in progress stops the first one from playing and starts the second song.

Creating a Song in RoombaComm

Say you want to define and play a song for song slot #1. The song is to be the three notes of a C-major chord (C-E-G) in the 4th octave, and each note should last half a second. You could do it by hand using RoombaComm.send() like so:

```
byte songcmd[] = {(byte)SONG, 1, 3, 48,32, 52,32, 55,32 };
roombacomm.send( songcmd );
byte playcmd[] = {(byte)PLAY, 1};
roombacomm.send( playcmd );
```

This is a little ungainly. It's also dangerous because the song length must be calculated accurately or the ROI will read too many or too few bytes and possibly end up interpreting your other commands as song notes. The above can be made a little better through some new methods added to RoombaComm, as in Listing 8-1.

Listing 8-1: RoombaComm createSong() and playSong()

```
public void createSong(int songnum, int song[]) {
  int len = song.length;
  int songlen = len/2;
  byte cmd[] = new byte[len+3];
  cmd[0] = (byte) SONG;
  cmd[1] = (byte) songnum;
  cmd[2] = (byte) songlen;
  for( int i=0; i < len; i++ ) {
    cmd[3+i] = (byte)song[i];
  }
  send(cmd);
}
public void playSong(int songnum) {
  byte cmd[] = {(byte)PLAY,(byte)songnum };
  send(cmd);
}
```

The createSong() method fixes the most dangerous problem by automatically computing song length. The process of defining and playing the song becomes:

```
int notes[] = {48,32, 52,32, 55,32};
roombacomm.createSong(1,notes);
roombacomm.playSong(1);
```

That's better, but it still isn't really conducive to composition, especially if one wants to play in real time. For that you need to create a way to play a single note.

Playing Single Notes

At the current stage of the ROI protocol, there is no way to play a single note. There may never be such a feature. However, with clever application of what you already know, a playNote() RoombaComm command can be created. Listing 8-2 shows one way to implement this. Both the SONG and PLAY commands are used, one after the other. The SONG command is used to define a single note song, and the PLAY command immediately plays it. The playNote() function uses song slot #15 to hold the note, so that slot cannot be used for normal songs.

One thing to note is that unlike MIDI, which uses separate note-on and note-off messages to indicate when a note starts playing and when it stops, the playNote() command necessarily must specify a duration. This puts some limitations on expression during real-time performance, but it's better than nothing.

Listing 8-2: RoombaComm playNote()

```
public void playNote(int note, int duration) {
  byte cmd[] = { (byte)SONG, 15, 1,
                 (byte)note, (byte)duration,   // define "song"
                 (byte)PLAY, 15 };             // play it back
  send(cmd);
}
```

Playing Roomba as a Live Instrument

Now that you can play single musical notes on Roomba, it should be easy to turn it into a musical instrument that can be played live. Although Java offers some amount of MIDI support, it doesn't offer the feature needed to turn the Roomba into a MIDI instrument: the creation of virtual MIDI destinations. Also, not everyone has a MIDI keyboard and a MIDI interface, but everyone should definitely be allowed to use their robotic vacuum cleaners to jam along with their favorite MIDI tunes.

The computer keyboard offers a workable replacement to a musical keyboard and because of the Roomba's limitations, using it doesn't affect the articulation of the Roomba as an instrument. The computer keyboard's simplicity of input mirrors the Roomba's simplicity of output.

Figure 8-4 shows the computer keyboard to musical keyboard mapping to be used in a Processing sketch called RoombaTune, shown in Listing 8-3. The a key becomes a C note, the w key becomes a C# note, the s key becomes a D note, and so on. The mapping allows only a little more than an octave, so the z and x keys are used to move the keyboard up and down an octave.

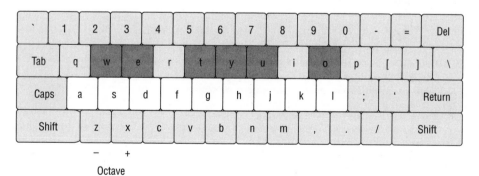

FIGURE 8-4: The RoombaTune computer keyboard mapping of musical keyboard

The core of RoombaTune in Listing 8-3 is the keyPressed() method. The switch(key){}
statement maps the computer keyboard to note number offset. Then if a note has been hit, the
note number offset is converted to the appropriate note number based on the current octave
setting and is played via the new playNote() command.

The RoombaTune sketch in Listing 8-3 is very minimal and not very Processing-like. Note
that it doesn't do any of the normal things in setup() like set the frame rate or sketch size. It
also has a blank draw() method. RoombaTune turns the computer keyboard into a musical
keyboard to play Roomba. One simple and fun change you could do would be to draw lines or
circles based on a note number. RoombaTune then becomes a music-activated screensaver.

To add even more musical control, you can add more functions to the unused keys of the computer keyboard to add turning on and off the vacuum motors, drive motors, and LEDs.

Listing 8-3: RoombaTune Live Instrument Processing Sketch

```
String roombacommPort = "/dev/cu.KeySerial1";   // or COM3, etc.

void setup() {
  if( !roombacomm.connect(roombacommPort) ) {
    println("couldn't connect. goodbye.");  System.exit(1);
  }
  println("Roomba startup");
  roombacomm.startup();
  roombacomm.control();
  roombacomm.pause(50);
}
void draw() {
}
void keyPressed() {
  int note=-1; // -1 means no note yet
  switch( key ) {
  // pseudo-keyboard to play notes on
  case 'a': note = 0; break;
  case 'w': note = 1; break;
  case 's': note = 2; break;
  case 'e': note = 3; break;
  case 'd': note = 4; break;
  case 'f': note = 5; break;
  case 't': note = 6; break;
  case 'g': note = 7; break;
  case 'y': note = 8; break;
  case 'h': note = 9; break;
  case 'u': note =10; break;
  case 'j': note =11; break;
  case 'k': note =12; break;
  case 'o': note =13; break;
  case 'l': note =14; break;
  case 'z': octave--; break;  // change octaves
  case 'x': octave++; break;  // change octaves
  }
  // we actually hit a note key
```

Listing 8-3 *Continued*

```
if( note >= 0 ) {
  octave = (octave < 2) ? 2 : (octave > 9) ? 9 : octave;
  note = note + ((octave+1)*12);
  roombacomm.playNote(note, 10);
  println("playing note: "+note+" (octave:"+octave+")");
}
}
```

Roomba Ringtones

The Roomba piezo beeper sounds a lot like the ringer of older mobile phones. Many of these phones enable you to customize the ringtone by inputting a string of text that describes the melody. This text is standardized and called the ungainly acronym RTTTL (Ringing Tones Text Transfer Language), also known as *Nokring* since Nokia was the first phone maker to provide it.

An example RTTTL format ringtone looks like this:

```
simpsons:d=4,o=5,b=160:c.6,e6,f#6,8a6,g.6,e6,c6,8a,8f#,8f#,8f#,
2g,8p,8p,8f#,8f#,8f#,8g,a#.,8c6,8c6,8c6,c6
```

The format has three parts: a name, a set of defaults, and the list of notes, all separated by colons. The defaults specify the duration of the notes, a default octave, and a speed in bpm (beats per minute). The note list is a comma-separated list of notes, or p for a rest (pause). A note letter can be preceded by a duration value and/or followed by an octave, if either differs from the default. The format isn't that interesting and its specification can be looked up online at www.mxtelecom.com/tech/sms/ringtone. In the RoombaComm API, there is a roombacomm.RTTTLParser class that parses the ringtone into a usable note list.

The important aspect of the RTTTL format is that there are thousands of free ringtones available on the Net. Do a search for monophonic ringtones or RTTTL ringtones to find them. Go out and find a handful of melodies you'd like your Roomba to play. Most of the sites enable you to audition them so you can get an idea of what they sound like.

Note The RTTTLParser class expects no whitespace in the RTTTL string, so be sure to delete any spaces or tabs in the ringtone.

Note Many RTTTL ringtones have speeds that are too fast for Roomba to play. The "b=" part in the default section specifies the beats per minute. Usually you should halve the original number. So if the original ringtone has "b=200", change it to "b=100".

Listing 8-4 shows a Processing sketch called *RoombaRing*. It takes any RTTTL formatted ringtone and plays it on the Roomba. Change the `rtttl` variable to be whatever RTTTL ringtone you like. And as always, change `roombacommPort` to reflect which serial port your Roomba is on.

After using RTTTLParser to create a `notelist` from an RTTTL string, RoombaRing follows one of two paths. If the length of the song is less than 16 notes, it uses `createSong()` and `playSong()` to let Roomba deal with the song's timing.

If the song is greater than 16 notes, it goes through each note, plays it with `playNote()`, and waits the appropriate amount of time for the note to play. Notice that there is a `fudge` factor added to the pause. This was determined experimentally when it was noticed that a pause of the exact right time wasn't long enough and the notes were getting cut off. Your Roomba and serial connection may have different characteristics and require a slightly different value of `fudge`.

Note When playing songs longer than 16 notes, timing between the computer and Roomba becomes important so a wired connection and not Bluetooth should be used.

Note There is also the command-line Java program `RTTTLPlay.java` that functions very similarly to Listing 8-4.

Tip Search for "monophonic ringtone" on Google to find websites with databases of RTTTL-formatted songs you can use with `RTTTLPlay.java`. One such site is www.handphones.info/ringtones/category/. Transcribing normal piano sheet music into an RTTTL song requires some experience in arranging for a monophonic instrument. If you can pick out melodies, try www.freesheetmusicguide.com and download some sheet music to find the notes of songs you can't find in RTTTL format.

Listing 8-4: RoombaRing Ringtone Player Processing Sketch

```
import roombacomm.*;

String rtttl = "tron:d=4,o=5,b=100:8f6,8c6,8g,e,8p,8f6,8c6,"+
               "8g,8f6,8c6,8g,e,8p,8f6,8c6,8g,e.,2d";
String roombacommPort = "/dev/cu.KeySerial1"; // or COM3, etc.
RoombaCommSerial roombacomm = new RoombaCommSerial();
if( !roombacomm.connect(roombacommPort) ) {
  println("couldn't connect. goodbye.");  System.exit(1);
}
println("Roomba startup");
roombacomm.startup();
roombacomm.control();
```

Listing 8-4 *Continued*

```
roombacomm.pause(50);

ArrayList notelist = RTTTLParser.parse(rtttl);
int songsize = notelist.size();
// if within the size of a roomba song,  make the song, then
play
if( songsize <= 16 ) {
  println("creating a song with createSong()");
  int notearray[] = new int[songsize*2];
  int j=0;
  for( int i=0; i< songsize; i++ ) {
    Note note = (Note) notelist.get(i);
    int sec64ths = note.duration * 64/1000;
    notearray[j++] = note.notenum;
    notearray[j++] = sec64ths;
  }
  roombacomm.createSong(1, notearray);
  roombacomm.playSong(1);
}
// otherwise, try to play it in realtime
else {
  println("playing song in realtime with playNote()");
  int fudge = 20;
  for( int i=0; i< songsize; i++ ) {
    Note note = (Note) notelist.get(i);
    int duration = note.duration;
    int sec64ths = duration*64/1000;
    if( sec64ths < 5 ) sec64ths = 5;
    if( note.notenum != 0 )
      roombacomm.playNote(note.notenum, sec64ths);
    roombacomm.pause(duration + fudge);
  }
}
System.out.println("Disconnecting");
roombacomm.disconnect();
```

RoombaMidi: Roomba as MIDI Instrument

If you have a MIDI keyboard or a MIDI sequencer and you want to use Roomba as a MIDI instrument, that's possible too. As mentioned earlier, Java provides a basic MIDI API, but it isn't well-developed across platforms. Fortunately Mac OS X does provide Java wrappers to its entire Core MIDI C language API, making programming MIDI applications in Java possible. Because of Core MIDI, these Java applications appear as just another MIDI device to other

MIDI-capable software and devices. When similar wrappers are created for other operating systems, then RoombaMidi could be easily ported to those operating systems, too. MIDI communication operates at essentially the device driver level, requiring operating system support and making a 100 percent Java solution not possible.

Since Mac OS X–specific Java libraries are required and a more standard GUI is desired, it doesn't make as much sense to use Processing. Instead, a standard technique of bundling Java JAR files into Mac OS X executables is used. This technique uses a free program that comes with Mac OS X called Jar Bundler. A similar free bundling tool exists for Windows called Launch4J (http://launch4j.sourceforge.net/).

Figure 8-5 shows what RoombaMidi looks like. At first glance it looks the same as Roomba CommTest with which you are already familiar. Like RoombaCommTest, RoombaMidi uses RoombaCommPanel for the Roomba GUI. RoombaMidi works just like RoombaCommTest and adds only a thin application wrapper, a MIDI parser, and a tabbed pane to hold multiple RoombaCommPanels. Figure 8-6 shows RoombaMidi as part of a larger setup including Ableton Live, a digital audio and MIDI workstation, and MidiKeys, a virtual MIDI keyboard.

Tip   You can download free MIDI sequencers (sometimes called *hosts*) and other MIDI programs for Mac OS X at http://xmidi.com/apps.html. Another good resource for free music programs for all operating systems is www.kvraudio.com.

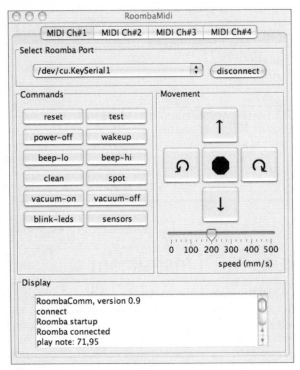

FIGURE 8-5: RoombaMidi running

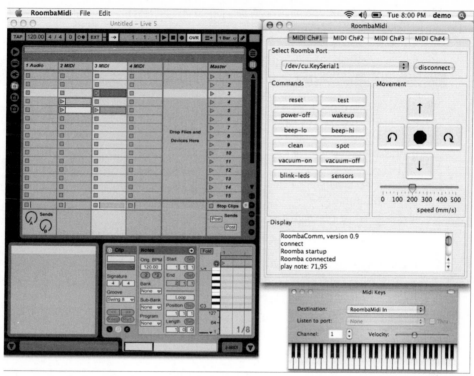

FIGURE 8-6: RoombaMidi as part of a virtual studio

Translating MIDI Notes into Roomba Actions

One part of RoombaCommPanel not mentioned earlier is the `playMidiNote()` method. It is shown in Listing 8-5. Structurally it is very similar to the `keyPressed()` method of the Processing sketches in this and the previous chapter. The method accepts two parameters: a MIDI note number and a note velocity. In MIDI, note velocity is basically how loud the note should be. Because Roomba can't vary note loudness, in RoombaMidi the note velocity is mapped to note duration: a high velocity makes for a longer note. Although MIDI defines an event for both note-on and note-off messages, note-on with a zero velocity is interpreted as note-off, too. The `playMidiNote()` method utilizes that fact to make implementation easier.

The first `if()` clause handles the case where the note falls within the range playable by the Roomba piezo beeper. The next `if()` clause turns the vacuum on or off if C0 is played. The one after that blinks all the LEDs if C#0 is played. The final two `if()` clauses spin Roomba left or right, for MIDI notes E0 or F0, respectively. The speed of the spin and the brightness of the Power LED are determined by the MIDI velocity.

The `playMidiNote()` method uses no Java MIDI classes, and it can be called like any other method. This is the manner in which a port to a non–OS X platform would occur.

Listing 8-5: playMidiNote() in RoombaCommPanel

```
public void playMidiNote(int notenum, int velocity) {
  updateDisplay("play note: "+notenum+","+velocity+"\n");
  if( !roombacomm.connected() ) return;

  if( notenum >= 31 ) {                    // G and above
    if( velocity == 0 ) return;
    if( velocity < 4 ) velocity = 4;    // has problems at lower
durations
    else
      velocity = velocity/2;
    roombacomm.playNote(notenum, velocity);
  }
  else if( notenum == 24 ) {             // C
    roombacomm.vacuum( !(velocity==0) );
  }
  else if( notenum == 25 ) {             // C#
    boolean lon = (velocity!=0);
    int inten = (lon) ? 255:128; // either full bright or half
bright
    roombacomm.setLEDs(lon,lon,lon,lon,lon,lon, velocity*2,
inten);
  }
  else if( notenum == 28 ) {             // E
    if( velocity!=0 ) roombacomm.spinLeftAt(velocity*2);
    else roombacomm.stop();
  }
  else if( notenum == 29 ) {             // F
    if( velocity!=0 ) roombacomm.spinRightAt(velocity*2);
    else roombacomm.stop();
  }
}
```

Working with Core MIDI

In the Mac OS X Core MIDI API, to participate in MIDI communications, you first create a MIDIClient object by calling new `MIDIClient()`, passing in the name you want your client to be called. Once you have a MIDIClient, you can create a virtual MIDI interface by calling `destinationCreate()` on it. This is shown in the `setupMidi()` method in Listing 8-6. Listing 8-6 shows the MIDI handling methods of RoombaMidi.java.

The second parameter of `destinationCreate()` determines what object to notify when MIDI data arrives for that MIDIClient, and it is set to the `RoombaMidi` object itself. This means that RoombaMidi must implement `MIDIReadProc`, and it must have an appropriate `execute()` method. Listing 8-6 also shows the very simple `execute()` method that goes through the incoming `MIDIPacketList` and processes each `MIDIPacket` with `processMIDIEvent()`. The `processMIDIEvent()` method parses the incoming MIDI data, looking for a note-on (0x90) or note-off (0x80) MIDI message. If it finds one of them, it extracts the entire message from the `MIDIData` buffer and invokes `doNote()` with it. Finally, `doNote()` finds the corresponding `RoombaCommPanel` based off of MIDI channel, and calls `playMidiNote()` to cause Roomba to sing or dance.

Listing 8-6: RoombaMidi.java MIDI Handling Methods

```java
public void setupMidi() {
  try {
    midiclient = new MIDIClient(new CAFString("RoombaMidi"),
null);
    midiclient.destinationCreate(new CAFString("RoombaMidi
In"), this);
  } catch(Exception e) {
    e.printStackTrace();
  }
}
public void execute(MIDIInputPort port, MIDIEndpoint
srcEndPoint,
                    MIDIPacketList list) {
  for(int i = 0; i < list.numPackets(); i++) {
    MIDIPacket packet = list.getPacket(i);
    processMIDIEvent(port, srcEndPoint, packet.getData());
  }
}
void processMIDIEvent(MIDIInputPort port, MIDIEndpoint
endpoint,
                      MIDIData data) {
  try {
    int startOffset = data.getMIDIDataOffset();
    int numMIDIdata = startOffset + data.getMIDIDataLength();
    int i = startOffset;
    do {
      if(i >= numMIDIdata)
        break;
      int commandByte = data.getUByteAt(i);
      int command = commandByte & 0xf0;
      int channel = commandByte & ~0xf0;
      if(command == 0x90 || command == 0x80) {  // NOTE_ON or
NOTE_OFF
          doNote(command, channel, data.getByteAt(i+1),
                                   data.getByteAt(i+2));
```

Continued

Listing 8-6 Continued

```
        i += 3;
      } else {
        i++; // otherwise skip over data we don't care about
      }
    } while(true);
  }
  catch(Exception cae) {
    cae.printStackTrace();
  }
}
void doNote(int command, int channel, int notenum, int
velocity) {
  if( channel > tabbedPane.getTabCount() )
    return;
  RoombaCommPanel roombapanel =
              (RoombaCommPanel)
tabbedPane.getComponentAt(channel);
  roombapanel.playMidiNote(notenum, velocity);
}
```

Summary

So Roomba is not a high-quality professional polyphonic synthesizer. But it does make some neat melodic and percussive noises. Having Roomba perform by itself is fun, but imagine getting your friends together to form a Roomba quartet.

Reusing the large collection of RTTTL ringtones as musical source material or using MIDI are just two methods of inputting musical information into Roomba. There are many others, and you may have a favorite one you'd like to add. Programming your own songs or algorithmic compositions for Roomba to play, whether long or short, can easily be accomplished.

Adding MIDI control to Roomba not only enables you to play the robot as a live instrument, but also to move it around. Although it only spins under MIDI control right now, you can make it perform any other movement triggered by MIDI, even complex actions. Using a simple MIDI sequencer, you can create elaborate song and dance routines for one or even several Roomba robotic cleaners.

Creating Art with Roomba

One of the first things people did after they discovered how to control Roombas remotely was to put two of them in a ring and have them fight. These Roomba Fights, as they've come to be called, are funny in part because Roomba is not designed as a battle bot. Upon reflection, it's incongruous in another way too. The Roomba's design is friendly and non-threatening. It performs a mundane task. Its movements while working shift between the random organic and the elegant geometric. It seems to be more of an artist than a fighter.

You've already started exploring the artistic capabilities of Roomba in the previous chapter. What can be accomplished musically with Roomba may be limited, but those very limitations enable new ways of thinking about music and what is musical. Music isn't constrained by a particular set of tools and is open to new expressive techniques. Witness the rise of the synthesizer in modern music in the last 40 years. It has gone from academic experiment to compositional requirement.

As in music, painting and drawing are open to new technologies and new expressive styles. The techniques range from cave drawings with charred sticks to mega-pixel Photoshop images. There is no best technique for producing art: all are valid. Paintings are sometimes judged on how photo-realistic they are. This is almost always beside the point. A painting may be about non-physical concepts, like belief or melancholy. Or it may explore how colors and shapes can evoke feelings. A single octagon may not evoke feelings but what about a red octagon? What about a red octagon with the word *stop* in white letters? Humans are pattern-matching creatures and even seemingly random combinations of color and shapes can provoke discussion and have meaning to those viewing it. Even abstract drawings like those produced by a spinning robot may be aesthetically pleasing.

As you've watched Roomba and experimented with its capabilities, you've seen how it can move. Using that knowledge, this chapter explores how a Roomba robotic vacuum cleaner can be used as a new way to make art. By virtue of the commands used to control the robot, it's natural to make it move in complex circular spirals. Such moves look similar to the curves made by a Spirograph, so the equations that produce such curves will be explored and you'll see how to implement a program that moves Roomba in a similar fashion.

in this chapter

☑ Add a paintbrush to Roomba

☑ Learn the math behind spiral curves

☑ Explore parametric curves

☑ Create art with Roomba

Can Robots Create Art?

The question as to whether or not robots (or computers in general) can make art comes up now and again. It is usually answered negatively against works of generative art: pieces that are generated, sometimes randomly, from an initial set of equations and heuristics. One could argue that humans make art through rules (equations) of perspective and color and use their past experiences (heuristics) as a basis for their choices. So really the question becomes one of degree. Both humans and computers/robots can make art; just the rule sets used by one are simpler. Also, a piece of art isn't purely the creation of the artist, it is embedded in the context of its environment and relies on observers to examine and appreciate it. Everyday objects, in the right context and with the right viewers, have become art.

Another way to consider the question is that for every computer program there is an author. Even if the program incorporates randomness, the amount of randomness is chosen by the author. Some generative art is based on evolutionary principles. In this case there may be no author, *per se*, and the resulting output may not be aesthetically interesting to humans. But the act of creating such art leads to discussion, which may be a point in and of itself.

The way this book approaches the question is based on a common accessible aesthetic: Does it look cool? Computers excel at mathematical equations; robots are how computers interact with the physical world. Programming the computer to run through variations of interesting equations and have the robot attempt to act out those equations can lead to some interesting images. Are these art? Try some of the techniques in this chapter, look at them, show them to your friends, and judge for yourself.

Note   There are several online galleries of and sites about generative/robot/AI art. Besides the Processing gallery (`http://processing.org/exhibition/`), Dataisnature is a news site (`http://dataisnature.com/`) showing upcoming generative art. Bogdan Soban's site (`www.soban-art.com/`) has some nice works and provides interesting essays and links about generative art. Of course, fractal math, the mathematical poster child for chaos theory, can result in beautiful images. The Infinite Fractal Loop webring (`www.fractalus.com/ifl/`) is a collection of websites with good fractal art. A really interesting example of robotic art is AARON, a set of artificial intelligence programs driving a large-format custom plotter to create paintings of people. AARON has been crafted by Harold Cohen over a period of 30 years. Its work is a common discussion point about who exactly is the author of the resulting works: the programmer or the program. A gallery of AARON's work and a downloadable program containing AARON's algorithms can be found at `www.kurzweilcyberart.com/aaron/static.html`.

Parts and Tools

In addition to the same basic RoombaComm setup as before, you'll need to stop by the craft store and the hardware store to pick up a few things. These supplies fall into two categories: things to draw on and things to draw with.

- Several colors of markers, non-toxic and washable preferred, like Crayola markers
- Sidewalk chalk

- Roll of craft paper (30″ wide or greater, 15′ long or greater)
- Large plastic tarp or old shower curtain liner
- Blue painter's tape
- Duct tape
- Small spring-loaded clamps
- Ruler
- Cable ties
- Scissors or utility knife to cut craft paper
- Drill with 1/4″ drill bit for wood
- 6′ × 6′ of open space on a hard floor

These particular items were chosen after some experimentation. It may be that you'll discover better supplies to use. Figure 9-1 shows examples of the supplies used to create the art in this chapter. The craft paper has been rolled out and the supplies are sitting on it.

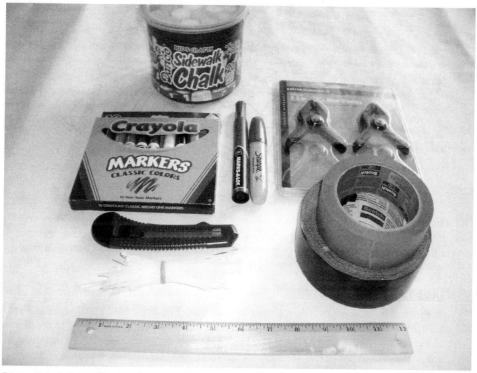

FIGURE 9-1: Art supplies used

Adding a Paintbrush to Roomba

The term *paintbrush* is being used lightly here. Because Roomba has no way to control the amount or color of pigment or even lift the brush, a real paintbrush would be difficult to use. And since paint takes time to dry, there's the possibility of Roomba tracking across wet paint. This might be neat artistically, but it makes for a messy robot.

Thus the paintbrushes used must be things that produce a mark when moved along a surface with little pressure and make a mark that dries quickly. Marker pens and chalk both meet these criteria. Normal pens or pencils could also be used, but the lines they create are thin and hard to see compared to the markers. A large mark makes a bold statement and also makes it easy to see what Roomba is doing.

Brush Types

The main type of brush used is a marker pen, the wider the better. Sharpie makes some huge markers, but the indelible ink used in them smells pretty toxic and permanently stains materials you may not want stained. Crayola makes some non-toxic, more water-soluble markers that produce reasonably thick lines and are easy to clean up in case of accidents. Also, the Marks-A-Lot and Sharpie markers are harder to use because they both have chisel tips. The conical tips of the Crayola markers make a more consistent line. Try out the chisel tips to see if you like the variation effect it produces as the robot spins.

Markers assume you have something to mark on. If you don't have big sheets of paper, there's still the option of using chalk on your driveway. The brightly colored sidewalk chalk available at most craft stores has the added benefit of being easily swept away with a push broom, making erasing quick and easy. The chalk has the downside of being less precise and you must constantly reposition the chalk as it is consumed. Plus, you can't easily save your masterpiece when done.

Attaching the Brush

The Roomba's smooth plastic surface doesn't lend itself to attachment devices that must touch the ground. One option is to take Roomba apart and find the appropriate places to drill holes for markers. This would enable you to create a very stable holder for the marker but would definitely void the warranty.

Whatever attachment method you use, it must press down a little on the marker so it makes a decent line. There's room for experimentation here. The following sections detail the three non-destructive techniques for attaching markers used in this chapter.

Duct Tape

Rumor has it the world is held together by duct tape. The embedded cloth fibers and strong adhesive of duct tape provide a decent anchoring for markers and enable you to position a marker virtually anywhere on Roomba. Duct tape only works on non-porous materials, however, so it will not hold chalk in place.

Figure 9-2 shows a Roomba with two markers attached to it in this fashion. Figure 9-3 shows an alternate attachment point. By removing the dirt bin, you get two vertical sides for attaching things. But without the bin, the Roomba is slightly unbalanced. This is usually an annoyance, because the Roomba will pop up on quick turns, but that can be used to lift the brush momentarily to create interesting drawing effects. If you don't want the popping behavior, either make all the turns the Roomba does be gradual and avoid drastic changes in directions or speed, or lay a weight on top of the Roomba. The Roomba is surprisingly strong — it can move around with a ballast consisting of a full gallon of paint resting on its top.

FIGURE 9-2: Duct-taped markers on the Roomba perimeter

Wedged Clamp

A small spring-loaded clamp holding a marker is big enough to fit in the space between the robot's upper shell and the plastic brush box that holds the main brush. The brush box moves up and down a little to accommodate different types of carpet. You can use this feature to hold a clamp. First remove the dust bin. Then pick up Roomba so the brush box drops down and insert the arms of the clamp. Put the robot back down to wedge the clamp. Figure 9-4 shows a marker held in place using this technique. The wedged clamp is the most accurate for the software featured later, which assumes an idealized marker at the very center of the Roomba. This technique is also very fast to add and remove, making it a great option for those quick art sessions.

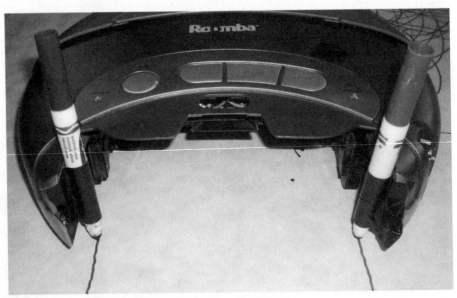

FIGURE 9-3: Duct-taped markers in the dirt bin area of Roomba

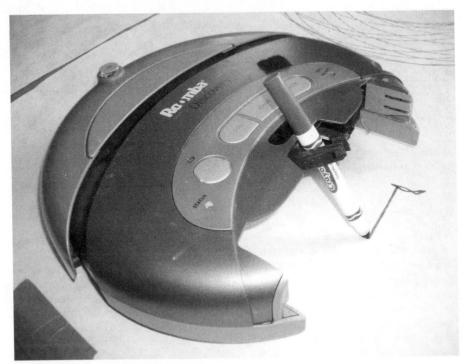

FIGURE 9-4: Roomba with wedged clamp

Ruler and Clamp

Both the duct tape and the wedged clamp suffer from an inherent unsteadiness. As the Roomba moves around, the pen shifts slightly. This results in either a wobbly drawing or the pen coming totally dislodged. The clamp holds the pen securely, so then the question becomes how to attach the clamp to Roomba more securely.

Take a standard wooden ruler and lay one side of a plastic clamp along it at the end. Drill holes on either side of the clamp handle and on one side of the clamp jaw. After the holes are drilled, run cable ties through the holes and around the clamp and cinch them tight. Figure 9-5 shows the result for the particular clamp used here. You may have a slightly different style of clamp and slightly different hole positions.

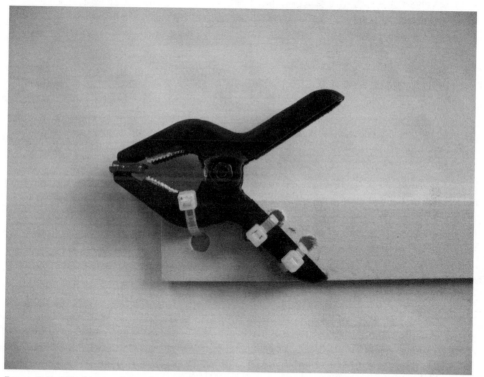

FIGURE 9-5: Attaching a clamp to a ruler

With the clamp securely fastened, the rest of the ruler is a good match for the large flat space on top of Roomba. Duct tape the ruler to the top of Roomba in whichever way you think would make interesting marks. Figure 9-6 shows one possible combination. It may not be aesthetically pleasing, but it does the job and can be removed in a snap with no permanent damage to Roomba.

FIGURE 9-6: Ruler and clamp attached to Roomba

The ruler and clamp method has the benefit of enabling you to switch colors quickly without messing up the position of the Roomba. It's also the easiest way to draw with the street chalk. The downside of this method is that it's the least accurate for the software.

Figure 9-7 shows another positioning of the ruler-clamp combination, this time with chalk. A piece of painter's tape is wrapped around the chalk before inserting it into the clamp so the clamp's teeth don't dig into the soft chalk.

Note Duct tape leaves residue and eventually your Roomba will start looking grungy. Use a damp paper towel with a little dish soap, and you can rub off the residue easily.

Laying Out the Canvas

Don't just let the Roomba start driving around with markers attached to it. It *will* stain floors. Besides, without a canvas, you can't save for later the great works of art you and your Roomba create.

FIGURE 9-7: Ruler and clamp with sidewalk chalk

The canvas space should be around 6´ × 6´ square. Lay out the plastic tarp or old shower curtain liner and tape it down with the painters tape. This forms the protective layer between the markers and the floor. Then roll out two 5´ sections of craft paper and tape them on top of the plastic. Figure 9-8 shows the result. Normally you will work within one of the sections and only use half of it. You can usually get four drawings with the arrangement of Figure 9-8. But by laying out two sections next to each other, you have the option of creating really large works and you save setup time. If the seams between sections of craft paper stick up and catch Roomba, they can be held down with more painters tape. If you want a truly invisible seam, use two-sided adhesive tape.

The plastic provides another benefit: when it's time to stop letting your Roomba be an artist, it's easy to roll up the plastic and paper assembly without disturbing the arrangement of the paper.

Note Do not use the duct tape to attach the plastic to the floor. It could damage the finish of the floor. The painter's tape is designed to leave no residue and not harm finishes.

FIGURE 9-8: Paper canvas laid out and secured

Trying It Out

With canvas and brushes in place, you can quickly test the mechanics of everything by telling Roomba to do a normal cleaning cycle. The Spot mode is most interesting as it immediately starts an expanding spiral. Figure 9-9 shows that spiral done twice, from slightly different starting positions. The little squiggle on the right side of Figure 9-9 is what it looks like when you keep tapping the bump sensor of Roomba.

You can read more about the Spot mode and other modes in Chapter 2.

Be sure to watch Roomba and make sure it doesn't stray off the paper. If it starts to, just pick it up and move it where you want it.

Tip If you want to ensure the Roomba stays within the canvas, make a frame out of lengths of cheap 2 × 4 lumber. You could nail or screw the 2 × 4s together or even just lay the loose 2 × 4s at the canvas edge. If you modify your code to watch for the bump sensors you could even have it bump of the frame in artistically pleasing ways.

Note To make Roomba more stable and less likely to smudge the line it's drawing, remove the Roomba brushes and brush guard.

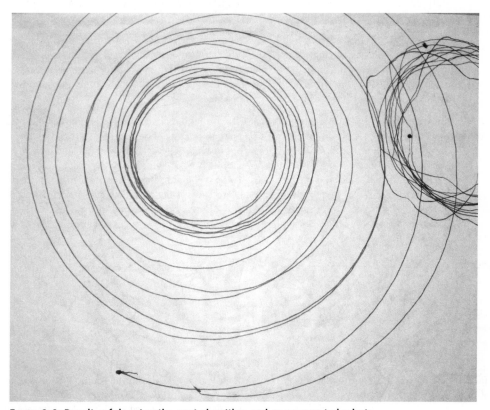

FIGURE 9-9: **Results of drawing the spot algorithm and an aggravated robot**

What Are Spiral Equations?

Now that you have a working robot artist, what are some good things to draw? Due to the nature of the artist, the subject matter should be:

- **Continuous:** Requiring no pen lifts
- **Bounded:** Existing only within a defined area
- **Smooth:** Having no sudden turns or inflections

If the preceding descriptions sound a little mathematical, that's because they are. Many mathematical curves can be quite beautiful and complex. Some of the easiest types to draw are the curves created by a Spirograph.

Spirograph is a toy by Hasbro that has been around for about 40 years. It consists of a set of plastic gears and plastic rings. By placing a smaller gear against a larger gear, inserting a pen into one of the gears, and moving the pen around according to how the shapes interact, you can produce neat geometric art. They are great fun. You should get one to play around with and experience the variety of curves it can produce.

SpiroExplorer is a Processing sketch that uses similar concepts to the ones employed by the Spirograph, albeit with modifications necessary to operate on a screen. Figure 9-10 shows a diagrammatic representation of a Spirograph. Instead of gears, imagine circles that move without slip against each other. The larger outer circle is fixed, with radius R. Inside of it moves a circle of radius r. Anchored to the moving circle is the pen, some distance d from the center of the moving circle. The pen distance d can be larger or smaller than the smaller circle r (although in the original Spirograph usually $d<r$ and $r<R$). By varying the values of R, r, and d, you can create a wide variety of interesting looking curves. Some of the types of curves obtained are shown in Figure 9-11.

Note   For another take on generating Spirograph-style art, see www.wordsmith.org/~anu/java/spirograph.html.

Parametric Curves

The curves a Spirograph produces can be described by a fairly simple set of equations. The simplest manner of representation is as parametric equations. A *parametric equation* is one that just takes in a couple of values (parameters) and outputs other values. One of the simplest parametric equations is for a circle. Any point x,y on a circle of radius r can be obtained with the equation:

$x^2 + y^2 = r^2$

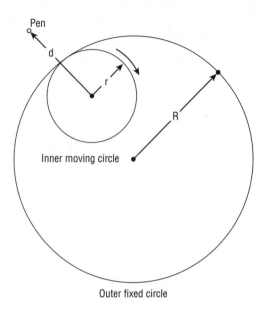

FIGURE 9-10: Spirograph diagram

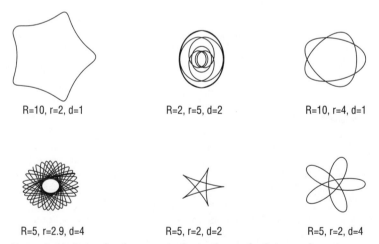

R=10, r=2, d=1   R=2, r=5, d=2   R=10, r=4, d=1

R=5, r=2.9, d=4   R=5, r=2, d=2   R=5, r=2, d=4

FIGURE 9-11: Example of curves similar to those of a Spirograph

Figure 9-12 shows such a circle, with the point x,y being angle *t* above the horizontal. Using standard trigonometric functions, you can solve for x and y to get:

$x = r\ cos(t)$

$y = r\ sin(t)$

The angle *t* is the same as the angle θ (theta) you used in Chapter 5 when you were calculating a radius for the DRIVE command. In parametric equations like the ones here, theta is usually replaced by just *t*. In Figure 9-12 the angle *t* represents which direction to go from the center of the circle to some point on its edge.

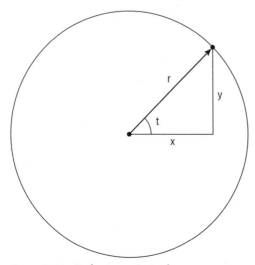

FIGURE 9-12: Circle parameterized

This is a parametric equation for a circle, and you could write a little Processing program to draw circles with it. If you think of *t* as time instead of an angle, then as time increases, the point *x,y* revolves around and around in an anti-clockwise motion, like a clock's second hand going backward. (Math isn't backward, our convention of clockwise is.) In Processing, *t* could be based off of the number of frames displayed, and you could make a little clock or a moving earth-moon diagram.

Figure 9-12 and the math above are derived from the *unit circle*, so called because the radius has been set to one. Analyzing unit circles is the basis for much of trigonometry. To learn more about the unit circle and general trigonometry, visit The Math Page website (http://mathpage.com/), particularly the page on unit circles at http://themathpage.com/aTrig/unit-circle.htm. Also, Wikipedia has a page with extremely useful unit circle diagrams at http://en.wikipedia.org/wiki/Unit_circle.

Hypotrochoid Curves

The equation that describes the curves a Spirograph makes is not that different from the one for the circle. However, since it is a moving circle r within a fixed circle R, the equation becomes a little more complex:

$$x = (R-r) \cos(t) + d \cos(((R-r)/r)t)$$

$$y = (R-r) \sin(t) - d \sin(((R-r)/r)t)$$

The types of curves produced by the above equation are called hypotrochoid curves. The Spirograph also makes another type, called epitrochoid curves. These differ in that the moving circle goes on the outside of the fixed circle instead of the inside. The epitrochoid curves are usually less interesting, so the hypotrochoids will be focused on.

Lissajous Curves

Another set of curves somewhat related to these curves are Lissajous curves. The general equation of them is:

$$x = A \sin(at + d)$$

$$y = B \sin(bt)$$

You've probably seen Lissajous curves. They are the moving curve shapes on computer displays in the background of old sci-fi movies.

SpiroExplorer

It can be difficult to get a feel for how the preceding equations result in different curves. All the equations take a set a parameters and spit out an x,y pair. Usually, the parameter t is varied while the other parameters like R, r, and d are kept constant.

Listing 9-1 shows a basic version of SpiroExplorer, a Processing sketch to experiment with different parametric curves. Figure 9-13 shows SpiroExplorer in action. The full SpiroExplorer sketch enables you to modify R, r, and d in real-time using keys on the keyboard. The update _xy() function is the heart of the sketch. It starts by saving the old values of x,y to xo,yo and then computes new values of x,y using whichever parametric equation you like. In Listing 9-1 the Java version of the hypotrochoid equation is being used. The update_xy() function also increments the angle t (which you can also think of as time here) by some incremental value called dt. dt is the step size you use to walk through the equation. When setting dt to a larger value, SpiroExplorer appears to move more quickly through the equation, whereas a smaller dt makes for a slower but smoother curve. The line() command draws each little bit of the line drawing from $xo.yo$ to x,y. You can modify update_xy() to use any function that sets x and y, like the Lissajous and circle equations (shown but commented out), or any other equation you can think of.

Listing 9-1: SpiroExplorer

```
float R,r,d;
float t,dt;
float x,y,xo,yo;
void setup() {
  size(400,400);
  framerate(15);
  background(127);
  strokeWeight(7);
  t = 0.0;    // time
  dt = 0.2;   // time increment
  R = 110;    // radius of outer fixed circle
  r = 64;     // radius of inner rolling circle
  d = 90;     // distance from center of rolling circle
  update_xy();
}
void update_xy() {
  xo = x; yo = y;   // save old values
  x = (R-r)*cos(t) + d*cos( ((R-r)/r)*t );    // hypotrochoid
  y = (R-r)*sin(t) - d*sin( ((R-r)/r)*t );    // hypotrochoid
  // x = (R+r)*cos(t) - d*cos( ((R+r)/r)*t ); // epitrochoid
  // y = (R+r)*sin(t) - d*sin( ((R+r)/r)*t ); // epitrochoid
  // x = R * sin(d*t);   // lissajous
  // y = R * sin(r*t);   // lissajous
  // x = r * cos(t);     // circle
  // y = r * sin(t);     // circle
  t += dt;
}
void draw() {
  update_xy();
  pushMatrix();
  translate(width/2,height/2);
  line(x,y, xo,yo);
  popMatrix();
}
```

Another way to explore these fun sets of curves is with a graphing calculator. Mathematica from Wolfram Research (available on all platforms) is the pinnacle of equation visualization but is expensive if you're not a student. Mac OS X comes with a free 2D/3D-graphing calcula-tor called simply Grapher that can graph extremely complex equations. It is easier to explore a much larger space of possible equations than SpiroExplorer, but it's not as interactive for a given equation. Figure 9-14 shows an example of it in use, graphing both a hypotrochoid and a Lissajous equation. For other platforms, there is the free online program GCalc at http:// gcalc.net/. It's a Java application or applet and will graph parametric equations. It's not as easy to use as Grapher, but it's a great free resource you can use from any computer anywhere.

Once you've explored how the equations change and have found some curves that look nice to you, it's time to try to draw them.

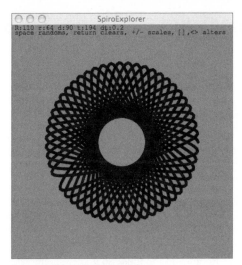

FIGURE 9-13: SpiroExplorer in use

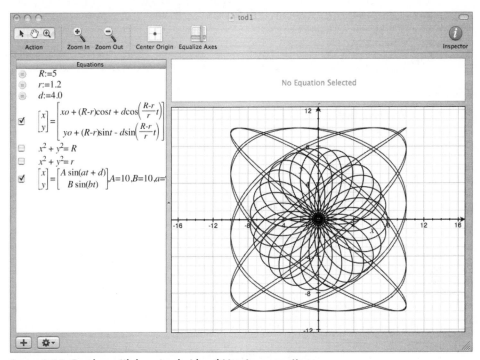

FIGURE 9-14: Grapher with hypotrochoid and Lissajous equations

Drawing Spirals with RoombaSpiro

Turning SpiroExplorer into a program that controls Roomba is pretty easy with the mathematical tools you already have available to you. The first step is to add the standard RoombaComm initialization and control statements to your Processing sketch, just like in Chapter 7. Copy and paste are your friend here. The next step is to figure how to transform the SpiroExplorer equations into Roomba commands.

For every time increment *dt* in SpiroExplorer, the line moves the old point *xo,yo* to the new point *x,y*. The resulting *dx,dy* incremental movement is actually a vector, and you can calculate the angle of that vector using the arctangent. From that angle you can find the radius of the circle to command Roomba with to follow that vector.

Listing 9-2 shows the updated `draw()` method with the radius calculation and the Roomba `drive()` command. Two different methods of finding the radius are shown. You can use either one as the radius for the `drive()` command. Note that the speed is constant since it represents time to the robot. These few extra lines turn SpiroExplorer into RoombaSpiro. In the initialization of RoombaSpiro, you may want to put Roomba in full mode so it keeps going when you pick it up.

The `rscale` value scales up the computed radius to a usable radius. Experiment with different `rscale` values to see how they affect the Roomba's behavior.

Listing 9-2: RoombaSpiro's draw()

```
float rscale = 4.0;
void draw() {
  pushMatrix();
  translate(width/2,height/2);

  float dx = (x-xo);
  float dy = (y-yo);
  float rth = atan2(dy,dx);
  float rx = ((dx/cos(rth))/dt) * rscale;
  float ry = ((dy/sin(rth))/dt) * rscale;
  if( roombacomm != null )
    roombacomm.drive(speed, rx);  // or, ry

  line( x,y, xo,yo);
  popMatrix();
}
```

The Result

Figure 9-15 is an example of the basic kinds of geometric art possible with Roomba and RoombaSpiro. It is a single color using a single set of parameters that stay constant. The marker was held in place by the ruler clamp and positioned in the space where the dirt bin would be. That configuration seems to give the most consistent and accurate results.

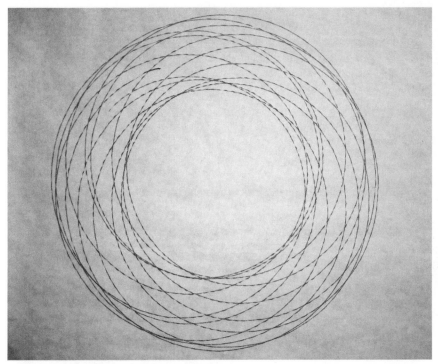

FIGURE 9-15: Drawing made with RoombaSpiro

Figure 9-16 shows the Roomba at work on a more complex piece that not only had two different colored markers but also had parameters that were being manipulated in real time. Figure 9-17 shows the result.

By letting Roomba operate for several minutes, dense, complex works can be created. The natural slippage of the wheels and pens guarantee that no two works will be the same.

Figure 9-18 shows a simpler work demonstrating some of these interesting slipping effects. It's done with two pens, one held by the wedge clamp and one taped. The outer taped one was purposefully allowed to be a little loose.

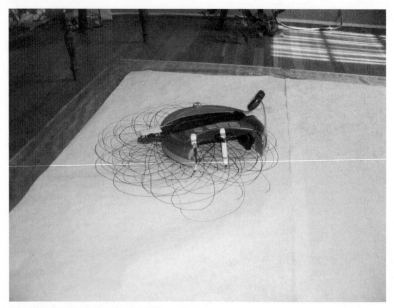

FIGURE 9-16: Roomba the artist in action

FIGURE 9-17: The resulting artwork

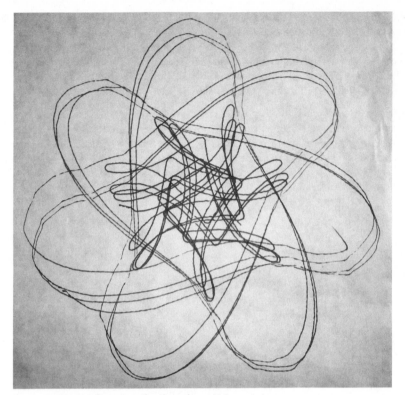

FIGURE 9-18: Another piece by the robo-artist

Drawing Other Types of Figures

Like SpiroExplorer, RoombaSpiro can use any parametric equation. In practice, the sudden turns caused by some equations (like most interesting Lissajous equations) make the Roomba jump around too much to make a consistent line. These jumps can be ameliorated somewhat by moving the Roomba at a very slow speed and by making the pen mount extremely secure. The smoothly changing hypotrochoid spirals are more forgiving of a loose pen.

Summary

The beginning of the chapter posed the question of whether robots can create art. In the case of physically making marks on paper, it turns out that your robot needs a lot of help from you to make it happen. But along with that technical help, you also make aesthetic judgments as to what parameters should be used by the equations used to control the Roomba. Art is a circular venture: it's created, observed, thought about, and created again using the ideas and opinions from previous art. Robots don't yet have opinions about art. When computers and robots begin to have an aesthetic sense, then we'll see true robotic art. This chapter is more about *robot-assisted* art. You are the creator of what Roomba draws. You make the color decisions, choose the algorithm and parameter values, and you determine how Roomba will hold its brush to maximize the effect you're after. The robot is simply a new tool to enable you to express yourself, one that takes a lot of the drudgery out of making certain types of art.

In fact, the techniques described in this chapter are just one interpretation, and you are likely envisioning improvements and changes that match your own sensibilities. Anywhere Roomba can drive can become a canvas for it to work upon. Any substance that makes a mark can be a brush for Roomba. Discover new ways of using Roomba artistically. The possibilities are limitless, even though Roomba itself is quite limited in its movements. Explore the space, make some art, take a picture of it, and upload it to the Roomba art gallery at roombahacking.com.

Using Roomba as an Input Device

The Roomba's sensors are designed to sense the world in very particular ways. Unlike our own "sensors" which have a wide sensing range and can be adapted for a variety of tasks, each of the Roomba's sensors are extremely limited. The limitation is partly due to cost reasons (this is a price-sensitive consumer device after all) and partly because creating high dynamic range durable sensors is hard. Current electronic vision sensors are low-resolution and require an enormous amount of space and computation when compared to even simple organic eyes. There has yet to be invented a touch sensor that responds as accurately and complexly as skin.

For now robots must make do with simple sensors that detect only a small bit of their environment. Such sensors are custom designed for a particular task and aren't meant for any other. But that doesn't mean *you* can't co-opt the sensors and make them do more.

The Roomba can act as a general-purpose input device. The sensors it normally uses to avoid obstacles and know its world can be turned upside down (literally as you'll see) and made to work as a multi-dimensional input device for whatever you can dream up. This chapter presents a few different examples of how to use the Roomba's inputs in ways its designers never imagined.

in this chapter

☑ Alternative uses for Roomba sensors

☑ Use Roomba as a mouse

☑ Make a theremin with Roomba

☑ Turn Roomba into an alarm clock

Ways to Use the Roomba's Sensors

As discussed in Chapter 7, Roomba has two different classes of sensors: internal and external. The internal sensors provide data about the internal Roomba state: how far it has gone, how much it has rotated, battery charge and drain, and so on. Another factor to consider is the sensor resolution. Most sensors are a single binary value: on or off, cliff detected or not, button pressed or not. A few Roomba sensors have greater resolution than one bit. It turns out that all the sensors with greater resolution than a single bit, except one (the dirt sensor), are internal sensors. Roomba needs accurate internal knowledge about its power system, so that makes sense. For the external sensors, it's usually easier to design a sensor for the physical world that unambiguously detects if a quantity is above or below a predefined value than it is to measure that quantity precisely.

Digital Sensors

All of the distance sensors Roomba employs are digital sensors. Whether it is the distance to the floor, to the wall, or from the wheels to the ground, all these distances are distilled down to a single Boolean value. Instead of "How far away is the wall?" the question is just "Is there a wall nearby?" The single-mindedness of these types of sensors makes them reliable but also harder to use for other purposes. However, by combining the readings from multiple sensors, or reading a single sensors multiple times, it may be possible to gather additional data. For example, by reading a button bit over time, you can determine how long the button was held down. A quick tap would mean one thing, but a longer hold would mean something else.

Analog Sensors

The only external sensor with a graduated value is the dirt sensor. This sensor doesn't seem as accessible to hacking because it appears to be tuned to the normal vacuuming environment (brushes moving, air moving past, and so on). It's difficult to get readings from the dirt sensor when Roomba is running its vacuum and brushes. The next most interesting analog sensor is the current drain value. By watching this value and the motor over-current sensor values, it may be possible to detect how hard Roomba is working as it moves its wheels. This could prove useful if the wheels are purposefully strained in a known way.

The distance and angle sensor values are a derived analog value from the digital sensors in the wheels. They offer high resolution but because they are "cooked" in a way that the other sensor values aren't, the distance values can sometimes be hard to use.

Using Roomba as a Mouse

The original computer mouse created in 1970 was a wooden box with two wheels mounted on its underside at right angles. When dragged along a desktop, one or both of the wheels would rotate in correspondence with the motion. The well-known ball mouse came soon after. The wheels were moved inside and a small rubber ball carried the mouse motion to the wheels.

Roomba has two wheels with sensors almost exactly like the sensors in a ball mouse. These sensors work and the data is available through the DISTANCE and ANGLE ROI commands even when Roomba isn't being driven. This means that the computeRoombaLocation() method used in several of your previous Roomba programs can be used verbatim. The difference in use now is that instead of using the *rx,ry* position pair from that function to represent the on-screen position of a controlled Roomba, you can use it as a virtual mouse pointer (in lieu of the *mouseX, mouseY* position pair) to represent how you are moving the robot.

Recall from Chapter 5 that Roomba only moves in straight or circular paths. This applies to it being either driven by its motors or positioned by you moving it. Figure 10-1 demonstrates some of the preferred ways Roomba moves. As discussed in Chapter 9, however, you can approximate almost any curve with many circle segments.

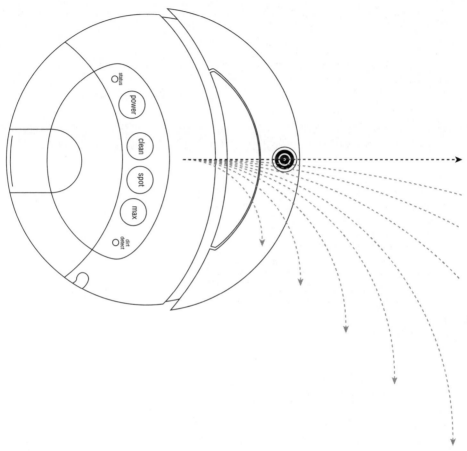

FIGURE 10-1: Roomba movements, right turns shown

Listing 10-1 shows the heart of RoombaSketch, a Processing sketch that turns Roomba into a mouse input for a vector drawing program. It has the following features:

- The distance and angle sensors become a virtual mouse pointer.
- The left and right bump sensors increment or decrement the drawing pen size.
- The Spot button becomes the main mouse button for starting or stopping drawing.
- The Clean button resets the cursor position to center of screen.
- The Power button quits the program.
- It draws a Roomba icon to show where the cursor is and how it's oriented.

The implementation is straightforward; the main hurdle is conceptual as you're using Roomba driving data without commanding the robot to move.

Instead of simply drawing lines as pixels onto the screen, an array of Line objects is created. Each Line holds an array of points that define the line. Each time draw() is called (determined by framerate), the current line is added if the Spot button is being held down. Each press and release of the Spot button creates a new Line object and thus a new line to be drawn.

Listing 10-1: RoombaSketch

```
Line[] lines = new Line[numlines];
int l = 0;
int strokeW = 5;
void draw() {
  computeRoombaLocation();   // same as before
  parseRoombaSensors();
  updateRoombaState();
  background(180);  stroke(0);
  for( int i=0; i<numlines; i++ )
    lines[i].draw();
  translate(rx,ry);
  rotate(rangle);
  image(rpic,-20,-20);
}
void parseRoombaSensors() {
  if( roombacomm.powerButton() ) {
      roombacomm.disconnect();
      System.exit(0);
  }
  if( roombacomm.cleanButton() ) {
    rx = width/2; ry = height/2;
    rangle = 0;
    strokeW = 5;
  }
  if( roombacomm.bumpLeft() ) {
    strokeW -- ;  if( strokeW<1 ) strokeW=1;
  }
  if( roombacomm.bumpRight() ) {
    strokeW++;  if( strokeW>100 ) strokeW=100;
  }
  if( roombacomm.spotButton() ) {
    if( drawing ) {
       if( rx != rxo && ry != ryo )
         lines[l].addPoint((int)rx,(int)ry,strokeW);
    }
    else {
      drawing = true;
      l++; l %= numlines;
      lines[l] = new Line();
```

Listing 10-1 *Continued*

```
      }
   }
   else if( drawing )
       drawing = false;
}
```

Figure 10-2 shows how you might hold Roomba and draw with it, while Figure 10-3 shows a drawing made with RoombaSketch. The ability to change the pen stroke width while drawing enables a much more fluid line than is possible with a normal mouse. You can create very organic drawings. Granted, as Figure 10-2 shows, using Roomba as a mouse requires a bit more physical movement than with a normal mouse, but with some people complaining that computer users don't get enough exercise, you can now point to the Roomba and say, "That's *my* mouse."

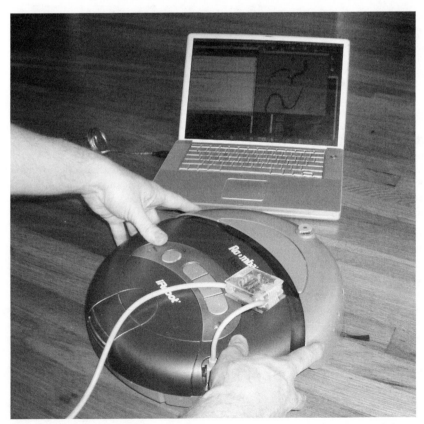

FIGURE 10-2: Using Roomba as a mouse

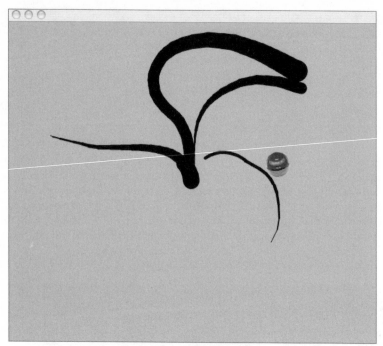

FIGURE 10-3: A drawing made with RoombaSketch

Using Roomba as a Theremin

The theremin is a strange and unique musical instrument. It's played without even touching it; placing your hands in front of it in particular ways adjusts its pitch and loudness. The Beach Boys song "Good Vibrations," old Star Trek episodes, and countless horror movies have used the theremin to good effect, so you have undoubtedly heard it. It produces a clear pure sine wave tone that glides between notes and sounds vaguely human-like.

Invented in 1919 by Leon Theremin, the theremin was the result of research into electrostatic proximity sensors. Figure 10-4 shows the inventor playing his instrument. The theremin consists of two antennae, one for controlling pitch and one for controlling volume. The pitch antenna is usually vertical and on the right-hand side of the player. The circuitry inside the theremin measures the varying capacitance between the player's body and the antennae and adjusts the loudness and pitch accordingly. If you've ever used your body to get better reception on your TV, you've experienced how the theremin works.

Making a Theremin Out of Roomba

The two Roomba dirt sensors are apparently capacitive-based and thus a likely candidate to use as a theremin input. Unfortunately, they do not seem to be tuned to picking up human-sized variations. But the other downward-facing sensors offer an alternative for the pitch control. By tilting Roomba left and right you can get a few bits of resolution by combining the cliff sensors with the wheeldrop sensors. Figure 10-5 shows how Roomba can be used to detect variations in tilt. The scale of the tilt is a bit exaggerated in the diagram to demonstrate the effect. Since the theremin glides from one pitch to the next, having relative pitch adjustments via tilting actually works pretty well.

FIGURE 10-4: Leon Theremin and his musical instrument

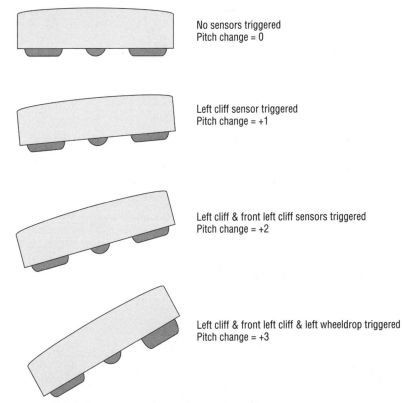

No sensors triggered
Pitch change = 0

Left cliff sensor triggered
Pitch change = +1

Left cliff & front left cliff sensors triggered
Pitch change = +2

Left cliff & front left cliff & left wheeldrop triggered
Pitch change = +3

FIGURE 10-5: Tilting Roomba for pitch control

Listing 10-2 shows the core of a Processing sketch called RoombaTheremin. As before, the draw() method calls parseRoombaSensors() and draws a growing vertical line showing pitch change. Figure 10-6 shows what the sketch looks like after running for a while and performing with it. Figure 10-7 shows the typical way you hold and use the robot when playing it with RoombaTheremin.

Listing 10-2: RoombaTheremin

```
int pitch = 70;
int dur = 6;
int pshift = 0;
void draw() {
  parseRoombaSensors();
  updateRoombaState();

  stroke(0);  strokeWeight(7);
  int x = pitch * 4;
  y = (y+1) % height;   // keep going down, but loop at height
  if( y==0 ) { background(180); }  // clear screen if back at
top
  line( x,y,  x,y+2);
}
void parseRoombaSensors() {
  if( roombacomm.cliffLeft() )        pshift++;
  if( roombacomm.cliffFrontLeft() )   pshift++;
  if( roombacomm.cliffRight() )       pshift -- ;
  if( roombacomm.cliffFrontRight() )  pshift -- ;
  if( roombacomm.wheelDropLeft() )    pshift++;
  if( roombacomm.wheelDropRight() )   pshift -- ;

  if( roombacomm.wheelDropLeft() &&
      roombacomm.wheelDropRight() &&
      roombacomm.wheelDropCenter() ) {
    ;     // all wheels down, do nothing
  }
  else { // otherwise, play
    pitch += pshift; pshift=0;
    roombacomm.playNote(pitch, dur);
  }
}
```

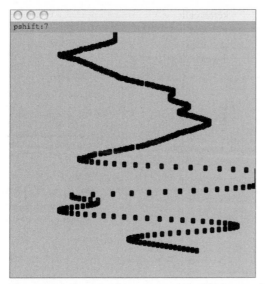

FIGURE 10-6: RoombaTheremin tracking tilt changes to modify pitch

FIGURE 10-7: The three main positions when performing with RoombaTheremin

Better Sound with the Ess Library

The necessary time delays between each note sent to Roomba makes for a less-than-convincing theremin simulation. Instead of a smooth tone characteristic of a theremin, the best that can be done on Roomba is a repetitive *beep-beep-beep* that is limited by the time it takes the robot to execute the SONG and PLAY commands.

An optional library for Processing called Ess makes it really easy to load, play, and manipulate sounds on your computer. It is built on JavaSound (the standard Java sound API), but is much easier to use than JavaSound normally is. For example, the following snippet is a complete Processing sketch that loads an MP3 file, pitch shifts it up five semitones, and plays it.

```
import krisker.Ess.*;
Ess.start(this);
AudioChannel myChannel = new AudioChannel("somesong.mp3");
myChannel.filter(new PitchShift(Ess.calcShift(7)));
myChannel.play();
```

You can use Ess instead of the `playNote()` command in RoombaComm. Granted then the sound is coming out of the computer instead of Roomba, but by using Ess's pitch shifting ability you can get a fairly smooth glissando from pitch to pitch.

Note   Ess is available at `www.tree-axis.com/Ess/` and, like any Processing library, is installed by unzipping its downloaded zip file into the `libraries` folder of the main Processing application. After restarting Processing, Ess is visible when you use the Sketch ⇨ Import Library menu command.

Touchless Sensing

Another way of triggering pitch changes that is perhaps more theremin-like is to flip Roomba over and use the cliff sensors as hand proximity sensors. Figure 10-8 shows how this works. By holding your hand over one or more of the cliff sensors, you un-detect a cliff. The cliff sensors are barrier sensors just like the wall sensor, but its logical sense is flipped and it's called a *cliff sensor* instead of a floor sensor. It's more logical (and marketable) to say cliff sensor instead of floor sensor.

But flipped over and used as a theremin, these cliff sensors work well detecting hands, as long as your hand has its fingers together and is mostly parallel to the surface of Roomba. Listing 10-3 shows a variation of the previous code, called RoombaThereminToo. It includes the Ess library method of making sound and has a modified version of `parseRoombaSensors()` to deal with the fact that it should change pitch when *not* detecting a cliff. The other features of that method are that by pressing both wheels down, it mutes the audio. Pressing the bump sensor starts the audio back up again.

You've probably noticed that `parseRoombaSensors()` doesn't actually change the pitch. It sets the `pshift` variable but doesn't act on it. That's because changing the pitch of a sound while it's playing causes an audible glitch. By waiting for the sound to finish playing through and changing its pitch before it loops again, the pitch transitions are smooth. In Ess, if the method `channelLoop()` is declared in your sketch, it will be called when a sound is finished playing. By adding that method and a check to see if `pshift` is actually set, the pitch (actually the `rate` here since it's a faster computation and yields the same result) can be changed seamlessly.

FIGURE 10-8: Alternate playing method, used in RoombaThereminToo

Listing 10-3: RoombaThereminToo, Using Ess and Hand-Waving

```
Channel mySoundA;
void setup() {
  // ... other standard setup
  Ess.start(this);
  mySoundA=new Channel();
  mySoundA.initBuffer(mySoundA.frames(250));
  mySoundA.wave(Ess.SINE,960,.75);
  pshift=0;
}
void parseRoombaSensors() {
  if( !roombacomm.cliffLeft() )        pshift++;
  if( !roombacomm.cliffFrontLeft() )   pshift++;
  if( !roombacomm.cliffRight() )       pshift -- ;
  if( !roombacomm.cliffFrontRight() )  pshift -- ;
  if( !roombacomm.wheelDropLeft() )    pshift++;
  if( !roombacomm.wheelDropRight() )   pshift -- ;

  if( !roombacomm.wheelDropLeft() &&
      !roombacomm.wheelDropRight() ) {
    pshift = 0;
    mySoundA.stop();
  }
  else if( roombacomm.bump() ) // bump to start playing
    mySoundA.play(Ess.FOREVER);
}
void channelLoop(Channel ch) {
  if( pshift != 0 ) {
   mySoundA.rate(1.0 + pshift*0.01);
   pshift = 0;
  }
}
```

Turning Roomba into an Alarm Clock

Roomba is now so familiar that it's almost like a pet. Why not have it sleep at your feet like a faithful dog? Except unlike a dog, with Roomba you can set the exact time when it will wake you up. And you can even make it turn on the radio for you.

Listing 10-4 shows the basics of the Processing sketch RoombAlarmClock, an alarm clock implemented on Roomba. It has the following features:

- **Alarm:** Beep and make noise at a particular wakeup time.
- **Snooze button:** Must hit both the left and light bump sensors.
- **Turn off alarm:** Pick up Roomba and press the Power button.
- **Turn on or off radio on iTunes:** Press the Clean button to play or pause.

The alarm and snooze times are determined by using Java `Date` and `DateFormat` objects. Alter the initial time for the alarm to go off (`wakeupTime`) and the length of the snooze (`snoozeSecs`) and run the sketch to start the alarm. When the alarm goes off, the method `playAlarm()` is called, which plays a tune and vibrates Roomba back-and-forth. Pressing both bumpers snoozes the alarm, while picking up Roomba and pressing the Power button turns off the alarm until the next day.

The radio is turned on and off with the `runRadioCmd()` method. This method uses `Runtime.exec()` to execute a system command. The particular command shown uses `osascript`, the Mac OS X command-line tool for running Applescript statements. The Applescript statement `tell app "iTunes" to playpause` will tell iTunes to play if it's paused or pause if it's playing. So before using RoombAlarmClock, set iTunes to the playlist or Internet radio station you'd like to wake up with. Then, when it's running, at any time press the Clean button to turn on the radio.

For other operating systems, change the `radioCmd` variable to any command-line command that you fancy. You could even add additional commands that are triggered for the other buttons or sensors. Perhaps the Max button runs a command that talks to an X-10 controller to turn on the lights. If you encapsulate these various commands into a shell script (or batch file on Windows), you can change it all you want without re-exporting the sketch in Processing.

In Chapter 13 you'll learn how to make Roomba fully autonomous. When fully stand-alone, having it be an alarm clock would be even more interesting. For example, it could run away from you as you try to turn off the alarm.

Note   If you use RoombAlarmClock to wake you up, don't put Roomba on your nightstand table. It may very well fall off. Roomba is quite sturdy, but why take the chance? Place Roomba on the floor.

Tip   You probably want to use the simpler serial tether rather than the Bluetooth adapter for the alarm clock. Some computers may power down their Bluetooth interface if it's idle. Regular serial ports don't have that problem.

Listing 10-4: RoombAlarmClock

```
String wakeupTime = "6/20/06 07:48 am";
Date wakeupDate = null;
boolean alarm = false;
int snoozeSecs = 60 * 9; // nine minutes
String radioCmd[] = {"osascript", "-e",
                     "tell app \"iTunes\" to playpause"};
void setup() {
  framerate(1);
  roombacommSetup();
  DateFormat
df=DateFormat.getDateTimeInstance(DateFormat.SHORT,

DateFormat.SHORT);
  try { wakeupDate = df.parse(wakeupTime); }
  catch( Exception e ) { println("error:"+e); }
  println("wakeupTime: "+wakeupDate);
 }

void draw() {
  if( roombacomm == null ) return;
  parseRoombaSensors();
  updateRoombaState();
  Date now = new Date();
  if( now.compareTo( wakeupDate ) > 0 )
    playAlarm();
}
void playAlarm() {
  alarm = true;
  println("playAlarm");
  int song[] = { 78,4, 77,4, 76,4, 75,4,
                 74,4, 73,4, 72,4, 71,4,
                 70,4, 69,4, 70,4, 71,4 };
  roombacomm.createSong(5,song);
  roombacomm.playSong(5);      // play rude song
  for( int i=0; i<5; i++ ) {  // and shudder a little
    roombacomm.spinLeftAt(75);  roombacomm.pause( 100 );
    roombacomm.spinRightAt(75); roombacomm.pause( 100 );
    roombacomm.stop();
  }
}
void runRadioCmd() {
  try {
```

Continued

Listing 10-4 *Continued*

```
        Process p = Runtime.getRuntime().exec(radioCmd);
        p.waitFor();
    } catch(Exception e) { println("exception:"+e); }
}
void parseRoombaSensors() {
    if( alarm ) {
        if( roombacomm.bumpLeft() &&
            roombacomm.bumpRight() ) {  // snooze
            wakeupDate.setTime(wakeupDate.getTime()+snoozeSecs*1000
);
            println("snooze until "+wakeupDate);
            alarm = false;
        }
        else if( roombacomm.wheelDropLeft() &&
                 roombacomm.wheelDropRight() &&
                 roombacomm.wheelDropCenter() &&
                 roombacomm.powerButton() ) {
            println("alarm off! (until tomorrow)");
            wakeupDate.setTime(wakeupDate.getTime() +
(60*60*24)*1000);
            alarm = false;
        }
    }
    if( roombacomm.cleanButton() )
        runRadioCmd();
}
```

Summary

Even though the Roomba's sensors are primitive, they can be put to some interesting uses. These uses need not be vacuum-related or even robotics-related. The cliff sensors are one of the best examples of this, becoming non-contact proximity sensors for hands or other movable objects when Roomba is turned upside down. While upside down, the wheeldrop sensors become buttons to trigger actions. You could even turn the wheels when upside down and register the movement as a variable function.

By combining the sensors into more complex aggregations, you can create a novel way to measure something Roomba normally cannot measure. Combining cliff and wheeldrop sensors gives a sense of tilt, but you can imagine other combinations as well. Perhaps a combination of buttons and driving motor over-current sensor could make a weight sensor and turn Roomba into a scale. Or using the distance and angle sensors and a little calibration could yield a Roomba yardstick. Try making Roomba a DJ input device and "scratch" audio files with it by rotating it back and forth like a record. The number of possible sensor combinations is huge and many of them produce useful results. Even those that produce non-optimal results are fun and instructive.

More Complex Interfacing

part

in this part

Chapter 11
Connecting Roomba to the Internet

Chapter 12
Going Wireless with Wi-Fi

Chapter 13
Giving Roomba a New Brain and Senses

Chapter 14
Putting Linux on Roomba

Chapter 15
RoombaCam: Giving Roomba Eyes

Chapter 16
Other Projects

Connecting Roomba to the Internet

The objects in our homes are becoming smart. Not just the obvious ones like the TV and stereo, but also the more mundane ones like the stove and vacuum cleaner. The "smart home" movement of a decade ago with its centralized house computer is giving way to the emergent phenomena of all the little parts of our homes becoming smart.

If you have a Roomba robotic vacuum cleaner, you're already aware of this. Roomba has more computing power than large corporations could afford in the 1960s. Imagine what bits of disposable computing will be present in our everyday devices a few decades from now.

A smart object is useful, but as anyone who has used the Internet can attest to, connecting smart objects with each other leads to entirely new and higher-level interactions. This network effect characteristic is so important, it's considered a field of study in and of itself. The effect has long been recognized (no one will go to a stock exchange with only a few traders, and a phone company with 10 users isn't nearly as useful as one with 10,000 users), but it took computers and the Internet to bring it into sharp focus. Network effects can apply to any aspect of a group that becomes more efficient or useful when a higher percentage of the group participates.

We do not yet live in a world where a large percentage of the objects in our lives are smart and networked, but we are on the brink. Often with network effects, once a critical mass of participants has been reached, all others in the group quickly follow suit. A recent example of this is e-mail: Before 1990 it seemed like no one had it, and then just a few years later suddenly everyone did.

Right now, only certain household objects are smart, and even fewer are networked. Recent advances in networking allow even the simplest smart device to communicate with others and to do it cheaply. This chapter discusses two of those types of devices that enable networking via Ethernet.

in this chapter

☑ About Ethernet

☑ Choosing the right embedded Net device

☑ Building an Ethernet adapter

☑ Using the SitePlayer Telnet

☑ Using the Lantronix XPort

☑ Updating RoombaComm for network use

Why Ethernet?

Both RS-232 and Bluetooth turn Roomba into a kind of networked object, but a subservient one. Objects communicating through simple serial protocols like those two require a computer translator to convert between the TCP/IP protocol used on the Internet and the simpler serial port protocol.

The object is beholden to a single computer and cannot function without it. Ethernet is the simplest and most pervasive physical protocol that can handle TCP/IP as a peer, rather than being subservient to a larger computer. However, even though it's the simplest, dealing with Ethernet and TCP/IP requires a great deal more processing power than a simple serial port.

What Is Ethernet?

You're likely very familiar with Ethernet as a user. The standard Ethernet jack is an RJ-45 connector that looks sort of like a fat 8-pin telephone jack. It is built into every computer now as the basic means of computer-to-computer connectivity. At its basic, Ethernet is a protocol for the physical transmission of serial data. It's different from RS-232 in many ways, but the two most notable are:

- Ethernet breaks all transmission streams into *frames*, a kind of data packet.
- Ethernet supports multiple devices on the same cable.

The packetization of data was an important and crucial step in the creation of the Internet. Before packets, computers were connected with serial cables or phone lines. These dedicated circuits joined just two computers together. If you wanted your computer to talk to another computer, you had to get another serial port or phone line. It wasn't a very scalable design, so Ethernet was created to allow multiple computers to use the same wire. In order to keep one computer from monopolizing the wire, data transmission was divided into frames and sent one after the other. Between frames other computers could interrupt to transmit their own data.

Ethernet is a physical layer protocol and thus is only valid on a local area network (LAN). To communicate between LANs, a higher-level protocol is needed, and the one everyone uses now is the Internet Protocol (IP). IP provides a common, open language for all networked devices and is the reason why the Internet works. If you then want to recreate a virtual direct connection between two computers on the Internet, you can use the Transmission Control Protocol (TCP), which exists on top of IP the way IP does on top of Ethernet. TCP and IP are so commonly used together that they're often called TCP/IP. TCP/IP doesn't require Ethernet to work. In fact, there are many alternatives to Ethernet. The most common alternative available to consumers is Wi-Fi (also known as 802.11b/g).

To create a device that exists on the Internet like any other, you need to implement TCP/IP and Ethernet (if using Ethernet). These are complex protocols, not something you can quickly whip up by hand. For Ethernet, there are several chips that provide that protocol's functionality. For TCP/IP, there are several tiny *stacks* (collections of interlocking software) that fit within modern microcontrollers and talk to these Ethernet chips.

In the open-source realm, one of the most popular TCP/IP stacks is uIP available from www.sics.se/~adam/uip/. This is a great stack that has been ported to many different types of microcontrollers. The downside of rolling your own embedded TCP/IP system is that the Ethernet chips are all surface-mounting and hard to solder for a hobbyist.

Fortunately, several companies have done the hard work of putting together an Ethernet chip, a microcontroller, and a TCP/IP stack and making it all function. To the user of these devices, they appear as a serial port on one side and an Ethernet port running a TCP/IP server on the other. The user hooks their circuit up to the serial port like any other and then connects to their device over the Internet.

Figure 11-1 shows a few of these embedded Ethernet devices designed to add TCP/IP and Ethernet capability to an existing device. The two devices focused on in this chapter are the SitePlayer Telnet by NetMedia and the Lantronix XPort. There are many others out there, and they're getting smaller all the time.

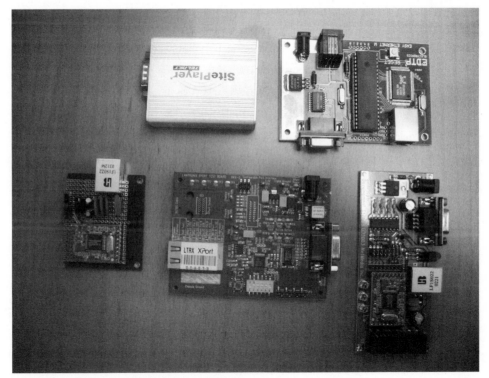

FIGURE 11-1: Various embedded Ethernet devices

Parts and Tools

In this chapter, I show you three ways to get your Roomba on the Net. Which way you choose affects which parts to get. The usual parts and tools described in Chapter 3 are also needed. Chapter 3 also covers voltage regulators, RS232 serial transceivers, and other sub-circuits discussed in these three projects.

To use your existing serial tether and a SitePlayer Telnet System box, you need:

- NetMedia SitePlayer Telnet System, NetMedia part number SPTS

And that's it. That isn't a very elegant solution, so if you want to build your own Ethernet-to-Roomba interface, the parts you need are:

- NetMedia SitePlayer Telnet Module, NetMedia part number SPT1
- LF1S022 10base-T filter, NetMedia part number FIL0011F
- Mini-DIN 8-pin cable, Jameco part number 10604
- 7805 +5 VDC voltage regulator, Jameco part number 51262
- 220-ohm resistor (red-red-brown color code), Jameco part number 107941
- Two 1μF polarized electrolytic capacitors, Jameco part number 94160PS
- 8-pin header receptacle, Jameco part number 70754
- Male snap-apart header, Jameco part number 160881
- General-purpose circuit board, Radio Shack part number 276-150
- Four 0.01μF ceramic disc capacitors, Jameco part number 15229

Finally, if you want to experiment with the Lantronix parts, the easiest path is to get the XPort evaluation kit:

- Lantronix XPort Evaluation Kit, Mouser part number XP100200K-03

NetMedia sells the SitePlayer modules and 10base-T filters direct from the company's website, at `http://siteplayer.com/`. The best way to get Lantronix parts is through Mouser, at `http://mouser.com/`.

SitePlayer Telnet

Several years ago NetMedia came out with the SitePlayer module. The SitePlayer is a small (about 1 sq. in.) and cheap ($29) website co-processor meant to integrate to an existing microcontroller. On the SitePlayer you could store a tiny dynamic website. The web pages on that site could change memory values in the SitePlayer or toggle pins, and the microcontroller could do the same. It gained renown among microcontroller hobbyists as the cheapest and easiest way to get a project on the Net. What it couldn't do was create a straight tunnel between the Net and a serial port.

In late 2004 NetMedia released the first version of the SitePlayer Telnet (SPT), the same hardware as a normal SitePlayer, at the same price, but specially configured to be an Ethernet-to-serial converter. It runs a Telnet server whose input and output go to a serial port. The initial versions of the SitePlayer Telnet were a little buggy. With the firmware released at the beginning of 2006, it became a robust and usable device. You can download all the documentation about the SPT, as well as read forums of others using the device, at `http://siteplayer.com/telnet/`.

Figure 11-2 shows what a SitePlayer Telnet module looks like. The two black legs are two 9-pin female headers that plug into a circuit you create. NetMedia also sells a ready-to-run configuration called the SitePlayer Telnet System, shown in Figure 11-3. Really it's just a few connectors on a simple circuit board surrounded by a small aluminum box, but having all that already done for you is a real time saver. Figure 11-4 shows what the system box looks like on the inside.

FIGURE 11-2: SitePlayer Telnet module

Notice that Figure 11-4 contains a 7805 5 VDC voltage regulator. Underneath the SitePlayer module is a MAX232 transceiver chip, which is also familiar. You could build an exact duplicate of this circuit if you wanted (in fact NetMedia provides the schematic for it). The only slightly oddball part is the LF1S022, which is a 10base-T filter and RJ-45 jack.

FIGURE 11-3: SitePlayer Telnet System box

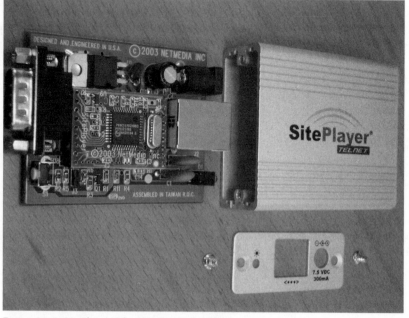

FIGURE 11-4: SitePlayer Telnet System internals

Creating Your Own SitePlayer Telnet Roomba Adapter

The SitePlayer Telnet box is very nice, but somewhat expensive. And if you don't need true RS-232 but instead need 0-5V TTL serial (like with the Roomba), it's wasteful. Figure 11-5 shows a schematic based on the box design, but simplified. The circuit consists of really just the SitePlayer Telnet module, the RJ-45 jack, a voltage regulator, and a Mini-DIN 8-pin cable. The RJ-45 jack needs a few resistors and capacitors to help filter out any noise picked up on the Ethernet cable, and that's about it.

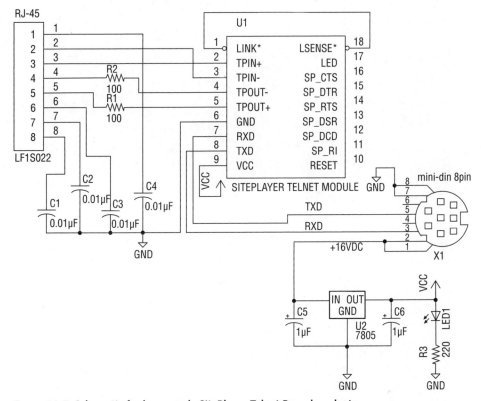

FIGURE 11-5: Schematic for homemade SitePlayer Telnet Roomba adapter

Figure 11-6 shows the constructed circuit. Construction of it is very similar to the serial tether, and all the same techniques and tools used for the serial tether should be used here. The SitePlayer Telnet module is attached by way of female header modules like those used for the Bluetooth module in Chapter 4.

Caution If you do build your own SitePlayer Telnet Roomba adapter circuit, be sure to thoroughly test it before plugging it in, and when plugging it in, use an Ethernet hub or switch. If anything goes wrong, it is better to destroy a $30 hub than the built-in Ethernet port on your computer's motherboard.

FIGURE 11-6: Finished SitePlayer Telnet Roomba adapter

LF1S022 RJ-45 Jack

The only odd part of the SitePlayer circuit is the RJ-45 jack. It is designed for 10base-T Ethernet and is available cheaply from NetMedia. Each row of pins has the standard 0.1″ spacing that matches breadboard spacing, but the rows are offset from each other by half a row. You can get around this by inserting the jack into a breadboard at an angle and then carefully rotating it to bend the pens slightly. Figure 11-7 shows what the pins look like after doing this. This weakens the jack pins and surely causes other problems, but for prototyping it works fine. If you're squeamish about doing this, SparkFun sells an RJ-45 breakout board for 95 cents.

Note You may be tempted to forgo the RJ-45 jack and wire an Ethernet cable directly to the circuit, as was done with the serial tether's DB-9 cable. Do not do this. Ethernet is a much faster protocol than RS-232 and more care must be taken with cable interconnects and signal filtering. The LF1S022 jack deals with all of that for you.

Roomba Prototyping Plug

The constructed circuit in Figure 11-6 shows a different method of connecting the Mini-DIN 8-pin Roomba cable. Instead of directly soldering the wires of the cable to the circuit board, you can install on the circuit board a female header like that used for the SitePlayer or

Bluetooth module. For the cable, solder header pins in a way that makes the most sense to you. Figure 11-8 shows one possible wiring of such a cable. For that cable, the power and ground pins are separated by one pin spacing from the data pins.

FIGURE 11-7: RJ-45 jack with the bent pins for breadboard insertion

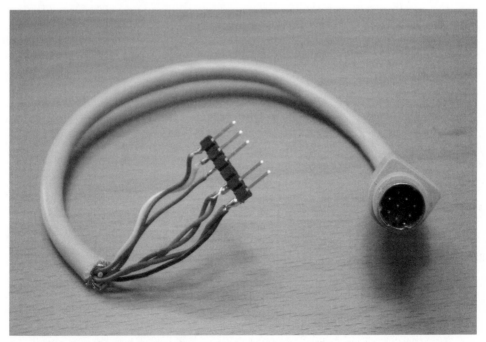

FIGURE 11-8: Roomba prototyping plug

Setting Up the SitePlayer Telnet

Since the SitePlayer Telnet silver box and the homemade SitePlayer Telnet Roomba adapter have the same SitePlayer Telnet module inside, configuration is identical for both of them. What is being configured is the SitePlayer Telnet module. The SitePlayer Telnet is configured entirely through its built-in web configuration page. The SitePlayer Telnet is easier to use because it supports ZeroConf/Rendezvous/Bonjour to announce its name and IP address to any compatible system. On the Safari web browser in Mac OS X, for example, looking in the Bonjour bookmarks list automatically shows the SitePlayer. For Windows, NetMedia provides a small Bonjour Browser tool to help with auto-discovery. On Linux, there are several tools, some built into web browsers. The SitePlayer also supports DHCP and will get a DHCP address from your DHCP server.

Note Bonjour (also known as Rendezvous or ZeroConf) networking is an open source standard for automatic discovery of network devices and services on those devices. It works by sending out multicast DNS announcement packets on a default link local IP address. To find out more about how Bonjour works and to get source code with which to experiment, see `http://developer .apple.com/networking/bonjour/`.

To get started configuring a SitePlayer Telnet device, plug it and your computer into the same Ethernet hub and plug the hub into your home network. That last step isn't strictly required for testing, but you'll want both your computer and the SitePlayer Telnet on the Internet eventually. Plug in the SitePlayer Telnet power adapter and in a few seconds it will be on your network.

Use either Safari (Mac OS X) or the Bonjour Browser (Windows) to watch for the SPT to show up. Its Bonjour name will be something like SitePlayer Telnet (7A12AD). The digits in parentheses are the last three digits of the SitePlayer Telnet MAC address. Select the SitePlayer Telnet and the browser will open showing the status page/homepage of the SitePlayer Telnet (see Figure 11-9). If you cannot find the SitePlayer Telnet on your network, you can use some of the debugging techniques mentioned in the "Debugging Network Devices" sidebar.

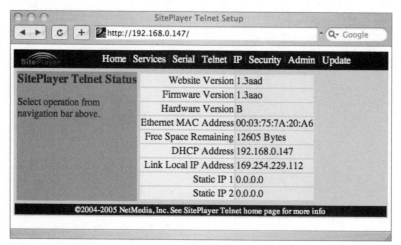

FIGURE 11-9: SitePlayer Telnet configuration home page

Configuring the SitePlayer Telnet is easier than the original SitePlayer because no tiny website needs to be built, compiled, and uploaded to the SitePlayer. The serial port parameters do need to be specified, however, to be compatible with the Roomba. Figure 11-10 shows the SPT serial configuration page with the parameters set correctly. Set the parameters to match and click Set Parameters.

If you're concerned about other people changing settings on your SitePlayer, you can change the username and password to something secret on the Security tab.

Note For a complete description of getting the SitePlayer Telnet on the network, see the SitePlayer Telnet manual.

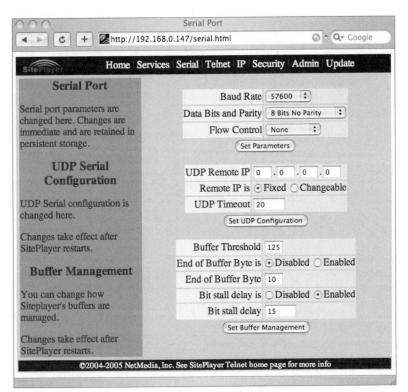

FIGURE 11-10: Configuring the SitePlayer Telnet serial port

Debugging Network Devices

Even though the SitePlayer Telnet makes an easy time of getting it on the Net, not all devices are so agreeable, and you may even run into problems where the SitePlayer Telnet appears to be acting up. In those cases, it's good to know how to debug network devices. Modern operating systems provide all the tools you need to debug most network problems.

First, note that there are usually three levels of addressing what happens on the Internet:

- **DNS name:** The human-readable name, like `www.yahoo.com`. It is sort of like a person's full name. The DNS name is made for humans; computers don't need it. A DNS server is a device that acts like a phone book and maps DNS names to IP address numbers. There are many giant DNS servers on the Internet, for the big namespaces like `.com` and `.net`, but your ISP also runs one for your neighborhood network, and your router may be running one too.

- **IP address:** The number obtained from the DNS name used on the Internet. An example is `66.94.230.42`. It is analogous to a person's phone number, except instead of being made up of seven single digits, it's made up of four 8-bit numbers, usually written with periods between them. Your router may have a DHCP server, which is a device that dynamically assigns IP addresses to new devices on your personal network.

- **MAC address:** The hardware address used by the Ethernet and Wi-Fi protocols. It is analogous to a person's home address, as it identifies the actual physical device, whereas IP addresses can be moved from device to device. Every network device contains its own mapping of IP addresses to MAC addresses for the addresses it has seen. This mapping is called an ARP cache. If a network device needs to use an IP address for which it doesn't know the MAC address, it announces that it doesn't through the Ethernet cable and hopes that someone responds.

The expansion of all the acronyms above isn't that important (and in fact can be counter-intuitive) compared to their functions. If you'd like to know more about the different protocols mentioned above, one of the best sources is Wikipedia (`www.wikipedia.org`).

Debugging DNS Names

DNS can be debugged with almost any network program, since they all perform a DNS lookup. The preferred tools of choice however are `host` and `dig`, with `nslookup` as the standby if those are not present. If your computer does not have the host program, search on the Net and get it as it's much easier to use than the others.

The fastest way to test whether a DNS name resolves to an IP address is to type **host** and the DNS name you're curious about. Here are two examples of the `host` command, one finding a DNS lookup and one not:

```
% host www.wiley.com
```

Debugging Network Devices *Continued*

```
www.wiley.com has address 208.215.179.146
% host www.wiley.cmo
Host www.wiley.cmo not found: 3(NXDOMAIN)
```

If you want to test a particular DNS server, append its IP address to your command. For example, if your router is at 192.168.1.1 and it runs a DNS server, typing the following will cause `host` to explicitly query that nameserver:

```
% host www.wiley.com 192.168.1.1
```

Debugging IP Addresses

You may not be using a DNS server on your home network, but you most certainly are using IP addresses. Every device must have an IP address to be a network citizen. Your home network is likely a private network with an address starting with `192.168` and your router translates requests between your private network and the Internet. This has the effect of giving your household one public address, but inside your house you can have as many networked gizmos as you like.

The first tool to turn to is `ping`. Sort of like a submarine's sonar, `ping` makes a special noise (an echo request packet) and listens for the echo. Network devices aren't required to respond to pings, but most do. If your router is at 192.168.1.1, then pinging it looks like:

```
% ping 192.168.1.1
PING 192.168.1.1 (192.168.1.1): 56 data bytes
64 bytes from 192.168.1.1: icmp_seq=0 ttl=64 time=2.125 ms
64 bytes from 192.168.1.1: icmp_seq=1 ttl=64 time=2.012 ms
64 bytes from 192.168.1.1: icmp_seq=2 ttl=64 time=1.787 ms
^C
--- 192.168.1.1 ping statistics ---
3 packets transmitted, 3 packets received, 0% packet loss
round-trip min/avg/max/stddev = 1.787/1.952/2.125/0.128 ms
```

Usually you either ping something you know about and want to see if it's still there (like the preceding example), or you want to see what is on your Internet. In the latter case, the `nmap` tool is a great resource. You could ping each address from 192.168.1.1 to 192.168.1.254, but that would take too long. Some operating systems and routers support a broadcast ping to the special broadcast address of 192.168.1.255, but it's not consistent. `nmap` solves the problem by pinging each IP address in turn for you and then summarizing the results. On a sample network (192.168.0.x), the network scan looks like the following; the notation 192.168.0/24 is how you specify the 192.168.0.x network, whereas -sP tells `nmap` to perform a ping scan:

```
% sudo nmap -sP 192.168.0/24
Starting Nmap 4.03 ( http://www.insecure.org/nmap/ ) at ↵
2006-06-28 11:42 PDT
```

Continued

Debugging Network Devices *Continued*

```
Host gw.home (192.168.0.1) appears to be up.
MAC Address: 00:06:29:15:F3:13 (IBM)
Host openwrt.home (192.168.0.8) appears to be up.
MAC Address: 00:16:B6:DA:91:2F (Cisco-Linksys)
Host nasty.home (192.168.0.9) appears to be up.
MAC Address: 00:0D:A2:01:04:70 (Infrant Technologies)
Host minimi.home (192.168.0.25) appears to be up.
MAC Address: 00:11:24:77:84:FA (Apple Computer)
Host 192.168.0.134 appears to be up.
Host 192.168.0.136 appears to be up.
MAC Address: 00:11:24:3F:3A:50 (Apple Computer)
Host 192.168.0.144 appears to be up.
MAC Address: 00:13:10:3A:16:17 (Cisco-Linksys)
Host 192.168.0.148 appears to be up.
MAC Address: 00:30:65:06:63:67 (Apple Computer)
Host 192.168.0.149 appears to be up.
MAC Address: 00:04:20:03:00:34 (Slim Devices)
Nmap finished: 256 IP addresses (9 hosts up) scanned in 7.459
seconds
```

To find errant network devices, first do a scan with the device disconnected; then do a second scan with it plugged into your network. There should be an extra host in the nmap list that will be your gadget.

nmap is capable of much more than simple ping scanning. It's an awesomely powerful network scanning tool that happens to be free, open-source, and available for Mac OS X, Linux, and Windows. It was also featured briefly in the second *Matrix* movie when Trinity used it to learn about an enemy computer system.

Debugging MAC Addresses

The most poorly understood of the three address layers are MAC addresses. It's not understood mainly because it works so well no one has to think much about it. The Ethernet hubs, switches, and routers used by everyone do the MAC routing for us. (The MAC acronym has nothing to do with Macintosh computers.) For a computer connected to the same hub as a networked gadget in question, that computer will store the gadget's MAC address in its ARP cache as soon as either of them try to talk to one another. The way you inspect this mapping of IP address to MAC address is with the arp command:

```
% arp -a
gw (192.168.0.1) at 0:6:29:15:f3:13 on en1 [ethernet]
nas (192.168.0.9) at 0:d:a2:1:4:70 on en1 [ethernet]
minimi (192.168.0.25) at 0:11:24:77:84:fa on en1 [ethernet]
? (192.168.0.131) at 0:2:2d:c:11:1 on en1 [ethernet]
? (192.168.0.139) at 0:4:20:5:2:d1 on en1 [ethernet]
```

Debugging Network Devices *Continued*

The `arp` command in the code is from a computer on the same network as the previous `nmap` command. Notice how the computer doesn't know the MAC addresses of every device on the network because it hasn't talked to all of them. When displaying ARP entries, the `arp` command does a DNS lookup of the IP address and displays the name if there is one. If there is no name, it displays `?`.

Other Tools

There are many other useful tools in addition to the above. The `ifconfig` (`ipconfig.exe` on Windows) command shows you what IP address your computer has. The `telnet` program is a great way to connect to arbitrary ports of any IP address. By telnetting to port 80, you can see what web servers and browsers are really saying to each other. The `traceroute` (`tracert.exe`) command will show the path through the Net that your packets take. The `netstat -r` command will show you what routers your computer knows about.

Testing SitePlayer Telnet

When your SitePlayer Telnet–based gadget is on the Net and configured correctly, you should test it before sticking on your Roomba. Fortunately, in the case of the SitePlayer Telnet System box, you can do a full end-to-end test using the equipment from previous chapters. The plan is to use your computer to go from Ethernet to SitePlayer Telnet System to USB-to-serial adapter and back to the computer.

Start with the Keyspan USB-to-serial adapter (or other serial port adapter) and plug it into your computer. Connect a DB-9 female to DB-9 female cable to the computer, and connect the other end of the cable to the SitePlayer Telnet System box. Plug an Ethernet cable into the SitePlayer Telnet System box and into an Ethernet hub. Finally, plug your computer into the Ethernet hub, too. Figure 11-11 shows the completed loop.

When everything is connected and turned on, open up both a serial port terminal program like ZTerm or RealTerm and a terminal window like Terminal or CMD.EXE. In the serial terminal, open the serial port like before. In the terminal window, telnet to the SitePlayer Telnet IP address. When it connects, anything you type in one terminal window shows up in the other. Congratulations, you've just created the world's smallest multi-protocol network (and perhaps the most pointless network for not testing serial-to-Ethernet devices).

Hooking the SitePlayer Telnet to Roomba

Now that you've fully verified the SitePlayer and know that it works from both ends, it's time to hook it up to your Roomba. The easiest way to do that is with the SitePlayer Telnet System box and the serial tether created in Chapter 3. Connect the serial tether as normal, plug the tether into the SitePlayer Telnet System box, and then plug the SitePlayer Telnet into an Ethernet hub. Figure 11-12 shows this setup.

FIGURE 11-11: Full end-to-end test using the SitePlayer Telnet System box

FIGURE 11-12: SitePlayer Telnet System hooked up to Roomba by way of serial tether

To power the SitePlayer Telnet box you can either plug it in normally with its wall wart or fashion a power plug from an old DC adapter. If you recall, the SitePlayer Telnet box uses a standard 7805 voltage regulator, so it can handle the 16 VDC from the Roomba. You could fashion a good reusable connector if you plan on using the SitePlayer Telnet System box in this configuration, or you could do what was in Figure 11-12 and create a simple power tap. If you've built your own SitePlayer Telnet Roomba adapter circuit, as in Figure 11-13, you don't have to worry about power problems because the circuit has its own voltage regulator and taps the Roomba battery in the standard way via the Mini-DIN 8-pin plug.

FIGURE 11-13: SitePlayer Telnet Roomba adapter hooked up to Roomba

If you know how to write network programs, you can immediately start sending ROI commands to Roomba by connecting to the telnet port (port 23) on the SitePlayer Telnet. As a quick experiment, try sending the byte sequence 0x80,0x82,0x86 (with 100msec pauses in between), and Roomba should start doing a spot clean. For example, if you have Perl and netcat (nc) on your system, you could do this:

```
% perl -e '@b=(0x80,0x82,0x86); for(@b){ printf("%c",$b);sleep(1);
}' \
    | nc 192.168.0.147 23
```

The little Perl program outputs the bytes with a one-second delay between them, and netcat connects the Perl program to your Roomba's SPT.

Lantronix XPort

Lantronix was one of the first companies to produce an embedded device server for serial devices, thus enabling those devices to be on the Internet. Their Cobox Micro was a module very similar in look to the SitePlayer, but included an Ethernet jack. The XPort is a miniaturization of the Micro. The entire device server fits inside of a slightly elongated Ethernet jack. Figure 11-14 shows what an XPort looks like not connected to anything. It is tiny. It's hard to believe there's really a computer in there.

FIGURE 11-14: Lantronix XPort

While the SitePlayer is very much aimed toward the hobbyist and includes such hacker-friendly things as standard breadboard pin spacing and 5V tolerant inputs, the XPort is aimed at the professional device integrator. It has 3.3V inputs and a high-density spacing. These two factors make it a little harder for the typical hacker to use. The XPort evaluation board, shown in Figure 11-15, converts it to a more hacker-friendly format. It still requires 5VDC power, so a small power supply to use the Roomba's battery is needed. Notice how much more complex it seems compared to the SitePlayer Telnet System box. This is partially because it is an evaluation board and so has extra parts to let engineers properly evaluate the device, but also because integrating the XPort just takes more infrastructure when dealing with the 5 VDC and RS-232 world of most hackers.

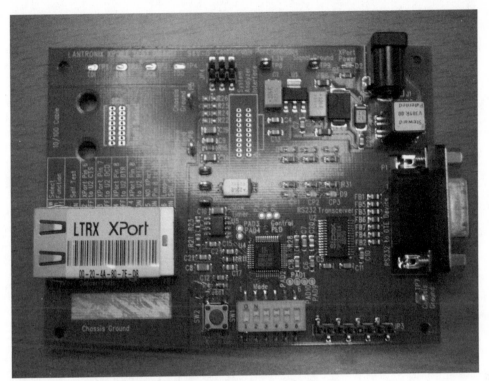

FIGURE 11-15: Lantronix XPort evaluation board

If you're designing a new system (which will thus likely run on 3.3V) and it needs to be small, the XPort is a great product. For hackers, or if you're adding to an existing 5V system, the SitePlayer or the Cobox Micro is better. Many prefer the SitePlayer because it has a slightly more modern configuration style.

Configuring the XPort

In its default configuration, the XPort responds on a single IP address and on a number of different ports:

- **Port 10001:** Serial-to-Ethernet gateway
- **Port 9999:** Text-based configuration
- **Port 80:** Web-based configuration

Unlike the SitePlayer, which uses the cross-platform ZeroConf/Bonjour/Rendezvous protocol to help you auto-discover it, the XPort assumes you will use their Windows-based DeviceInstaller. If you use Windows, go ahead and use that. If you don't have Windows, you can still configure the XPort using the networking debugging techniques mentioned in the "Debugging Network Devices" sidebar earlier in this chapter.

The XPort will ask your network's DHCP server for an IP address so that you can connect to it. One method of finding its IP address is to do one `nmap` network scan before plugging the XPort in and then another `nmap` network scan after plugging it in. The extra IP address that appears is the XPort.

Somehow the slightly annoying Java applet that is the XPort's web interface manages to be both simpler and more confusing than the SitePlayer (see Figure 11-16). Thankfully, the only parameters that you *need* to change are at the very top. Change the serial parameters to be Roomba compatible (57600 bps 8N1). First, make sure Serial Protocol is set to RS232. Next, set the Speed setting to 57600, the bits per second speed the Roomba expects. Then, set Character Size to 8, for 8-bit bytes, and the Parity to None. Finally, change the Flow Control setting to 2, which means no flow control to the XPort. Leave all other settings alone. Click Update Settings to complete the changes. You could also use the simpler Telnet configuration interface to do the same thing. The Lantronix WiMicro Wi-Fi module in Chapter 12 is configured almost exactly the same as the XPort and you'll use Telnet to configure it.

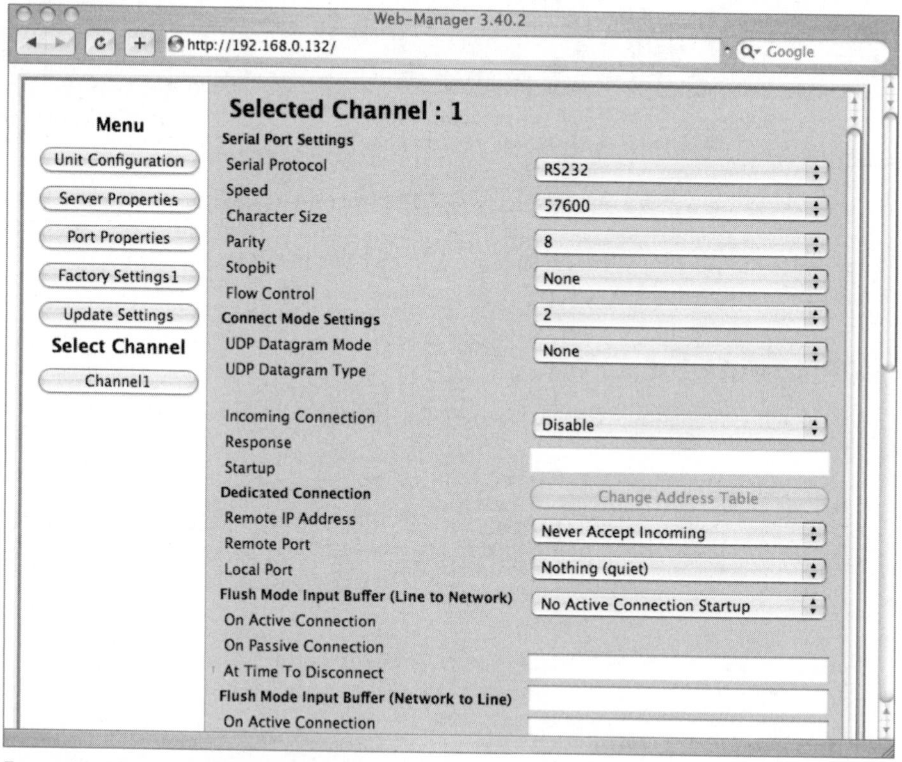

FIGURE 11-16: XPort web interface

There is no username and password combination on either the web interface or the Telnet interface. You can enable a password on the Telnet configuration interface or disable the Telnet interface entirely. This doesn't affect the web interface, which has no password protection, but you can also disable that interface too.

Note Be sure to read the XPort documentation thoroughly about configuration, especially if you're using Windows.

Using the XPort

When the XPort is configured, using it is just like using the SitePlayer Telenet. The only difference is the Telnet port. SitePlayer Telnet uses the standard port 23, while XPort uses port 10001.

In lieu of creating a custom circuit board for the XPort, the evaluation board is small enough to be mounted on the top of Roomba. To supply power, either bring out the 5V from the power supply in the serial tether and attach it to the 5V input of the evaluation board or build a small 5V power supply.

Going Further with XPort

Lantronix did a smart thing in making their line of embedded device server products similar to each other. If you choose not to use the XPort, the Micro or Mini modules might be appropriate for you. If you want Wi-Fi connectivity instead of Ethernet, the WiPort or WiMicro modules are replacements for their Ethernet cousins. The WiPort will be covered in detail in the next chapter.

Modifying RoombaComm for the Net

You now have Roomba on the Net, but all the code you've created thus far has been designed for the serial port. In Java, as in most modern languages, dealing with serial ports or Ethernet ports is fairly similar. The RoombaComm library uses that fact to create a new subclass of the RoombaComm base class that knows how to deal with the TCP telnet port that both the SitePlayer Telnet and XPort produce. This new subclass is called RoombaCommTCPClient, and most of it is shown in Listing 11-1.

Listing 11-1: RoombaCommTCPClient

```
package roombacomm.net;
import roombacomm.*;
public class RoombaCommTCPClient extends RoombaComm
{
    String host = null;
```

Continued

Listing 11-1 *Continued*

```java
int port = -1;

Socket socket;
InputStream input;
OutputStream output;

public RoombaCommTCPClient() {
    super();
}
// portid is "host:port"
public boolean connect(String portid) {
    String s[] = portid.split(":");
    if( s.length < 2 ) {
        logmsg("bad portid "+portid);
        return false;
    }
    host = s[0];
    try {
        port = Integer.parseInt(s[1]);
    } catch( Exception e ) {
        logmsg("bad port "+e);
        return false;
    }
    logmsg("connecting to '"+host+":"+port+"'");
    try {
        socket = new Socket(host, port);
        input  = socket.getInputStream();
        output = socket.getOutputStream();
    } catch( Exception e ) {
        logmsg("connect: "+e);
        return false;
    }
}
public boolean send(int b) {  // will also cover char
    try {
        output.write(b & 0xff);   // for good measure
        output.flush();
    } catch (Exception e) {
        e.printStackTrace();
        return false;
    }
    return false;
}
// ...other methods to implement RoomabComm...
}
```

You should note two key differences in `RoombaCommTCPClient`. First, notice that the `String` argument to `connect()` goes from being a serial port name to being a host:port combination. The host is the IP address of the Ethernet-to-serial device and the port is either port 23 (for SitePlayer) or port 10001 (for XPort).

The second thing to note is that Java uses the exact same objects (`InputStream` and `OutputStream`) to represent reading and writing data over a network as it does for communicating over a serial line. This means that most of the code like `send()` can be almost exactly the same. For network devices, the `Socket` object provides `InputStreams` and `OutputStreams`; for serial ports, the `SerialPort` object does.

The `updateSensors()` and associated code to read data back from Roomba aren't shown, but they are largely the same. Unlike `SerialPort`, which runs a separate thread and provides an EventListener interface, Java's `Socket` doesn't. So a standard EventListener-like `thread` is created to periodically look for input. When information arrives, the EventListener buffers it and calls an internal event method to deal with the data, just like `RoombaCommSerial.serialEvent()`.

All of the example programs in this book thus far have explicitly created `RoombaCommSerial` objects. This was done to make things more obvious, but all of the example programs can be quickly changed to use another subclass of `RoombaComm`. Listing 11-2 shows a version of the familiar SimpleTest example program, very slightly modified to use `RoombaCommTCPClient`. In fact, the only modification necessary is changing what type of RoombaComm object is instantiated and to remove the serial-specific parameter setting.

Similarly, all of the Processing sketches can quickly be modified to use `RoombaCommTCPClient` instead.

Listing 11-2: SimpleTest, for Networked Roombas

```java
package roombacomm.net;
import roombacomm.*;
public class SimpleTest {
    static boolean debug = false;

    public static void main(String[] args) {
        String portnamem = args[0];
        if( !roombacomm.connect(portname) ) {
            System.out.println("Couldn't connect to
"+portname);
            System.exit(1);
        }

        System.out.println("Roomba startup on port "+portname);
        roombacomm.startup();
        roombacomm.control();
        roombacomm.pause(50);
```

Continued

Listing 11-2 Continued

```
        System.out.println("Checking for Roomba... ");
        if( roombacomm.updateSensors() )
            System.out.println("Roomba found!");
        else
            System.out.println("No Roomba. :(  Is it turned
    on?");

        System.out.println("Playing some notes");
        roombacomm.playNote(72,10);   // C
        roombacomm.pause(200);
        roombacomm.playNote(79,10);   // G
        roombacomm.pause(200);
        roombacomm.playNote(76,10);   // E
        roombacomm.pause(200);

        System.out.println("Spinning left, then right");
        roombacomm.spinLeft();
        roombacomm.pause(1000);
        roombacomm.spinRight();
        roombacomm.pause(1000);
        roombacomm.stop();

        // ...and so on...
        System.out.println("Disconnecting");
        roombacomm.disconnect();
    }
}
```

Summary

Getting a Roomba on your LAN is pretty easy with the right tools. Now anyone on your network can access Roomba and run the programs you write for it. No special serial port drivers are needed, just an Internet connection. The Ethernet tether turns out to be a pretty good replacement for the serial tether because dealing with Ethernet, as a user, is just simpler. The Ethernet module and your computer do all the hard work. Ethernet has the added benefit of giving you much longer cable lengths, up to 100 meters (325 feet). For the next chapter, the Ethernet tether can function in a similar support role for the Wi-Fi adapter as the serial tether did for the Bluetooth adapter: providing a known-good interface that is the same in all ways except one is wired and the other wireless.

Both the SitePlayer and the XPort are good embedded device servers. For Roomba hacking, the SitePlayer is a bit more appropriate, but the XPort is more useful if you're trying to add network capability to devices with less available space. For example, if you wanted to put your coffee maker or alarm clock on the Internet, the extra space savings the XPort affords could be critical.

Modifying RoombaComm to use a networked version of Roomba was easy. And while modifying your existing programs and sketches to use the new, networked RoombaComm is a little clunky, no doubt you have some ideas on how to make it work. Java has some patterns for dealing with this situation, and they're easy to add. Both the SitePlayer and the XPort support a UDP mode instead of the Telnet-like TCP. UDP is connectionless, making you deal with packets of data instead of streams. For most cases, TCP is preferred, but you may like dealing with Roomba (or other networked objects you create) in a packetized fashion.

Going Wireless with Wi-Fi

In a few short years wireless Internet connectivity has gone from research project to required computer feature. All new laptops have wireless capability built in and many desktops do, too. USB adapters to add wireless to existing computers can be had for under $20. It seems we hardly know how we ever lived without wireless Internet. And that's the interesting thing. Being free from a physical cable has changed how we interact with our computers. Laptops are outselling desktops. The dedicated computer nook is giving way to computer use any time, anywhere. You can surf the Net (often for free) in public places like coffee shops, airports, and hotels around the world. Cities are rolling out metro-wide Wi-Fi access for all, partly as a way to seem progressive, but also as a valid and inexpensive way of providing a critical resource to its citizens. The computer is becoming less of a destination and more of a companion. The Net is now the destination, and it must be available for use wherever people want it. Both new cell phones and Skype phones have Wi-Fi built-in, and they are both able to forgo the standard cellular network for an Internet connection. Everything that can is going Wi-Fi.

In the previous chapter, you saw how to add an embedded Internet device server to an existing system. Adding Internet connectivity to stationary domestic objects with Ethernet is relatively cheap and simple. Everyone should experiment with putting his or her coffee maker on the Net. The tools and techniques learned for a wired network adapter carry over to a wireless one.

For a mobile device like Roomba, it makes less sense because the cable gets in the way. It's more of a test device and a stepping-stone to Wi-Fi. Like the RS-232 serial tether as a debugging tool for the Bluetooth adapter, a wired version of a network adapter complements a wireless one. This chapter shows how to build the Wi-Fi version of a network adapter. It's currently much more expensive to add Wi-Fi in a manner similar to the Siteplayer, but having a Wi-Fi Roomba is really cool.

# chapter 1

in this chapter

- ☑ Understand and debug Wi-Fi

- ☑ Use Lantronix WiMicro with Roomba

- ☑ Try SitePlayer with a wireless bridge

- ☑ Build a Wi-Fi Roomba

- ☑ Control Roomba with a Web page

- ☑ Control Roomba with PHP

Understanding Wi-Fi

In everyday use, Wi-Fi is wireless Internet connectivity. More specifically, Wi-Fi is a marketing term for a class of wireless IEEE standards for connecting computers in an Ethernet-like way via microwave radio transceivers. The most common of these standards are:

- **802.11a:** Up to 54 Mbps data rate on the 5 GHz unlicensed ISM band
- **802.11b:** Up to 11 Mbps data rate on the 2.4 GHz unlicensed ISM band
- **802.11g:** Up to 54 Mbps data rate on the 2.4 GHz unlicensed ISM band

These are all updates to the original 802.11 standard, which had a maximum data rate of 2 Mbps on the 2.4 GHz band. There's a new standard emerging called 802.11n that promises speeds up to 540 Mbps.

The 2.4 GHz and 5 GHz radio bands are unlicensed, meaning that you do not need a permit to operate a radio transmitter on these frequencies. These frequencies are two of the several industrial, scientific, and medical (ISM) bands that have this freedom. Bluetooth is also in the 2.4 GHz band, as are microwave ovens, cordless phones, wireless video cameras, home automation protocols, and just about anything else you can think of. It's a noisy region, but engineers have found ways to sidestep most of it. Of course, if you have an old-style 2.4 GHz cordless phone, chances are it'll cause your Wi-Fi to go down when you use it.

Sometimes the Wi-Fi standards are called Wireless Ethernet, and this is a very apt description. The designers of Wi-Fi wanted the same sort of simple connectivity and configuration that Ethernet affords. To connect to an Ethernet network, you simply plug in a cable. To disconnect, you unplug the cable. The hardware and software attached to the cable automatically figure out the details of setting up and tearing down pathways for data. To arbitrarily connect and disconnect was quite a novel concept when Ethernet was invented, but we've come to expect it in every communication bus we use. Wireless Ethernet works in almost the same way, but actually a bit better. Figure 12-1 shows typical wired and wireless networks, showing their topological similarity from the users' perspective. In both cases there's a resource shared by multiple computers. In the wired case it's the Ethernet hub; with wireless it's the access point. Often the functionality of hub + router and access point + router (or all three) are combined into a single unit.

The problem on a cable that people experience is the same as when they use walkie-talkies or CBs: No two people can talk at the same time. A CB channel may have many people participating, but that problem is dealt with by using a protocol of adding "breaker," "over," and "out" to conversations. Ethernet solves the problem in a similar way with a technique called CSMA/CD: Carrier-sense multiple access with collision detection. This sounds complex, but you intuitively know how it works. *Carrier sense* means listen for others before talking. *Mutiple access* means there are more than just two devices on the wire. *Collision detect* means if someone starts talking at the same time as you, stop and wait a bit before trying again. It's a simple and elegant solution.

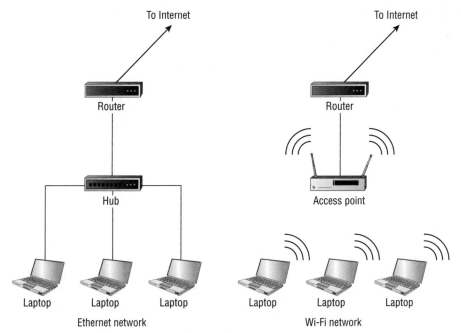

To Internet

Router

Hub

Laptop   Laptop   Laptop

Ethernet network

To Internet

Router

Access point

Laptop   Laptop   Laptop

Wi-Fi network

FIGURE 12-1: Wired versus wireless networks, the topology is the same

Wi-Fi modifies the algorithm a bit and is instead CSMA/CA; the CA stands for collision avoidance. Unlike on a physical cable, with radio it's hard to tell if someone else is transmitting while you're transmitting. So in CA, when a device wants to talk, it will first send a jam signal. This is sort of like blowing a whistle or clearing your throat: it gets through when normal conversation tones wouldn't. The jam signal puts all other devices in a listen state for a while as they wait for the transmitting device.

One problem results with both CD and CA algorithm: Data transmission in both degrades sharply once a critical mass of devices get on the shared medium (wire or radio channel). With many devices on the same Ethernet with CD, collisions happen so frequently hardly any data is sent. In Wi-Fi with CA, jam signals are frequent as devices start to talk; stopping all transmissions and again hardly any data is sent. In both cases the solution is to reduce the number of devices. In Ethernet that means moving devices to a different hub; in Wi-Fi that means moving them to a different access point. Hubs or access points are then connected together and talk among themselves as if they're each a single entity.

Roaming and Disconnects

In Ethernet, when you disconnect the cable and move a device to a different location and hub, the hub can tell when the cable is unplugged and plugged in. These events are used to update the internal ARP tables used by all devices to route Ethernet traffic. The result is that you're back on the network seamlessly. In Wi-Fi there are analogous events of connecting and disconnecting from an access point, but there are two complications:

- A Wi-Fi device can get turned off without properly disconnecting.
- A Wi-Fi device can roam to a new access point.

Both situations happen regularly. The former is dealt with by regular keep-alive signals. These signals approximate the electrical cable detection of Ethernet. The access points deal with the latter transparently. When a Wi-Fi node detects an access point on the same network but with a stronger signal, it sends a request to that access point to switch to it, and the access points communicate so that the old one gives up ownership and the new one takes it over. This takes a little time so if you're unlucky enough to be in a situation where your computer thinks two access points are about the same strength, you may bounce back and forth between them.

Power Consumption Concerns

Wi-Fi is inherently more power hungry than Ethernet. This is mostly due to the radio transceiver that must be powered all the time. Bluetooth is a great improvement in terms of power consumption, because the protocol allows devices to power down their radios for small amounts of time. Wi-Fi devices must be on all the time or they'll be disconnected from their access point. The 802.11 standard does define an optional power save mode which allows the device to periodically power down and the access point will buffer data until it wakes. Getting into and out of this mode requires extra logic on the parts of both the device and the access point.

Most devices support power management, but most don't enable it because it drastically affects real-time behavior. When in power save mode, the device wakes up only once every 100 ms or so. For web browsing, 100 ms isn't noticeable, but for embedded device control it can make the difference between a robot falling down the stairs or not. Even if the Wi-Fi radio could micro-sleep the way the Bluetooth radio can, the Wi-Fi data protocol and TCP/IP require a certain level of computational complexity that prevents truly tiny low-power devices. As embedded systems get more powerful, this issue will be addressed, but for now it means Wi-Fi embedded systems draw about two to three times more power than equivalent Ethernet systems.

Debugging Wi-Fi networks

Debugging network problems on a wireless network is almost entirely the same as debugging a wired network, as discussed in Chapter 11, in the sidebar "Debugging Network Devices." The same tools (`ping`, `arp`, `nmap`, `traceroute`) can be used. You should have additional tools and techniques at your disposal, too, such as a stumbler, described below, and a wireless access point or router you can control. If you don't have a wireless access point of your own yet, Chapters 14 and 15 describe two different inexpensive ones you can use and then later reconfigure for use with Roomba.

The first thing to do when debugging any network is to reduce the number of variables to make things as simple as possible. If you can create a private network with just the device under test and the computer you use to test with, you don't have to worry about getting confused by data from other network devices. On an Ethernet network this means using a hub with only the two devices plugged in. On a Wi-Fi network this means configuring an access point with a different SSID name and only configuring the devices you want to connect to it.

Also, simplify your test Wi-Fi networks by turning off all security and authentication features. When you have everything working, you can turn it back on, but they just get in the way when you're testing.

Note

On Windows, the default PING.EXE program may not give you the expected results. You may see either no response or responses with the broadcast address. This is wrong and partly due to the Windows implementation of TCP/IP. Using Cygwin and its ping package helps a little, but Windows machines may still be invisible to broadcast pings. In such cases, you can use nmap -sP in place of ping.

Stumblers

Your operating system has a rudimentary means of detecting Wi-Fi networks, but it reports only what it has noticed in a small window of time. Stumbler applications continuously scan for wireless networks and provide a historical view of the observed networks and their signal strength, usually in graphical form. For Windows there is NetStumbler (http://netstumbler .com/), the progenitor of the stumbler moniker. For Mac OS X you can use iStumbler (http://istumbler.net). For Linux, the built-in system command-line programs iwlist and iwspy coupled with a few simple shell scripts give you the same information, and there are several GUI programs available.

Low-Level Debugging

If you need to debug at an even lower level than what a stumbler provides, and look at the raw Wi-Fi data emitted from both access points and wireless clients, then Kismet (http:// kismetwireless.net/) is for you. Kismet is an open-source tool for Linux to passively scan Wi-Fi networks by putting a computer's wireless adapter in promiscuous mode. It is a very powerful tool used by network administrators to perform intrusion detection, detect unauthorized access points, and do accurate site surveys of their facilities' wireless networks. If you're unsure if a wireless device is transmitting at all, Kismet can detect if it is emitting any information.

If you suspect interference of a sort not identifiable by even Kismet, then you need a spectrum analyzer. They examine a frequency spectrum you're interested in and display it graphically. Normally, these devices are extremely expensive: many thousands of dollars for a basic one. The 2.4 GHz spectrum is full of chatter, not just Wi-Fi. The clever geeks at Metageek (www .metageek.net/) have created a spectrum analyzer for $99 that plugs into your USB port and analyzes just the frequencies of interest around 2.4 GHz for Wi-Fi, Bluetooth, cordless phones, microwave ovens, and so on. It's perfect when you've exhausted all your ideas as to why your wireless connection has problems.

Parts and Tools

To build the Wi-Fi Roomba adapter, you'll need the following parts:

- Lantronix WiMicro, Mouser part number 515-WM11A0002-01
- Mini-DIN 8-pin cable, Jameco part number 10604
- 7805 +5 VDC voltage regulator, Jameco part number 51262
- Two 1μF polarized electrolytic capacitors, Jameco part number 94160PS
- Two 8-pin header receptacle, Jameco part number 70754
- General-purpose circuit board, Radio Shack part number 276-150

Except for the WiMicro, all the parts you've seen before from previous projects, and the circuit you'll be constructing is very similar to previous ones, so you'll need all the same tools.

To aid in setting up, it would be helpful to have a dedicated wireless access point not connected to your home network. In a pinch you can use your computer in either ad-hoc or Internet sharing mode.

Lantronix WiMicro

Lantronix makes the XPort seen in the previous chapter. The wireless brother of the XPort is the WiPort. It is a tiny silver box with a pigtail antenna lead coming out of it and a high-density set of pins on its bottom. Like the XPort, it's not very hacker-friendly because of its pinout and 3.3V power requirements.

Lantronix also makes the WiMicro (shown in Figure 12-2), which contains a WiPort and some support circuitry. It is mostly a drop-in upgrade for their Micro (also known as Cobox Micro) Ethernet board. Like the Micro, it is more hacker-friendly with a standard 0.1″ spacing header connector on its underside and is driven by +5 VDC (see Figure 12-3). The pinout of this header is the same as the Micro and is shown in the diagram on the right side of Figure 12-4. The left side of Figure 12-4 shows the location of the header on the WiMicro board when looking at it from the top. Connecting the WiMicro to your circuit is easy: hook up the +5V and GND for power and TXA and RXA to serial in/out.

The WiMicro board is not cheap. One board costs $165 from Mouser. Like all Lantronix devices, it can be configured through web-based interface, telnet, or serial line. It has two serial ports called A and B, 0 and 1, or Channel 1 and Channel 2, depending on the Lantronix document you're reading. This project shows use of the first serial port (also known as A or 0), but you can use either. In fact if you have some other device (like a microcontroller controlling other sensors or actuators as described in the upcoming chapters), you can communicate with it through the second serial port.

FIGURE 12-2: Lantronix WiMicro board, with WiPort attached

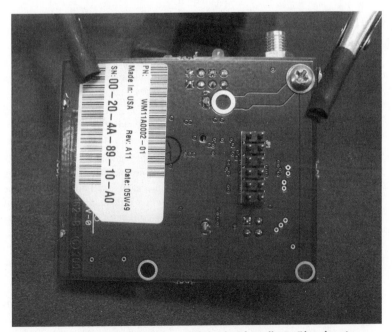

FIGURE 12-3: Underside of WiMicro with hacker-friendly 0.1″ header pins

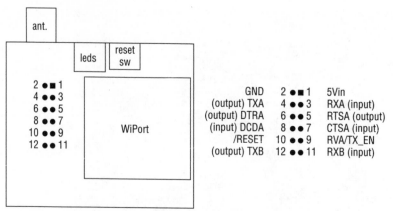

FIGURE 12-4: Pinout and location of WiMicro header (top view)

Note

Lantronix doesn't produce a single document on how to use and configure the WiMicro. Instead you must download several different documents, most for the WiPort. Particularly useful is the WiPort User Guide (WiPort_UG.pdf), which describes how to configure the device, and the WiMicro Addendum (WiMicro_Addendum.pdf), which goes over the layout of the WiMicro board.

DPAC Airborne and Other Wireless Device Servers

There aren't as many Wi-Fi device servers as there are Ethernet ones. This is unfortunate since Wi-Fi is much more interesting. The devices that do exist are bulky like the WiMicro. For example, another company called Digi (www.digi.com/) makes the Wi-ME (www.digi .com/products/embeddedsolutions/digiconnectwime.jsp). The Wi-ME is very similar in terms of usage, power, and cost to the Lantronix WiPort or WiMicro. There's not as much hacker-friendly information about it on the Internet, however, so using the WiMicro is easier in that respect.

An interesting device that is just now making its way into the hobbyist space (thanks to SparkFun) is the DPAC Airborne embedded server. This is a truly tiny device, at only about an inch square (see Figure 12-5). In large quantities they cost about $80 a piece. In small quantities you can get them for $125 from SparkFun. SparkFun also makes them a little easier to interface to by offering a $15 breakout board that converts the high-density connector to a useful 0.1″ header. To properly experiment with the Airborne module, you get the evaluation kit. This kit is large not only in price ($479) but also in size (see Figure 12-6). It's unfortunate that DPAC doesn't do what Lantronix did with the WiMicro, which is just a packaging of the WiPort. A small, experimenter-friendly drop-in board would make it easier to use.

FIGURE 12-5: DPAC Airborne module, with standard resistor for scale

FIGURE 12-6: DPAC Airborne evaluation kit

The Airborne is more than a Wi-Fi-to-serial converter, however. It is a complex programmable system with its own command-line interface and has eight analog I/O and eight digital I/O. It's meant to be more of an integral part of a product rather than just a wireless bolt-on. The power consumption of the Airborne is a little higher than the WiPort, perhaps to match its slightly greater capabilities.

The Airborne is an interesting device and you'll likely see it being used in hacks as people become more familiar with it. Come back to this book's web site for updates to see if it ever is used in Roomba.

Reusing SitePlayer

It seems a shame that the work from the previous chapter can't be reused. A wired network and a wireless network don't seem that different. It is possible. There are devices called wireless bridges that act as a client on an existing Wi-Fi network and have an Ethernet jack to plug in an Ethernet-based gadget. Wireless bridges are usually marketed toward gamers who want to hook their game consoles up to Wi-Fi. A few companies produce portable wireless access points that can also operate in a wireless bridging mode. One of the most notable of these is the AirPort Express with its built-in AC adapter and audio interface that works with iTunes.

Figure 12-7 shows an Asus WL-330g portable access point connected to the SitePlayer circuit from the previous chapter. The WL-330g comes with a spare DC adapter plug for use with a USB port, perfect for you to use as a power jumper cable. Cut off that USB side of the plug and solder the cable to the SitePlayer board to tap off its 5 VDC power the WL-330g requires. A short Ethernet patch cable completes the wiring.

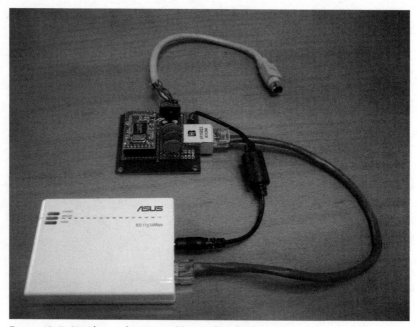

FIGURE 12-7: SitePlayer plus a portable wireless bridge

This combination is easy to get going and works okay. The combined cost is about $120 ($70 for WL-300g, $30 for SitePlayer, $20 for miscellaneous parts), which makes it a little cheaper than the WiMicro.

The biggest downside is the power consumption. The WL-330g consumes approximately 400 mA by itself. Add in the 150 mA drawn by SitePlayer, and that makes for 550 mA total. The WiMicro draws around 320 mA when used with Roomba. The difference could mean an extra hour of run time. You may save some money by using a wireless bridge, but you give up something else.

Thus reuse of SitePlayer is possible with a wireless bridge, and if you already have one around the house, try it out. The WiMicro is a nice integrated alternative and the rest of the chapter will focus on it.

Building the Roomba Wi-Fi Adapter

Due to the WiMicro board, building a Wi-Fi adapter for Roomba is easy. The only additional part needed beyond the connectors is a 7805 +5V voltage regulator. Figure 12-8 shows the schematic for the Wi-Fi adapter carrier board that the WiMicro will plug into. All the components are familiar and the construction is similar to other carrier boards you've created like the ones for the Bluetooth or Ethernet adapters.

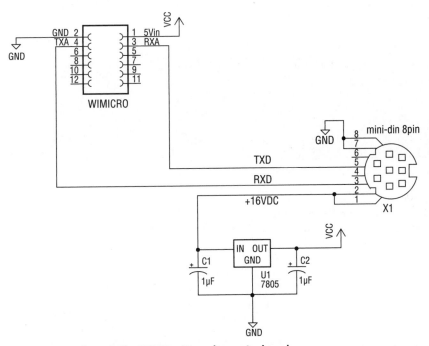

FIGURE 12-8: Schematic for WiMicro Roomba carrier board

You can see in Figure 12-9 the construction of the carrier board is quite simple. The dual female headers are made from two single headers used previously, cut down to the right size and soldered next to each other. Instead of soldering the Roomba Mini-DIN cable directly to the board, another single header is used to accept the Roomba prototyping plug from the previous chapter. (If you build a lot of interfacing circuits, you may want to construct a couple prototyping plugs.) The header for the plug is mounted horizontally so it fits underneath the WiMicro board. A little hot glue on the edges of the header secures it.

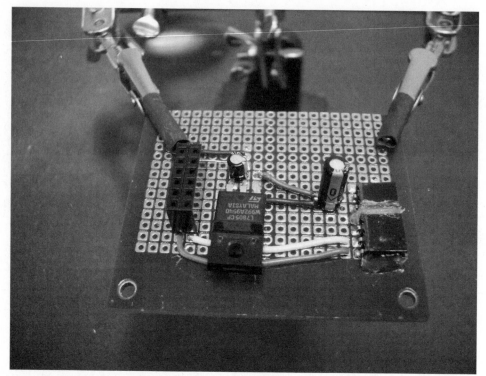

FIGURE 12-9: Carrier board wired up and continuity tested

To offer some mechanical support to the side of the WiMicro opposite its connector, a physically larger capacitor was used and placed at the edge of the board near the Roomba plug connector. The WiMicro board then sits on both the dual header connector and the top of the capacitor. This takes the strain off the dual header. If you aren't able to use a capacitor in this way, you can create a bumper by building up a few layers of hot glue.

WiMicro Configuration

After you've constructed the carrier board and tested the continuity to ensure you have no short circuits, plug in the WiMicro board and feed in DC power from a 9V wall wart. The WiMicro board should power up. Figure 12-10 shows the WiMicro powered up. Notice the lower-right power LED. If the power LED doesn't immediately come on, remove power quickly and look over your carrier board again. Be sure to screw on the included antennae if you haven't already. The upper-left LED indicates that the first serial port is active.

FIGURE 12-10: Configuring WiMicro and carrier board, powered by DC wall wart

Finding the WiMicro on the Net

Just like finding any new device on your network, you can use a broadcast ping to find it. With Wi-Fi there's the additional issue of getting both the device in question and your computer on the same wireless network. When the WiMicro first powers up, it creates an ad-hoc wireless network called LTRX_IBSS. (For Lantronix IBSS, IBSS is the technical abbreviation for an ad-hoc network.) Configure your computer to join that network. When you connect, only you and the WiMicro will be on that network. This makes finding it a bit easier. The IP address range used on this network is the default one used when there's no DHCP server or static IP address configured: the 169.254/16 network. /16 means the broadcast ping address is 169.254.255.255 and either device's IP address must start with 169.254.

When you're on the LTRX_IBSS net, the steps to find the WiMicro address are:

1. Determine your own IP address.

2. Broadcast ping 169.254.255.255.

3. Look at responses and pick the one that isn't yours.

Listing 12-1 shows an example of doing this on a Unix-like OS. The ifconfig command gives the IP address of the computer called demo: 169.254.86.198. The ping command then shows two devices responding, the demo computer and something else at 169.265.78.119. That's the WiMicro. From this point you can use the web interface or telnet interface to configure it.

Note If you don't have a Wi-Fi access point but do have a computer with wireless access, you can keep the WiMicro in ad-hoc mode and connect directly to its LTRX_IBSS wireless network. This enables you to experiment with a Wi-Fi Roomba without needing the Wi-Fi router normally required.

Listing 12-1: Finding the WiMicro with ping

```
demo% ifconfig en1
en1: flags=8863<UP,BROADCAST,SMART,RUNNING,SIMPLEX,MULTICAST>
mtu 1500
        inet6 fe80::20d:93ff:fe86:fb49%en1 prefixlen 64 scopeid
0x5
        inet 169.254.86.198 netmask 0xffff0000 broadcast
169.254.255.255
        ether 00:0d:93:86:fb:49
        media: autoselect status: active
        supported media: autoselect
demo% ping 169.254.255.255
PING 169.254.255.255 (169.254.255.255): 56 data bytes
64 bytes from 169.254.86.198: icmp_seq=0 ttl=255 time=0.240 ms
64 bytes from 169.254.78.119: icmp_seq=0 ttl=64 time=2.208 ms
(DUP!)
64 bytes from 169.254.86.198: icmp_seq=1 ttl=255 time=0.247 ms
64 bytes from 169.254.78.119: icmp_seq=1 ttl=64 time=2.304 ms
(DUP!)
^C
-- - 169.254.255.255 ping statistics -- -
2 packets transmitted, 2 packets received, +2 duplicates, 0%
packet loss
round-trip min/avg/max/stddev = 0.240/1.250/2.304/1.007 ms
```

WiMicro Telnet Configuration

Now that you have the IP address of the WiMicro, telnet to its configuration menu on port 9999, just like you can with the XPort. You should see something like:

```
demo% telnet 169.254.78.119 9999
Trying 169.254.78.119...
Connected to 169.254.78.119.
Escape character is '^]'.

MAC address 00204A8910A0
Software version V6.0.0.1 (050412)
AES library version 1.8.2.1

Press Enter for Setup Mode
```

Upon pressing Enter, the WiMicro will dump its current configuration and then display a setup menu:

```
Change Setup:
  0 Server
  1 Channel 1
  2 Channel 2
  3 E-mail
  4 WLAN
  5 Expert
  6 Security
  7 Factory defaults
  8 Exit without save
  9 Save and exit          Your choice ?
```

Configuring Baudrate

To make the WiMicro work with Roomba, you need to adjust the settings of the first serial port (Channel 1). All that needs to change from the default is to change the baudrate to 57600. Press 1, then Enter, and step through the configuration like this:

```
Baudrate (9600) ? 57600
I/F Mode (4C) ?
Flow (00) ?
Port No (10001) ?
ConnectMode (C0) ?
Auto increment source port  (N) ?
Remote IP Address : (000) .(000) .(000) .(000)
Remote Port  (0) ?
DisConnMode (00) ?
FlushMode   (00) ?
DisConnTime (00:00) ?:
SendChar 1  (00) ?
SendChar 2  (00) ?
```

Configuring Wi-Fi

If you want the WiMicro to join your network, you need to configure that too. In the following example, the network is called todbot and has no security. From the setup menu, press 4, then Enter, and then do this:

```
Enable WLAN (Y) ?
Topology 0=Infrastructure, 1=AdHoc (1) ? 0
Network name (SSID) (LTRX_IBSS) ? todbot
Security 0=none, 1=WEP, 2=WPA (0) ? 0
Data rate, Only : 0=1, 1=2, 2=5.5, 3=11 Mbps or
        Up to: 4=2, 5=5.5, 6=11 Mbps              (6) ?
Enable power management (N) ?
```

Configuring IP Address

Finally, to make it easier to find the WiMicro in the future, assign it an unused static IP address on your wireless network. In this example, the unused address is 192.168.0.90. From the setup menu, press 0, then Enter, and then do this:

```
IP Address : (000) 192.(000) 168.(000) 0.(000) 90
Set Gateway IP Address (N) ?
Netmask: Number of Bits for Host Part (0=default) (0)
Change telnet config password (N) ?
```

Saving the Configuration

With those three sets of changes made, it's time to save the configuration. Press 9 and Enter, and the WiMicro saves the configuration and resets itself to load the new settings. You'll find yourself disconnected from the LTRX_IBSS ad-hoc network. Connect to your wireless network as usual, and you should be able to ping and telnet to the IP address you assigned the WiMicro.

```
Change Setup:
  0 Server
  1 Channel 1
  2 Channel 2
  3 E-mail
  4 WLAN
  5 Expert
  6 Security
  7 Factory defaults
  8 Exit without save
  9 Save and exit              Your choice ? 9

Parameters stored ...
Connection closed by foreign host.
```

Note If you ever lose the WiMicro through a configuration mishap, you can reinitialize it to factory defaults by holding down the reset button for seven seconds.

Testing It All

If you can `ping` and `telnet` to the WiMicro with its new IP address, remove the wall wart power and plug in the Roomba prototyping plug. Then plug the assembly into the robot as shown in Figure 12-11. The WiMicro Power LED should light as soon as you plug it into the robot and within a few seconds it should be ping-able again. Turn on Roomba by pressing the Power button, and start sending ROI commands to it through port 10001 of the WiMicro.

FIGURE 12-11: Testing the bare WiMicro

As in the previous chapter, you can use the `roombacomm.net.SimpleTest` program to test Roomba. All the `roombacomm.net` API will work with a Wi-Fi Roomba as well as it does with an Ethernet Roomba. In fact, try out `roombacomm.net.SimpleTest` now. From the settings used above, invoking the test program would look like this:

```
demo% cd roombacomm
demo% ./runit.sh roombacomm.net.SimpleTest 192.168.0.90:10001
```

Controlling Roomba through a Web Page

Roomba is now on a wireless network, but to make it usable, it needs something external to send it proper ROI commands. You could use RoombaComm as you've done previously, but network accessibility opens up an entirely new space to play in. On the Internet there are programming languages that normally don't touch hardware, but now you control Roomba with them. Some of those languages are part of the LAMP web application platform. You can create a dynamic web page to control Roomba and have that page live anywhere on the Internet that has a LAMP-compatible server.

LAMP

LAMP is a suite of open source tools connected together to create dynamic web sites. It usually stands for Linux, Apache, MySQL, PHP, but P can also stand for Perl or Python. The great thing about LAMP is that it's more of a methodology than a specific set of technologies. Different pieces can be added or swapped around as the need arises. In fact, the LAMP philosophy isn't bound to Linux. There is also WAMP for Windows and MAMP for Mac OS X. For the following purposes, PHP on Apache will be used as the basic arrangement. The term LAMP will be used to refer to all similar setups, unless there are some WAMP- or MAMP-specific configuration issues.

Note The components of LAMP are part of every modern Linux distribution. To get WAMP for your Windows box, visit www.wampserver.com/. To get MAMP for Mac OS X, visit www.mamp.info/.

Both PHP and Apache will tell you when something is amiss by writing the errors to special log files. When debugging web programs, one of the easiest and most instructive techniques to find problems is to watch these files. This is sometimes called *tailing the logs* after the Unix practice of typing `tail -f logfile`. The `tail -f` command shows the last few lines of a file and will follow the end of the file as more is written to it. In LAMP environments the two most important logs are:

- **error_log**: The Apache error log. This shows web site configuration and page access errors.

- **php_error_log**: The PHP error log. This shows syntax and execution errors for PHP pages.

In MAMP systems both of these logs are located in `/Applications/MAMP/logs`. In WAMP systems both are located in `C:\WAMP\logs`. Various Linux distributions put the logs and other components in different locations.

Writing a PHP Web Page

On a LAMP system, writing a dynamic web page with PHP is as simple as creating a text file that ends with .php. A very simple PHP program is:

```
<html>
<?php
  print "<h1> hello world </h1>";
?>
</html>
```

Save this file as helloworld.php in the document root directory of your LAMP server. You can then test the file by loading it to the corresponding URL in your browser.

Note For MAMP, the document root is /Applications/MAMP/htdocs and the URL is http://localhost:8888/helloworld.php. For WAMP, the document root is C:\WAMP\www and the URL is http://localhost/helloworld.php.

Listing 12-2 shows roombacmd.php, which is the initial work of a PHP web page to control a network-connected Roomba. It is divided into four main parts:

- **Configuration:** Setting $roomba_host and $roomba_port to point to your Roomba.
- **Roomba functions:** A set of functions to replicate RoombaComm functionality.
- **HTML interface:** The dynamic HTML that makes up the actual user interface.
- **Command action:** Acting on the commands issued by the user through the user interface.

Listing 12-2: PHP Roomba Control with roombacmd.php

```php
<?php
// config: edit these two lines to point to your Roomba
$roomba_host = "192.168.0.90";
$roomba_port = 10001;

$cmd = isset($_GET['cmd']) ? $_GET['cmd'] : '';
$vel = isset($_GET['vel']) ? $_GET['vel'] : 200;

function roomba_stop() {
    roomba_drive(0,0);
}
function roomba_go_forward($velocity) {
    roomba_drive($velocity, 0x8000);
}
function roomba_go_backward($velocity) {
    roomba_drive(-$velocity, 0x8000);
}
```

Continued

Listing 12-2 *Continued*

```php
function roomba_spin_left($velocity) {
    roomba_drive($velocity, -1);
}
function roomba_spin_right($velocity) {
    roomba_drive($velocity, 1);
}
function roomba_drive($velocity,$radius) {
    $vhi = $velocity >> 8;
    $vlo = $velocity & 0xff;
    $rhi = $radius   >> 8;
    $rlo = $radius   & 0xff;
    print "vhi:$vhi, vlo:$vlo, rhi:$rhi, rlo:$rlo\n";
    roomba_send_cmd(pack("C*", 137, $vhi,$vlo, $rhi,$rlo));
}
function roomba_init() {
    roomba_send_cmd(pack("C", 128));   // START
    usleep(100000);   # wait 100 ms
    roomba_send_cmd(pack("C", 130));   // CONTROL
    usleep(100000);   # wait 100 ms
}
function roomba_send_cmd($cmd) {
    global $roomba_host, $roomba_port;
    $fp =
fsockopen($roomba_host,$roomba_port,$errno,$errstr,30);
    if (!$fp) {                  // couldn't connect
        echo "$errstr ($errno)\n";
    }
    else {
        fwrite($fp, $cmd);
        fclose($fp);
    }
}

function roomba_read_sensors() {
    global $roomba_host, $roomba_port, $roomba_sensors;
    $fp =
fsockopen($roomba_host,$roomba_port,$errno,$errstr,30);
    if (!$fp) {                  // couldn't connect
        echo "$errstr ($errno)\n";
    }
    else {
        fwrite($fp, pack("C", 142)); // SENSORS
        fflush($fp);
        $raw = fread($fp, 26);
        fclose($fp);
        $sensors = unpack("C*", $raw);
        return $sensors;
    }
}
?>
```

Listing 12-2 *Continued*

```
<html>
<head>
<title> Roomba Command </title>
<style> table,td { text-align: center; } </style>
</head>
<body bgcolor=#dddfff>

<h1> Roomba Command </h1>
<form method=GET>
<table border=0>
<tr><td><input type=submit name="cmd" value="init"></td>
    <td> </td>
    <td><input type=submit name="cmd"
value="sensors"></td></tr>
<tr><td> </td>
    <td><input type=submit name="cmd" value="forward"></td>
    <td> </td></tr>
<tr><td><input type=submit name="cmd" value="spinleft"></td>
    <td><input type=submit name="cmd" value="stop"></td>
    <td><input type=submit name="cmd"
value="spinright"></td></tr>
<tr><td> </td>
    <td><input type=submit name="cmd" value="backward"></td>
    <td> </td></tr>
<tr><td> velocity: </td>
    <td><input type=text size=5 name="vel"
                         value="<?echo $vel?>"></td>
    <td/> </td></tr>
</table>
</form>
<pre>
<?php

if( $cmd ) print "cmd:$cmd\n";

if( $cmd == 'init' ) {
    roomba_init();
}
else if( $cmd == 'stop' ) {
    roomba_stop();
}
else if( $cmd == 'forward' )  {
    roomba_go_forward( $vel );
}
else if( $cmd == 'backward' ) {
    roomba_go_backward( $vel );
}
else if( $cmd == 'spinleft' ) {
    roomba_spin_left( $vel );
```

Continued

Listing 12-2 *Continued*

```php
    }
    else if( $cmd == 'spinright' ) {
        roomba_spin_right( $vel );
    }
    else if( $cmd == 'sensors' ) {
        $sensors = roomba_read_sensors();
        $c = count($sensors);
        for( $i=1; $i<=$c; $i++ ) {
          printf("sensors[$i]: %x\n", $sensors[$i]);
          }
    }
    else if( $cmd == "drive" ) {
        $vel = $_GET['velocity'];
        $rad = $_GET['radius'];
        roomba_drive( $vel, $rad );
    }

?>
</pre>
</body>
</html>
```

Figure 12-12 shows what the script looks like in the browser when run.

FIGURE 12-12: Using the PHP Roomba program roombacmd.php

As you can see, the Roomba functions seem very similar compared to the RoombaComm functions. The only strange statements are the pack() statements, like:

```
pack("C*", 137, $vhi,$vlo, $rhi,$rlo)
```

In PHP, in order to send raw bytes with no string or numeric translation (which is normally what you want), you need to pack the bytes into a variable according to some packing rules. In the preceding statement, C* means to pack a series of unsigned bytes. The asterisk means an unknown number of bytes. In the preceding case you could use C5 since you're packing five bytes. When receiving data from Roomba, you need to use a corresponding unpack() method.

If you expand roombacmd.php to more fully control Roomba, the first thing to do is to move the Roomba functions into their own file to build a Roomba PHP library called, for example, roombalib.php. You can then use this separate file with include 'roombalib.php' at the top of your main PHP file.

The HTML user interface is an example of using multiple Submit buttons for a single form. Each button sends a different value for cmd. You could have just as easily created multiple links and set the cmd query argument yourself.

Putting It All Together

Now that you have a working PHP page to control Roomba, you can run it anywhere you can find a working LAMP system. This includes virtually all web hosting services on the Internet. In order to run roombacmd.php from one of them, you'll need to poke a hole in your firewall to allow computers outside your home network to connect.

Most home routers have a special configuration page just for such changes. Your home network looks like a single IP address to the rest of the Internet, but that IP address can have many different ports. What your router can do is translate one IP:port pair on the outside part of your network to another IP:port pair on the inside of your network. This is called *port mapping* or *port forwarding* and is often used to allow online gaming or P2P file sharing to work, so you may already have some port maps in place.

For example, if your external IP address to the Internet (WAN IP address) is 12.34.56.78 and the IP address of your Roomba is 192.168.0.1, then you'd configure your router to map port 10001 from one IP address to the other. This can be written in documentation as 12.34.56.78:10001 ---> 192.168.0.1:10001.

Note If you don't know your external WAN IP address, go to http://whatsmyip.org — you should be able to see it.

A final task is to make an enclosure for the WiMicro and carrier board. Having an exposed circuit board looks cool but as the WiMicro is expensive it's a good idea to protect it. Any small enclosure of the right size can be used. Figure 12-13 shows one example using a small project box. Small cut-aways on the sides are made where the antenna and Roomba cable come out. Figure 12-14 shows the finished enclosure mounted on the Roomba.

FIGURE 12-13: Enclosure for WiMicro and carrier board

FIGURE 12-14: The completed Wi-Fi Roomba

Going Further with LAMP

The M part of LAMP hasn't been touched on yet. The preceding PHP script shows the basics of how to receive sensor data, but the script doesn't do anything with the data besides printing it. With PHP and MySQL it's pretty easy to record the sensor data to a database table. With enough sensor data you can do an analysis on it to figure out things like:

- When is the house the dirtiest? Compare historical dirt sensor readings with time and date.

- Approximately how big or cluttered is the living room? Average out time between bump sensors to determine the mean free path of your rooms.

- How long are my walls? Analyze the wall sensor readings to find walls; then create a histogram to find the different wall sizes.

- How long does it take for the dustbin to fill up? Average the time from when the Roomba starts to when the vacuum motor over-current bit is set.

Collect enough data, and you can determine a great many things about the environment within which Roomba moves. Roomba may not be a very accurate sensor platform, but its sensors cover the same area again and again. Good data can often be recovered from noise with repeated measurements.

Summary

A Wi-Fi Roomba coupled with a LAMP application is pretty powerful. You can create not just web pages to control the robot, but dynamic database-driven web pages. Having the ability to tie a database into the robot enables you to create for an entirely different level of interaction as Roomba can now have a huge historical memory. You could create complex robotic logic based on this elephantine memory located on a server anywhere and remotely command one or several Roombas simultaneously. Or you could create a social networking site for Wi-Fi Roomba users and collect real-time Roomba stats from all over the world.

As the price of the Wi-Fi modules continues to fall due to cell phones and portable game systems having Wi-Fi, making a Wi-Fi Roomba will become even more attractive. To make the extra cost of the current devices more palatable, you can hook up the extra serial port offered to the microcontrollers discussed in the next chapter. Then you can get the best of both worlds: localized computation backed with a huge storehouse of previous knowledge.

Giving Roomba a New Brain and Senses

The projects presented so far have all been based on having the new brains of Roomba located remotely to it. From an RS-232 serial tether to a Wi-Fi connection, in all cases, Roomba is more of a *telepresence* device than a true robot. The term *robot* has come to have a flexible definition with the rising popularity of various telepresence/remote-control devices. The most egregious use of robot to mean remote-controlled machine are the battle bot programs on television. The contestants are billed as robots, but they're all controlled by humans through radio-control transmitters. Perhaps in a few years we'll have true robot battle shows where the machines are programmed, given a strategy, and let loose to fend for themselves.

This chapter shows a few ways to sever the connection with a controlling computer and instead replace the microcontroller brain of Roomba with one you can program. You'll turn your Roomba into a truly autonomous re-programmable robot. Like all projects in this book, this chapter doesn't harm Roomba and its original brain is available when you remove the new one you create.

In theory you could reprogram the microcontroller chip inside the Roomba, as that's what the OSMO//hacker device does to upgrade a Roomba's firmware. iRobot has yet to release specifications on the protocol used to upgrade firmware. They likely never will. Allowing anyone to re-program Roomba is a sure way to brick it, resulting in unhappy customers. There has been some effort in reverse engineering the upgrade protocol, but with the release of the ROI it's become less of an issue, since it addresses most of what was wanted once you add your own microcontroller that speaks the ROI protocol.

in this chapter

☑ Understand microcontrollers

☑ Build circuits without soldering

☑ Make a stand-alone robot out of Roomba

☑ Use the Basic Stamp with Roomba

☑ Use Arduino with Roomba

☑ Build a RoombaRoach

☑ Build a mobile mood light

Microcontroller vs. Microprocessor

A microcontroller is simply a small, integrated computer system designed for a particular purpose. Microcontrollers are pervasive in your life, yet almost totally invisible. You may be unfamiliar with the term and never realized there is a substantial difference between the computer in your laptop and the one in your coffee maker.

In the 1970s, when computers started to become small and cheap enough for everyday use, engineers realized that the modular architecture of microprocessor computer systems was too complex and cumbersome for designs that didn't need all the processing power and input/output of a normal computer. There were many tasks that needed computation but didn't need a powerful microprocessor, let alone a screen, keyboard, or printer. So smaller, less powerful CPUs were joined with a few simple I/O devices onto a single chip. These new creations were called microcontrollers.

Figure 13-1 is a diagrammatic view of the difference between a normal computer system and a microcontroller system. In a standard computer setup, different physical chips provided different functions: computation, memory, timing, input, and output. These separate devices were joined together by a collection of wires called a bus. The modern PCI bus is a descendant of this common system bus. In a microcontroller, reduced-functionality versions of the distinct devices were placed inside a single physical chip and connected by an internal bus that was usually not accessible to the outside. The lack of this bus may seem like a limitation, but in fact it means getting a microcontroller system working is much easier due to the so many fewer connections needed.

With a single small device containing everything needed to run algorithms, microcontrollers were embedded in all sorts of devices. This practice has come to be called *embedded systems*, and designing such systems is an engineering discipline in its own right. Embedded systems are what is responsible for all the exciting advances in our everyday objects becoming smart objects.

Modern home computer systems are a kind of hybrid approach. The CPU is still separate, but instead of many separate chips serving different functions, the functionality is consolidated into one or two large bridge chips. Similarly, some microcontrollers are so advanced now as to be able to run modern operating systems like Linux.

 Note The best magazine about embedded systems is *Circuit Cellar*, available at `http://circuitcellar.com/`. It is written by engineers for engineers and can get complex quickly for the uninitiated. But if you want to learn about the newest developments in tiny computational devices, it is a great resource.

Parts and Tools

There are many microcontroller development kits available that can be used with Roomba. The first shown here, the Basic Stamp, is a standard, easy-to-use, and well-known device. The second, Arduino, is a newcomer to the scene, but is based on standard parts, is becoming as easy to use as the Basic Stamp, and is much cheaper and more powerful than the Stamp. If you have the time, try out both microcontroller styles.

For both styles of projects, you'll need:

- Mini-DIN 8-pin cable, Jameco part number 10604
- Several 220-ohm resistors (red-red-brown color code), Jameco part number 107941
- Jumper wire kit, Jameco part number 19289
- Several LEDs, Jameco part number 156961 or similar

Typical Computer Architecture

Typical Microcontroller Architecture

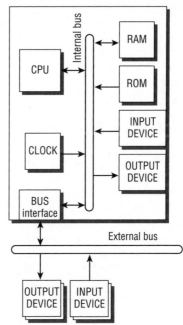

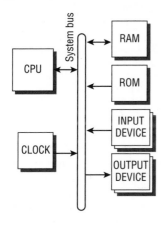

FIGURE 13-1: Typical computer architecture vs. microcontroller architecture

For the Basic Stamp projects, you'll need:

- Basic Stamp Board of Education Full Kit, Jameco part number 283101
- Two 0.1 μF capacitors, Jameco part number 545561
- Two photocells, Jameco part number 202403

For the Arduino projects, you'll need:

- Arduino USB board, Sparkfun part number Arduino-USB
- 8mm Red LED, Quickar part number 8R4DHCB-H or similar
- 8mm Green LED, Quickar part number 8G4DHCB-H or similar
- 8mm Blue LED, Quickar part number 8B4DHCB-H or similar
- Two 8-pin header receptacles, Jameco part number 70754
- Single line wire wrap socket, Jameco part number 101282
- Small solderless breadboard, Digikey part number 923273-ND or Sparkfun PType-BBM
- General-purpose circuit board, Radio Shack part number 276-150

Additionally, you'll need the various tools described in Chapter 3. For the Basic Stamp projects, you'll need the USB-to-serial adapter used in the previous projects.

Note Quickar Electronics (`http://quickar.com`) is one of several small companies that specialize in high-power LEDs and also sell in small quantities to hobbyists. Besides PC-board mount LEDs, they stock the blindingly bright LEDs used in flashlights and traffic lights as well as a wide variety of surplus components.

Solderless Breadboards

The previous construction projects were all interfaces of one sort or another and thus benefited from the level of sturdiness that you get from soldering a circuit. However when prototyping, it's usually quicker to sketch out circuit ideas on a solderless breadboard before committing them to solder. If you're unfamiliar with solderless breadboards, Figure 13-2 shows a typical one with parts plugged in and a kit of pre-cut jumper wires behind it. In a solderless breadboard, groups of five holes are connected together. In the figure, each vertical line of five is connected together, and at the top and bottom board between the red and blue strips, each horizontal line of five is connected. The holes are like the holes of an IC socket or header in that they have a spring-loaded metal connection that grabs a wire or part when inserted. And like a socket, repeated insertions wear it out. If you prototype a lot using solderless breadboards, you eventually collect them as your old ones become flaky.

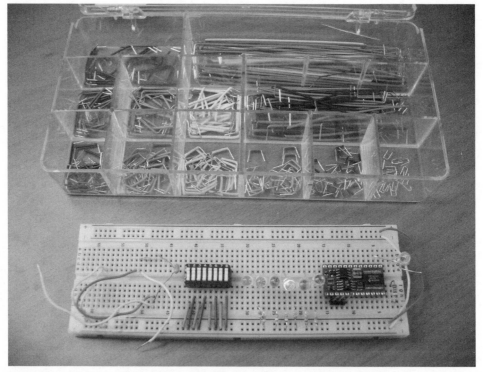

FIGURE 13-2: Medium-sized solderless breadboard with pre-cut jumper wire kit

It's tempting to skimp and not get a box of pre-cut jumper wires. Don't do it. It's possible to cut and strip wires yourself (and necessary if you run out of pre-cut wires), but using just any sized wire makes for an ugly and hard-to-read circuit. The ill-fitting wires move around, putting stress on the breadboard holes. Making jumper wires takes a lot of time, which is kind of beside the point of using a solderless breadboard. Get a box of pre-cut jumper wires and you'll be a much happier hacker.

Alternatives

Although the circuits presented in this chapter are both simpler and easier to build than the previous projects, even easier ways to re-program your Roomba are emerging. RoombaDevTools.com is working on two boards that are functionally very similar to the Basic Stamp and Arduino boards. They are:

- **RooStamp:** A carrier board for Basic Stamp 2 that interfaces to Roomba. It is Bluetooth-optional with servo ports and GPIOs exposed to the user, and it is powered by the Roomba battery.

- **RooAVR:** A high-end 16-bit Atmel AVR processor with plenty of I/Os, Bluetooth onboard, A2Ds, URART, JTAG, and servo ports, also powered by the Roomba.

Both of these boards can also be powered externally if your project needs to draw more current than Roomba can supply. The RooAVR uses the same type of microcontroller as Arduino, so it should be possible to use the Arduino programming environment with it.

If wiring up a microcontroller to your Roomba is not your cup of tea and you want to get started immediately making an autonomous robot, check out RoombaDevTools.com for these devices. But hooking up a little computer to Roomba isn't so hard really, as you'll see.

Adding a New Brain with the Basic Stamp

The Basic Stamp was created by Parallax in the early 1990s to address a persistent problem of embedded systems design: making it easy to use. Because microcontrollers are so self-contained, you couldn't just add a monitor and keyboard to them to see what's going on. Instead you needed some sort of development system, usually costing thousands of dollars. When you got the development system, learning the strange non-standard languages used to program the microcontroller was daunting. It could be months before you get a working system. And even if you could get all that working, acquiring the microcontrollers themselves was quite difficult unless you were a large company that wanted to buy 10,000 of them.

Parallax solved these problems by taking a common microcontroller, placing it on a hobbyist-friendly carrier board and writing a BASIC language interpreter for it. Now anyone who could program in BASIC could buy a single Basic Stamp and get a tiny stamp-sized microcontroller running a program they wrote, and do it in an afternoon. It was a revolutionary way of thinking about microcontrollers and for the past decade the Basic Stamp has been the standard in microcontrollers for hackers.

About the Basic Stamp 2

There have been a few variations of the Basic Stamp module as Parallax improved and expanded their product line. The most common is the Basic Stamp 2, also known as BS2, shown in Figure 13-3. The module is a small PC-board containing a specially programmed Microchip PIC microcontroller and a few support components. It is in the form of a wide DIP chip, so it's immediately familiar in shape to anyone who's built circuits before, and it plugs easily into a socket or solderless breadboard. It provides 16 input/output pins that can be used to read buttons or knobs, power LEDs or motors. The PBASIC version of the BASIC language it uses contains special keywords to help with these common tasks.

FIGURE 13-3: Basic Stamp 2 module

The BS2 can run directly off a 9V battery and a few of its pins can connect directly to a PC RS-232 serial port to serve as a means of both programming it with BASIC code and watching debugging output. Grab a solderless breadboard, some wire, a few other common parts and you can whip up a Basic Stamp programming board in a few minutes.

Many people, however, opt for getting one of the several Basic Stamp carrier boards that Parallax makes. One of the best (and cheapest) is the Board of Education, shown in Figure 13-4. It has jacks for power and RS-232 and has a small solderless breadboard for quickly creating projects. Alongside the breadboard, all the pins of the BS2 are brought out as headers for easy connecting. The Board of Education and BS2 are so popular that Parallax sells a kit containing both along with a serial cable and power adapter, which is exactly the one listed in the "Parts and Tools" section in this chapter.

Note In addition to the great products Parallax has created, they also have some of the best documentation about microcontrollers in general and Stamps in particular. The documentation covers everything from what a microcontroller is to a multitude of interfacing techniques for sensing and controlling the environment. All are available as free PDF downloads. Visit `http://parallax.com`, click Downloads, and look at the Documentation and Tutorials.

FIGURE 13-4: Basic Stamp Board of Education development board, with BS2 inserted

The Basic Stamp Environment

Figure 13-5 shows the Parallax Basic Stamp development environment with a typical "Hello World!" type of program for microcontrollers. Of course, the first program you write on a normal computer that verifies everything is working is to print out "Hello World!" to the screen. With a microcontroller, instead of saying "Hello World!", the first program you create verifies functionality by blinking an LED.

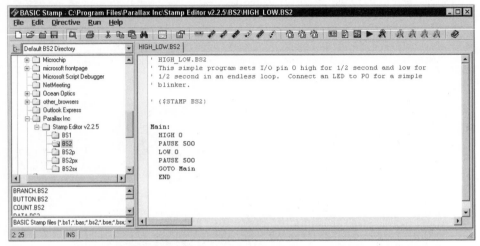

FIGURE 13-5: Basic Stamp editor in Windows

In PBASIC, the HIGH and LOW commands cause a pin to go to 5V or 0V, respectively. In Figure 13-5, pin 0 (P0) is brought high for half a second, then low for half a second, and then the program loops. Figure 13-6 shows the MacBS2 environment for programming Basic Stamps on Mac OS X with the same sort of "Hello World!" program but with a slightly different style. MacBS2 isn't as feature-rich as the official Parallax Basic Stamp IDE, but it is very easy to use and works very well.

Note MacBS2 was written by Murat N. Konar and not Parallax. It is free to use and can be downloaded from www.muratnkonar.com/otherstuff/macbs2/.

Note If you use Linux and want to program Stamps, check out the Basic Stamp Tools for Linux at http://bstamp.sourceforge.net/.

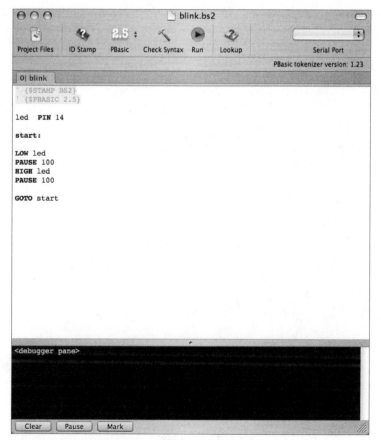

FIGURE 13-6: MacBS2 for Mac OS X

After you've downloaded one of the programming environments, getting it running and connected to the Basic Stamp is easy. Plug a 9V battery or power adapter into the Board of Education, connect the Board of Education serial port to your USB-to-serial adapter, and click the Identify/ID Stamp button. The program should beep and tell you it found the Basic Stamp. Hook up the familiar LED and resistor circuit to a pin on the Stamp and click Run. The program will be tokenized and uploaded to the Stamp and start running. Figure 13-7 shows a typical setup and a running blink program (although the light looks like a simple white circle when shown in a black-and-white photograph).

Note   If the Stamp isn't detected, make sure the 0-1-2 power switch on the Board of Education is set to position 1 or 2.

Figure 13-7: Typical setup when programming a Basic Stamp with a Board of Education

Hooking Up a Basic Stamp to Roomba

Since any pin on the BS2 can do serial input or output and the BS2 uses the same 0–5V signaling that the Roomba does, the circuit is really simple and can be done in a number of ways. One possibility is shown in Figure 13-8. To implement this circuit, the easiest thing to do is create (or re-use) a Roomba ROI prototyping plug (as shown in Figure 11-8). You may want to add a little bit of hot glue around the edges of the plug where the solder is to act as strain relief, since you'll likely be inserting and removing the plug a lot.

Tip   When wiring up Roomba to a microcontroller, the most common mistake is to accidentally swap the RXD and TXD lines. If your code compiles and runs yet doesn't control Roomba, try swapping those wires. If it still doesn't work, verify the robot still can receive ROI commands by using the serial tether from Chapter 3.

The BS2 can be run from a 9V battery, but you can also power it directly from Roomba. The Stamp and the BoE have standard 7805 voltage regulators that you're familiar with from earlier projects. They'll take the Roomba Vpwr (approx. 16VDC) line and regulate it to 5 VDC.

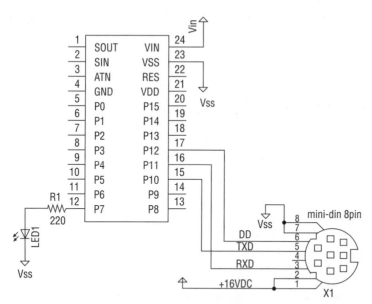

FIGURE 13-8: Schematic for Basic Stamp Roomba

Figure 13-9 shows a Board of Education wired up as in the schematic, using the ROI prototyping plug and with a few extra components from another experiment. For now, ignore the parts below the LED. Thanks to the solderless breadboard, the Board of Education, and the Roomba prototyping plug, wiring up this circuit takes just a few jumper wires.

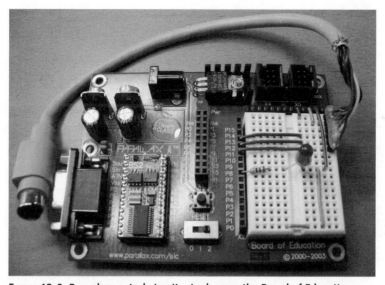

FIGURE 13-9: Roomba control circuit wired up on the Board of Education

Note The Basic Stamp nomenclature uses Vss to mean Gnd and Vdd to mean Vcc or +5V. The Vdd/Vss terms are more accurate but less well known.

Making BumpTurn with a Basic Stamp

Now that you have a new brain ready for Roomba, it's time to create some Roomba code. The BumpTurn functionality from Chapter 6 is a good starting point as it shows how to send commands to Roomba as well as get sensor data out of it. Recall the BumpTurn algorithm had only three actions:

- Go forward

- If bumped on the left, turn 90 degrees to the right

- If bumped on the right, turn 90 degrees to the left

Implementing BumpTurn in PBASIC will be as easy as implementing the Java version from Chapter 6. The main differences will be in the setup of the BS2 itself and setting the Roomba serial speed to be slow enough for the BS2.

The first step in writing a Roomba (or any) PBASIC program is to define which pins are being used and what direction the data is moving on those pins. These statements go at the beginning of any PBASIC program. For the above circuit, the pin definitions would be (using the ROI names):

```
DD          PIN 12
RXD         PIN 11
TXD         PIN 10
INPUT TXD
```

Downshifting the Serial Speed for BS2

The Basic Stamp 2 cannot send serial data at the standard Roomba speed of 57.6 kbps. Fortunately, the Roomba ROI designers allow you to downshift to 19.2 kbps by toggling the Roomba DD line three times in a row. In PBASIC this looks like:

```
FOR i = 1 TO 3
    LOW DD
    PAUSE 250
    HIGH DD
    PAUSE 250
NEXT
```

Although Roomba can send data at 19.2 kbps, it can't really receive data at that speed, so you need to downshift the Roomba communication speed again to 2400 bps with the ROI BAUD command.

Sending Serial Data with the BS2

In PBASIC, to send serial data out on a pin, use the SEROUT command with a special baudmode value that is defined in the Stamp reference manual. To set up the ROI as is standard, but at a 2400 bps speed, the PBASIC commands are:

```
' baudmode 32 == 19200 8N1, not inverted
BAUDFAST    CON 32
' baudmode 396 == 2400 8N1, not inverted
BAUDSLOW    CON 396
SEROUT RXD, BAUDFAST, [R_START]
PAUSE 100
SEROUT RXD, BAUDFAST, [R_BAUD, 3] ' 3 == 2400 bps
PAUSE 100
SEROUT RXD, BAUDSLOW, [R_START]
PAUSE 100
SEROUT RXD, BAUDSLOW, [R_CONTROL]
PAUSE 100
```

At this point you can send commands to Roomba like normal, but at a low level like in Chapter 3. For example, to tell Roomba to drive forward, the PBASIC command is:

```
SEROUT RXD,BAUDSLOW, [R_DRIVE,$00,$c8,$80,$00]
```

The R_DRIVE, R_START, R_CONTROL, and so on are constants similar to what's in RoombaComm. In general, creating constants like that are great mnemonics and make typos easier to catch. Otherwise you'd have to keep remembering that 137 meant the DRIVE command.

Note

PBASIC uses the dollar sign ($) to indicate hexadecimal characters. A single quote (') indicates the start of a comment.

Receiving Serial Data with the BS2

The Basic Stamp 2 has no serial input buffer, so as soon as the SENSORS command is sent, your code must wait around until all the data is received. Instead of getting all 26 bytes of sensor data, only data packet 1 (containing 10 bytes) will be requested. The SERIN PBASIC command takes a list of variables to fill out with serial data as its final arguments, making the code to get sensor data look like:

```
SEROUT RXD, BAUDSLOW, [R_SENSORS, 1]
SERIN  TXD, BAUDSLOW, 2000, No_Data,
        [SENS_BUMPWHEEL,SENS_WALL,

        SENS_CLIFF_L,SENS_CLIFF_FL,SENS_CLIFF_FR,SENS_CLIFF_R,
        SENS_VWALL,SENS_MOTOROVER,SENS_DIRT_L,SENS_DIRT_R]
```

The SERIN command optionally takes a timeout and GOTO label in case not all the data can be fetched. In the above code snippet, SERIN is to wait two seconds and then jump to No_Data if all the variables can't be filled out.

Putting It All Together: BumpTurn

Listing 13-1 shows the complete BumpTurn.bs2 program in PBASIC. Structurally it's not much different from the BumpTurn.java you're already familiar with from Chapter 6. Most of the BS2 code is spent just setting up Roomba to talk at the slower speed needed by the Stamp. If you don't need to get sensor data, you can omit the second downshift to 2400 bps and send commands at 19.2 kbps.

To use this program, just program the Stamp like normal and then plug it into Roomba. Roomba will wake up and start driving around, avoiding any obstacles it finds. One of the nice things about the Board of Education is that you can keep it connected through the serial cable while it's plugged into Roomba.

Listing 13-1: BumpTurn.bs2

```
' {$STAMP BS2}
' {$PBASIC 2.5}

' what pins are being used, named as in the Roomba ROI
DD          PIN 12
RXD         PIN 11
TXD         PIN 10
LED         PIN 7
LSRB        PIN 2
LSRF        PIN 1
INPUT TXD

' baudmode 32 == 19200 8N1, not inverted
BAUDFAST    CON 32
' baudmode 396 == 2400 8N1, not inverted
BAUDSLOW    CON 396

' define some Roomba constants
R_START     CON 128
R_BAUD      CON 129
R_CONTROL   CON 130
R_FULL      CON 132
R_POWER     CON 133
R_SPOT      CON 134
R_CLEAN     CON 135
R_DRIVE     CON 137
R_MOTORS    CON 138
R_SONG      CON 140
R_PLAY      CON 141
R_SENSORS   CON 142

' sensor bytes for data packet 1
SENS_BUMPWHEEL      VAR BYTE
SENS_WALL           VAR BYTE
```

Listing 13-1 *Continued*

```
SENS_CLIFF_L       VAR BYTE
SENS_CLIFF_FL      VAR BYTE
SENS_CLIFF_FR      VAR BYTE
SENS_CLIFF_R       VAR BYTE
SENS_VWALL         VAR BYTE
SENS_MOTOROVER     VAR BYTE
SENS_DIRT_L        VAR BYTE
SENS_DIRT_R        VAR BYTE
i VAR BYTE          ' counter

BUMP_RIGHT   VAR SENS_BUMPWHEEL.BIT0
BUMP_LEFT    VAR SENS_BUMPWHEEL.BIT1
WHEELDROP_C  VAR SENS_BUMPWHEEL.BIT4

' wake up robot
LOW DD
PAUSE 100
HIGH DD
PAUSE 2000

' pulse device detect 3 times to shift down to 19.2kbps
FOR i = 1 TO 3
    LOW DD
    LOW LED
    PAUSE 250
    HIGH DD
    HIGH LED
    PAUSE 250
NEXT

' start up ROI, then downshift again to 2400 bps for basic
stamp
SEROUT RXD, BAUDFAST, [R_START]
PAUSE 100
SEROUT RXD, BAUDFAST, [R_BAUD, 3] ' 3 == 2400 bps
PAUSE 100
SEROUT RXD, BAUDSLOW, [R_START]
PAUSE 100
SEROUT RXD, BAUDSLOW, [R_CONTROL]
PAUSE 100

Main:
    DEBUG "at the top",CR
    GOSUB Update_Sensors

    IF BUMP_RIGHT THEN
        GOSUB Spin_Left
```

Continued

Listing 13-1 *Continued*

```
            PAUSE 1000
        ELSEIF BUMP_LEFT THEN
            GOSUB Spin_Right
            PAUSE 1000
        ENDIF

        GOSUB Go_Forward
        PAUSE 100

        IF WHEELDROP_C THEN
            GOTO Done
        ENDIF
        GOTO Main

Done:
        PAUSE 1000
        SEROUT RXD, BAUDSLOW, [R_POWER]   ' turn off
        STOP

Spin_Left:
        SEROUT RXD,BAUDSLOW, [R_DRIVE,$00,$c8,$00,$01]
        RETURN
Spin_Right:
        SEROUT RXD,BAUDSLOW, [R_DRIVE,$00,$c8,$ff,$ff]
        RETURN
Go_Forward:
        SEROUT RXD,BAUDSLOW, [R_DRIVE,$00,$c8,$80,$00]
        RETURN
Go_Stop:
        SEROUT RXD,BAUDSLOW, [R_DRIVE,$00,$00,$00,$00]
        RETURN

' Get sensor packet 1, which is 10 bytes
Update_Sensors:
        DEBUG "update sensors",CR
        SEROUT RXD, BAUDSLOW, [R_SENSORS, 1]
        SERIN  TXD, BAUDSLOW, 2000, No_Data,
            [SENS_BUMPWHEEL,SENS_WALL,

            SENS_CLIFF_L,SENS_CLIFF_FL,SENS_CLIFF_FR,SENS_CLIFF_R,
            SENS_VWALL,SENS_MOTOROVER,SENS_DIRT_L,SENS_DIRT_R]
        RETURN
No_Data:
        DEBUG "no data!",CR
        RETURN
```

Making a Robot Roach

The many competing behavioral drives of a subsumption architecture robot are discussed in Chapter 1. The BumpTurn program already exhibits a few simple behaviors, but why not add an appropriate behavior in honor of the bugs that inspired the subsumption architecture idea? Most insects are sensitive to light in some way. The moth flies toward light, while the cockroach runs from it. Since Roomba can't fly (yet), maybe it can be made to act more like a roach. Having a robot that runs away from bright lights and scurries to the dimmest corner of the room is not only a great example of real working artificial intelligence, but is also comical to watch.

Figure 13-10 shows the schematic of an updated Roomba BS2 interface. The additions are photocells. Photocells are light sensitive resistors. Standard cheap photocells (even available at Radio Shack), range from 50 Ohms in very bright light to 500 kOhms in darkness. These values vary greatly and are not calibrated. But with such a wide dynamic range, it's at least easy to tell bright from dark. Figure 13-11 shows the circuit implemented on the Board of Education. Notice how each photocell is pointing in a different direction. The one pointing toward the RS-232 port will be the front sensor and the other the back sensor. With two sensors you can tell which direction the bright light is coming from.

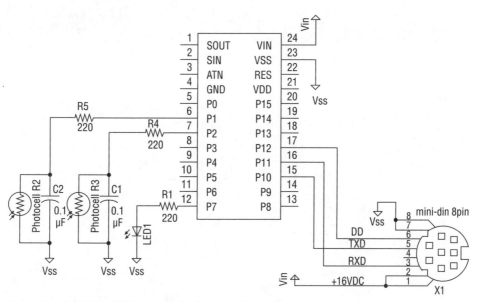

FIGURE 13-10: Adding light sensors for the RoombaRoach

Figure 13-12 shows the Board of Education plugged into Roomba and positioned correctly for the front and back orientation of the photocells. To anchor the Board of Education and keep it from sliding around, you can attach a little bit of Velcro between the Board of Education and Roomba.

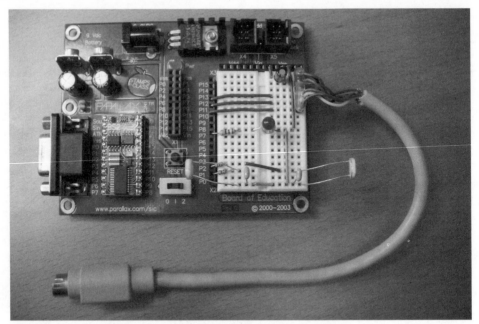

Figure 13-11: The RoombaRoach brain built on the Board of Education

FIGURE 13-12: RoombaRoach scurrying for the dark

In order to measure a variable resistor like a photocell or potentiometer, you would usually create a voltage divider with another fixed resistor and feed the output of that voltage divider to an analog input of your microcontroller. The BS2 has no analog inputs, but it does have an interesting function called RCTIME that records the amount of time it takes a capacitor to charge or discharge through a resistor. The charge time to go from a digital LOW to HIGH is related to the resistance value. The capacitors in parallel with the photocells and the resistors leading to the BS2 pins are part of this timing circuit RCTIME needs. Because capacitors, resistors, and photocells all have widely varying tolerances and are temperature sensitive, the values returned from RCTIME cannot be used as absolute values, but rather as a mark of relative change. For details on how RCTIME works, see the Basic Stamp language reference available from the Parallax site.

Listing 13-2 shows how to use RCTIME to measure the photocells. Only the front photocell measurement is shown, but the back one is just the same. The code spins Roomba around every so often, checking light levels. When it finds a darker area, it drives straight toward it. Even with such a simple algorithm, Roomba will almost consistently find the darkest area of a room. It never stops moving though, and still runs the BumpTurn algorithm so it avoids obstacles.

As a final bit of fun, you can now chase Roomba around with a flashlight.

Listing 13-2: Additions to BumpTurn to Make RoombaRoach

```
' the pin where our light sensitive resistor is
LSRF        PIN  1

Main
' ...everything else as in BumpTurn, then...

' check light levels, find the dark
    FOR i = 1 TO 4
        oldresF = resF
        GOSUB Read_LightF
        IF resF > oldresF+10 THEN ' its getting darker
            GOSUB Go_Forward
        ELSE
            GOSUB Spin_Left
            PAUSE 500
        ENDIF
    NEXT
    GOTO Main

Read_LightF:
    HIGH LSRF  ' charge cap
    PAUSE 1    ' for 1 ms
    RCTIME LSRF, 1, resF
    DEBUG DEC ? resF  ' print out value to debug port
RETURN
```

Adding a New Roomba Brain with Arduino

The Basic Stamp is an amazing tool and fun to play with, but it suffers from a few problems for the serious hacker:

- **It's expensive.** Each BS2 costs $49. If you want to start adding microcontrollers to lots of things, it gets expensive. And that's assuming you don't fry a BS2 or two while experimenting. What usually ends up happening is that you buy one BS2 and then re-use it, destroying previous projects to make future projects. Unless you're really good at documenting (and who is?), this cannibalization destroys the history of your previous learning experiences. You can't cheaply make several iterations on an idea.

- **It's limiting.** PBASIC provides many useful functions, but if you cannot do what you want using the PBASIC functions, you're stuck. There's no way to extend PBASIC to add any functions you might need.

- **It's closed.** PBASIC works well, but maybe you could make it work better. Or perhaps you'd like to get rid of the PBASIC functions you don't need to make space for your own code. Or maybe you'd just like to learn how PBASIC works. All of that is closed to you because the PBASIC interpreter is not open source. It's hard to fault Parallax for this, because they've done great things for the hobbyist community, but it is frustrating for the hacker who wants to learn.

- **It's too slow.** Your PBASIC programs run in a PBASIC interpreter on the BS2 PIC microcontroller. They aren't compiled down to machine code. This extra level of indirection makes PBASIC programs of any complexity run very slowly. This is likely the issue you'll run into first as you gain experience with the Basic Stamp.

Solutions exist here and there to these problems. Some companies offer BASIC compilers with a language very much like PBASIC. These compilers compile your program to machine code that can be burned on to cheap PIC microcontrollers. These compilers are usually several hundred dollars and are not open source. And then you need to purchase a PIC programmer to burn the chips. And even when you have it all hooked up, the cycle time between write-compile-burn-use is long because it doesn't work as seamlessly as the Basic Stamp does. The Basic Stamp environment is so easy to use, why would anyone want to bother with the hassle of compiling microcontroller code?

Interpreter vs. Compiler

With a microcontroller, interpreted code is almost always more problematic than compiled code. In a microcontroller, you usually attempt to maximize the speed of execution and minimize the memory footprint. An interpreted system necessarily results in slower code that takes up more space. Figure 13-13 shows how your blinking LED source code file gets turned into executable code for both an interpreted system like the Basic Stamp and a compiled system like Arduino.

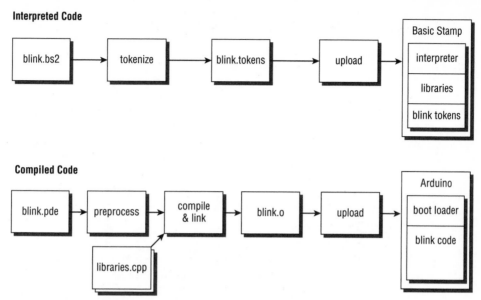

FIGURE 13-13: Interpreted vs. compiled code flows for the blink program

Both systems take the source code file, process it, and upload it to the microcontroller, which then executes it. In both cases, libraries of common routines (like how to toggle a pin to allow blinking) are used by your code. In the Basic Stamp case, the libraries are fixed, frozen in the firmware of the BS2. In the Arduino case, only those libraries that are needed are linked in before being uploaded. If you don't use the serial port, your executable code in the microcontroller won't contain the serial libraries. By compiling only those libraries you need, a compiled system will have more space available in the microcontroller for your code.

In the interpreted system, what is uploaded to the microcontroller are tokens: a compressed version of your PBASIC program. If you have a loop in your code, the PBASIC interpreter parses the loop's token each time, figuring out what each token means and turning it into machine code, each time. The compiled system does this translation once, at compile time. The benefit is two-fold: no interpreter is needed on the microcontroller (more space again for your code), and your code executes much faster.

About Arduino

Arduino (http://arduino.cc/) is several things. At first glance it is a microcontroller board, as in Figure 13-14. It is also a standardized way of generating compiled programs for the Atmel AVR series of microcontrollers, like the one in Figure 13-15. Viewed another way, Arduino is an environment and style of programming any type of microcontrollers. In many ways it is the spiritual successor to the Basic Stamp. It attempts to maintain the same ease-of-use pioneered by the Basic Stamp while giving people the power to write compiled C code.

FIGURE 13-14: The standard Arduino board

FIGURE 13-15: The standard Arduino microcontroller, Atmel AVR ATmega8

In short, Arduino is an open source environment for creating smart objects. The initial team who created and promoted the Arduino concept consisted of Massimo Banzi, David Cuartielles, Tom Igoe, David Mellis, and Nicholas Zambetti. This group of physical computing researchers and enthusiasts wanted to make creating hardware as easy and friendly as creating software has become.

The Arduino hardware addresses the cost issue of the Basic Stamp by using inexpensive parts and having an open-source design. Anyone can build an Arduino board like in Figure 13-14 or build an entirely different-looking one that still abides by the Arduino standards. If you do want to purchase a pre-built board, they only cost $29. This is for an entire board, not just a chip. An Arduino-capable chip, like the one shown in Figure 13-15, is a standard Atmel AVR ATmega8 and is about $3.60. A common practice is to buy one Arduino board and a handful of ATmega8s and build many Arduino circuits.

The Arduino Environment

Figure 13-16 shows the Arduino programming environment. If you think it looks a lot like the Processing environment introduced in Chapter 7, you're right. Arduino is based on Processing and adopts many of its idioms. Like Processing, Arduino runs on Mac OS X, Windows, and Linux. And like Processing, programs written with Arduino all have a similar code structure, consisting of two functions:

- `setup()`: Code that's run once, on startup
- `loop()`: Code that's run over and over, after `setup()`

Typically you put code to configure the Arduino board in the `setup()` function, much like you'd put graphics initialization code in the `setup()` of a Processing sketch. However, although Processing's `loop()` function is repeatedly executed at a specific framerate, the Arduino `loop()` is run as fast as possible. If you want to emulate a Processing-like periodic execution of `loop()`, add a `delay()` to the end of your `loop()` function.

Beyond that similarity and the familiar programming environment, Arduino code is completely different from Processing code. Processing code is just thinly wrapped Java code, whereas Arduino code is thinly wrapped AVR-GCC C code. In a similar way that Processing attempts to make Java a bit easier to deal with by creating many convenience functions, so Arduino does with C.

The idea of creating an embedded microcontroller version of Processing did not originate with Arduino. That honor goes to Wiring (http://wiring.org.co/) by Hernando Barragán. The Wiring I/O board has enormous amounts of I/O, almost like having two or three Arduino boards in one. It's priced to match, at around $80 from SparkFun. Arduino is based on Wiring but aimed more at being a low-cost solution. Wiring is open-source, and Arduino inherits much from its bigger brother.

The Arduino environment is not bound to the Arduino board, however; that is simply its default. Arduino supports different target types and through text file configuration changes, you can have Arduino program a Wiring board, a generic ATmega8 board, or any board with any CPU supported by AVR-GCC. You could create your own circuit using any Atmel AVR chip and use Arduino to program it.

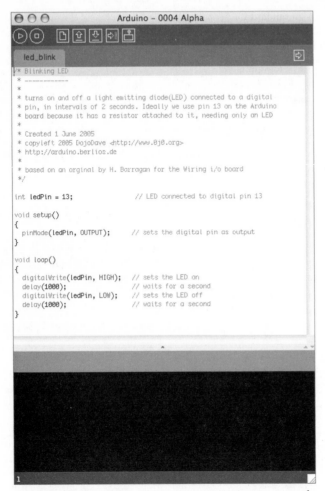

FIGURE 13-16: The Arduino programming environment, same for all OSes

GCC is the GNU Compiler Collection (formerly GNU C Compiler) toolset, the standard compiler used to create operating systems like Linux and Mac OS X, applications like Firefox, and embedded systems like wireless routers and smartphones. It is the most successful open-source project you've never heard of. Unlike user-focused software like Firefox, GCC is used only by engineers and hackers, and yet it is pervasive and critical to the functioning of the Net. Yahoo!, Google, ebay, and Amazon.com all use GCC on a daily basis. The philosophies and design methodologies encapsulated by GCC were a beacon to other software developers: Here was a best-of-breed toolkit given away for free and worked on by hundreds of people worldwide. And it all started in the late 1980s.

Today GCC can compile code for just about any microprocessor or microcontroller. The variant for the Atmel AVR is called AVR-GCC and is tuned to produce very tiny code while still letting the developer write standard C. (Most versions of C for microcontrollers for programmers tend to use very non-standard syntax.)

Every C compiler needs a library of basic functions. AVR-Libc is a mini-version of the standard C library that every C program is familiar with. This library contains functions like `printf()`, `toascii()`, and `malloc()`, as well as functions for manipulating AVR-specific registers.

For more complex I/O, Arduino also includes the Procyon AVRLib written by Pascal Stang. This library contains some helper functions to deal with the AVR chip, but also has functions to control LCDs, servos, compact flash disks, and even Wi-Fi cards. Using these functions isn't as easy as using the basic Arduino functions, but as Arduino grows it may find ways of encapsulating more of AVRLib.

Getting Started with Arduino

Getting an Arduino board up and running is almost as easy as getting Processing running. The most up-to-date information is available from the Arduino website, but when you have an Arduino board and have downloaded the Arduino software, the most common thing to do is:

1. Install the USB-to-serial driver the Arduino readme says is needed.

2. Connect an LED to the digital pin 13 socket and Gnd socket on the Arduino board.

3. Plug in the Arduino board to a USB port. Its green LED power light should go on.

4. Run the Arduino IDE. Select the serial port of the Arduino board. (It will be the serial port created by the USB-to-serial driver.)

5. Load up or write an LED blink program as shown in Figure 13-16 and compile it by clicking the Verify/Compile button. It should say Success.

6. Press the reset button on the Arduino board and click the Upload to I/O board button. While it's uploading, you can watch the TX/RX LEDs on the Arduino blink. The IDE will report Done when finished.

In a few seconds the Arduino board will reset to run your program, and your LED will start blinking. Figure 13-17 shows the result of the preceding steps. (The bright white dot at the bottom of the board is the power LED. The blinking LED is on the right toward the top, inserted into the header socket.) You don't need a resistor like you normally would for the LED, because the pin 13 socket already has a resistor on the board, just for this kind of test.

Note   The bootloader on the Arduino board is only active for the first three seconds after the board is reset, so right before uploading your code to the board, press the reset button.

Tip   The Arduino board contains a USB-to-serial adapter and so requires a driver to be installed on your OS before you can use it. You may already have this driver, but if you don't, the Arduino folder contains the correct driver for your OS.

FIGURE 13-17: Programming the Arduino board

Note Another inexpensive and interesting approach to programming AVR microcontrollers is the Atmel AVR Butterfly, a $20 credit-card sized board with a high-powered AVR controller and lots of I/O like a beeper and an LCD display. SmileyMicros (http://smileymicros.com/) produces a great beginner's book titled C *Programming for Microcontrollers* that uses WinAVR and the Butterfly.

Hooking Arduino Up to Roomba

Like the Basic Stamp, the Arduino board operates at 5V logic levels and has a built-in 7805 voltage regulator that can take the Roomba Vpwr line and convert it to 5V. Figure 13-18 is the schematic of a basic Roomba-to-Arduino connection. The Arduino serial TX/RX lines (digital pins 0 and 1) are connected to the Roomba serial lines. Figure 13-19 shows the circuit wired up on a breadboard that plugs into the Arduino header sockets. It's very straightforward.

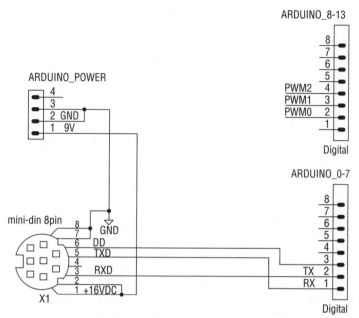

FIGURE 13-18: Schematic for Arduino-to-Roomba connection

FIGURE 13-19: Arduino Roomba circuit wired up on a small solderless breadboard

Making an Arduino Prototyping Shield

Unlike the Basic Stamp Board of Education, the standard Arduino board doesn't come with a solderless breadboard attached. This is okay, since they tend to wear out. In fact, it would be nice to keep a circuit you created by just unplugging it from the microcontroller board when the inspiration strikes to try something else.

You could use a stand-alone breadboard like in Figure 13-2, and run wires between it and Arduino. But since this is supposed to go on top of Roomba, it would be better if there was a way to make a sturdy connection between the two. The Arduino designers have thought of this and have created a set of shields that sit on top of the Arduino board, plugging in to its header sockets. This is a great solution since you can mix and match various shields to solve common problems. Some of the shields that have been created are motor controllers, bio-sensors, RFID readers, LCD displays, and prototyping. For an up-to-date list of shields, go to www.arduino.cc/en/Main/ArduinoShields.

The prototyping shields are the most interesting ones for these projects. These were created by Tom Igoe of NYU's ITP program. You can download his designs for the shields from www.tigoe.net/pcomp/code/archives/avr/arduino/.

You also can build your own Arduino prototyping shield out of a Radio Shack circuit board, a small solderless breadboard, and a few headers. It's a quick project that will only take a few minutes. It doesn't offer all the capability of Tom's shields, but it works in a pinch. The steps to make one are:

1. Take the wire wrap socket and break off an 8-pin chunk.

2. Push it through and solder it down to the edge-most pads.

3. Take an 8-pin header socket and solder it down immediately next to it. Notice that since the pads are joined, the connection between the two is made for you (see Figure 13-20). If the header socket has plastic flanges that make it hard to fit, snip them with cutters.

4. If you want the second set of digital pins, repeat Step 3, but lined up with the second Arduino header socket. Note that since the second socket isn't exactly inline with the Radio Shack circuit board, you'll have to bend the wire wrap pins a little.

5. Use the double-sided tape that came with the solderless breadboard and stick it to the top of the circuit board (see Figure 13-21).

Making an Arduino Prototyping Shield *Continued*

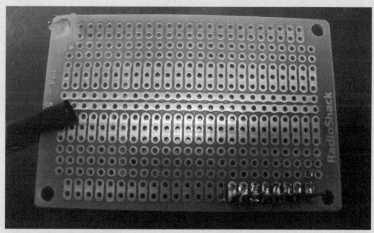

FIGURE 13-20: Soldering down the posts and header socket

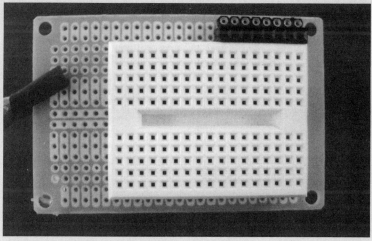

FIGURE 13-21: Attaching the solderless breadboard

Continued

Making an Arduino Prototyping Shield *Continued*

6. Plug it into the Arduino board and start using it (see Figure 13-22).

FIGURE 13-22: Finished shield plugged in to Arduino

7. You may want to dab a blob of hot glue to the opposing corner of the circuit board to keep the board level when plugged into the Arduino board (see the top left of Figure 13-20).

Switching between Coding and Running

The standard Arduino board has two issues that make it a little harder to use with Roomba compared to the Board of Education. The first is power. Most of the time when you're using the Arduino board, it's getting its power from the USB port. This is normally a great time-saver and enables you to try out new ideas fast without having to remember where you put the wall wart or finding a place to plug it in. But if you want to power Arduino from an external power source, you must disconnect the USB power and connect the external power input. This is done through the SV1 power switch. Move it to position 2-3 (toward the USB port) to enable USB power and move it to position 1-2 (toward the external power input) to enable external power. Figure 13-23 shows the SV1 switch in the external power position.

FIGURE 13-23: Switch SV1 in external power mode

The second issue is the serial port. The ATmega8 has a hardware serial port that can communicate at the fast 57.6 kbps speed that Roomba speaks. However, this hardware serial port is used by Arduino to load new programs. This means that you cannot have the USB cable hooked up and getting debug information at the same time as you're talking to the Roomba. You can use any of the normal I/O pins as a software serial port the way the Basic Stamp does, but that has the same speed problems as the Basic Stamp serial and in Arduino it's currently not as elegant.

Since you'll be using the same serial lines to load code into the board and control Roomba, you'll need to move a jumper. To switch from coding to running, the following steps should be performed:

1. Upload the program to the Arduino board using Arduino IDE.

2. Unplug the Arduino board from the USB.

3. Move the power switch SV1 from position 2-3 to position 1-2.

4. Plug the Arduino board into Roomba.

To switch back to coding, reverse the steps. There is a danger of ruining an Arduino board if it is plugged into Roomba and a USB port when the SV1 switch is set to the USB position. Just make sure both Roomba and the computer are not plugged into Arduino at the same time.

Coding BumpTurn with Arduino

After the experience of coding RoombaComm, working in Processing, and writing BumpTurn in PBASIC, seeing BumpTurn in Arduino is bound to trigger all sorts of déjà vu. Because the hardware serial port is used, Arduino can talk to Roomba at its default 57.6 kbps speed, so all the bending over backward to get Roomba to speak at a speed the BS2 could deal with is not needed.

The code should be fairly obvious. The Arduino function `digitalWrite()` is used to set the state of a pin, like the PBASIC `HIGH`/`LOW` commands. The `Serial.print()` function sends data out the serial port. That function can format different types of data before it is sent, so a BYTE type is specified, which does no formatting. To read Roomba sensor data, the `serialAvailable()` and `serialRead()` Arduino functions are used. Listing 13-3 shows the Arduino version of BumpTurn.

Because you're familiar now with both the ROI and how BumpTurn works, the friendly mnemonics like R_CONTROL and R_DRIVE aren't used here just to make the code a little more visually compact. If you do a lot of Roomba programming with Arduino, you should create those constants and use them. They'll save you time and headache in the long run. Arduino enables you to create complete libraries so you could implement all of the RoombaComm API in C for use with Arduino.

When you have this code loaded in the Arduino IDE, upload it to the Arduino board, unplug the board, and plug it into Roomba. In a few seconds it should light its LED and start driving around.

 Note If you've set Roomba to baud rates other than 57.6 kbps from previous projects, pull out its battery and put it back in to reset the Roomba ROI to its default speed.

Listing 13-3: BumpTurn with Arduino

```
int ddPin = 2;
int ledPin = 9;
char sensorbytes[10];
#define bumpright (sensorbytes[0] & 0x01)
#define bumpleft  (sensorbytes[0] & 0x02)

void setup() {
```

Listing 13-3 *Continued*

```
    pinMode(ddPin,  OUTPUT);    // sets the pin as output
    pinMode(ledPin, OUTPUT);    // sets the pin as output
    Serial.begin(57600);
    digitalWrite(ledPin, HIGH); // say we're alive

    // wake up the robot
    digitalWrite(ddPin, LOW);
    delay(100);
    digitalWrite(ddPin, HIGH);
    delay(2000);
    // set up ROI to receive commands
    Serial.print(128, BYTE);    // START
    delay(50);
    Serial.print(130, BYTE);    // CONTROL
    delay(50);
    digitalWrite(ledPin, LOW);  // say we've finished setup
}

void loop() {
    digitalWrite(ledPin, HIGH); // say we're starting loop
    updateSensors();
    digitalWrite(ledPin, LOW);  // say we're after updateSensors
    if( bumpleft ) {
      spinRight();
      delay(1000);  // spinning @ 200 for 1 sec == ~90 degrees
    }
    else if( bumpright ) {
      spinLeft();
      delay(1000);
    }
    goForward();
}

void goForward() {
    char c[] = {137, 0x00,0xc8, 0x80,0x00};    // 0x00c8 == 200
    Serial.print( c );
}
void goBackward() {
    char c[] = {137, 0xff,0x38, 0x80,0x00};    // 0xff38 == -200
    Serial.print( c );
}
void spinLeft() {
    char c[] = {137, 0x00,0xc8, 0x00,0x01};
    Serial.print( c );
}
void spinRight() {
    char c[] = {137, 0x00,0xc8, 0xff,0xff};
    Serial.print( c );
}
```

Continued

Listing 13-3 *Continued*

```
void updateSensors() {
  Serial.print(142, BYTE);
  Serial.print(1,   BYTE);   // sensor packet 1, 10 bytes
  delay(100); // wait for sensors
  char i = 0;
  while( Serial.available() ) {
    int c = serialRead();
    if( c==-1 ) { // error
      for( int i=0; i<5; i ++ ) {   //  blink 5 times on error
        digitalWrite(ledPin, HIGH); delay(50);
        digitalWrite(ledPin, LOW);  delay(50);
      }
    }
    sensorbytes[i++] = c;
  }
}
```

Making a Mobile Mood Light

Roomba is not very visible at night when operating. Sometimes people operate their Roomba in darkened rooms and don't want to accidentally step on it. Maybe you just want a mobile source of mood lighting. With the vacuum motors turned off and moving at a moderate speed, Roomba isn't very noisy and so having it slowly drive around, creating splashes of colored light on the ceiling and walls, is pretty relaxing. Between the random motions of Roomba and the rotating color cycling from the code below, a darkened room becomes a pretty groovy place.

The schematic in Figure 13-24 shows a modification to the previous circuit to add three high-output, high-efficiency RGB LEDs to the Roomba Arduino board. These LEDs are so bright they are painful to look at directly and can fill up a room with light. By having an RGB trio hooked up to the PWM outputs of Arduino, you can mix and match the RGB components to make any color you choose.

The blue and green LEDs have a different voltage drop than the red ones and so will need different resistor values. If you don't feel like calculating optimal values, use the ones in the schematic. The schematic's resistor values are purposefully larger than normal to keep the LEDs' brightness down. From the Ohm's Law discussion in Chapter 3, you'll recall the resistor controls the current to an LED and that, for a given voltage, a larger resistance gives a lower current. Since an LED's brightness is directly proportional to the current it's given, if you find you'd like the LEDs brighter, lower the resistor values. Be sure to know what the maximum current is from your particular LED's specification sheet and don't exceed it.

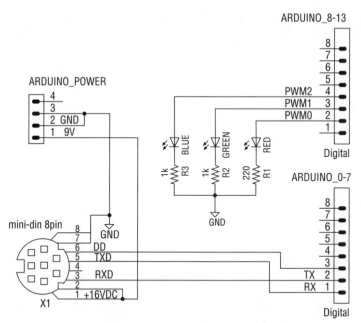

FIGURE 13-24: Adding RGB lights to the Roomba Arduino interface

Listing 13-4 shows the additions to BumpTurn needed to control the LEDs. The updateLEDs() function added causes slow color fades to occur over time. It uses the analogWrite() Arduino function to set a brightness value for each LED. A value of 0 turns off the LED and a value of 255 sets maximum brightness. The analogWrite() function doesn't truly write an analog value out but instead sets a pulse-width modulation (PWM) that when averaged over time looks like a varying voltage. Combining the LED functions with the BumpTurn functionality demonstrates how you can perform rudimentary types of multitasking with microcontrollers.

Note   There are a lot of useful code samples available on the main Arduino site and in the forums where the Arduino community swap code snippets. The RGB LED code additions shown here were adapted from code available there.

Figure 13-25 shows the LEDs mounted on the Arduino prototyping shield, with the shield reconfigured to cover the Arduino board. Notice how the component's leads have been trimmed down compared to the arrangement in Figure 13-18. When using solderless breadboards, keeping components closer to the board generally makes for a more well-behaved circuit.

Figure 13-26 shows the board plugged into Roomba and ready to go. The combination Arduino and shield make for a compact setup.

FIGURE **13-25**: Arduino and prototyping shield reconfigured with RGB LEDs added

FIGURE **13-26**: Arduino plugged into Roomba and getting moody

Finally, to give a slightly different effect (suffusive glow instead of a point light), you can stick the Arduino + prototyping shield into one of those tap lights that everyone seems to have and use its diffuser to better mix the colors, as shown in Figure 13-27. You'll have to snip out a small section of the edge of the tap light to make room for the ROI cable. Of course, the inner workings of the tap light have been removed so the tap function no longer works, but it wouldn't be hard to add back in and trigger some sort of action when the light is tapped by hooking up the original tap light switch to a digital input on the Arduino board and using digitalRead() on the input.

FIGURE 13-27: Arduino Roomba interface inserted into a tap-light for diffuse color

Listing 13-4: Additions to BumpTurn into MoodLight

```
int redPin   = 9;    // red LED
int greenPin = 10;   // green LED
int bluePin  = 11;   // blue LED

int ledi = 0;        // led loop counter
int redVal   = 255;  // brightness values to send to LEDs
int greenVal = 255;  // start with all at full bright(==white)
int blueVal  = 255;

void setup() {
  pinMode(redPin,   OUTPUT);
```

Continued

Listing 13-4 *Continued*

```
    pinMode(greenPin, OUTPUT);
    pinMode(bluePin,  OUTPUT);
    // ...setup continues as before...
}
void loop() {
  updateLEDs();
  // ...loop continues as before...
}
void updateLEDs() {
  ledi++;
  if( ledi < 255 ) {
    redVal   -= 1;   // red down
    greenVal -= 1;   // green down
  } else if( ledi < 255*2 ) {
    redVal   += 1;   // red up
    blueVal  -= 1;   // blue down
  } else if( ledi < 255*3 ) {
    greenVal += 1;   // green up
    blueVal  += 1;   // blue up
  } else {
    ledi = 0;  // reset
  }
  analogWrite(redPin,   redVal);  // write brightness vals to
LEDs
  analogWrite(greenPin, greenVal);
  analogWrite(bluePin,  blueVal);
}
```

Summary

Adding a brain to Roomba is one of the most fun things you can do to it. When you have a microcontroller on it that you can reprogram and add new sensors and actuators to, the Roomba becomes a truly interesting robotics platform. Both the Basic Stamp and Arduino enable you to read and control not only digital transducers like switches and lights, but analog ones as well. If you've looked over the Stamp and Arduino web sites, you've no doubt seen many of the other interesting devices you could hook up to Roomba. Add servos to Roomba to give it arms, or add microphones to give it ears. Make it come to you when you whistle for it. The add-on possibilities are limitless.

If you've not had much experience with microcontrollers before this, you now know about two of the friendliest yet most powerful embedded development systems out there. The Basic Stamp has a great library of tools and techniques developed from over a decade of work by enthusiastic hackers, all available for you to use for your own ideas. Arduino offers a gateway into a high-performance but low-cost system for when the BS2 is not appropriate. You can mix and match which microcontrollers you use depending on your needs or whims.

Adding a brain to Roomba is just the start. You can also make smart objects out of other everyday objects in your house. Of course, you're probably already considering adding the Ethernet or Wi-Fi adapters from the previous two chapters. Both are very possible. Try it out, and create your own programmable, networked smart objects.

Putting Linux on Roomba

Adding a microcontroller to Roomba is the small step that enables a near infinity of capabilities. The microcontrollers are cheap and power-conscious and hint at a world we'll soon be in, full of objects imbued with simple, ubiquitous intelligence. But their very design limits their usefulness for the more complex tasks we might call upon for a robot.

One of the main differences between the programs run on a microcontroller and those run on a PC (be it Windows, Linux, or Mac OS X) is the inability to run multiple programs concurrently. That is, the microcontrollers you've used so far don't have an operating system: a master program that abstracts the hardware with software libraries, allows multiple programs to run at once, and protects programs from one another. There are some real-time executive programs for some microcontrollers, which provide a degree of hardware abstraction and multitasking, but these are no better than DOS compared to Windows.

The tiny microcontrollers considered in Chapter 13 have small RAM and ROM footprints (on the order of several kilobytes), obviating the ability to run complex programs or store large amounts of data. There are a few exceptions to this rule, with some having the ability to talk to large flash ROM disks, but those are special cases and the code for such things eats up most of the precious memory space.

The smaller microcontrollers make it a snap to interface simple peripherals like LEDs and buttons, but complex peripheral standards like hosting Bluetooth or USB are out of reach. Interfacing something as seemingly simple as a USB mouse is not possible with small microcontrollers.

Between the tiny microcontroller chips and the large motherboards of modern PCs lies a realm of embedded boards that are small yet have the same power as desktop PCs of a decade ago. With such power, these embedded boards can run real preemptive multitasking operating systems. In the two architectures presented in Chapter 13 (in Figure 13-1), these boards exist in the middle: some memory and peripherals are on-board the processor, but many are external.

This chapter focuses on using such an embedded board, with Linux as the operating system running on it. Since you already have the ability to add a processor and communicate through Wi-Fi, this new system should be able to do at least that. And it goes without saying that it should be able to be powered by Roomba.

# chapter 14

in this chapter

☑ Understand embedded Linux

☑ Pick the right embedded Linux system

☑ Install and configure OpenWrt

☑ Control Roomba from the command line

☑ Control Roomba from Perl

☑ Make a battery pack for your embedded Linux box

Linux on Roomba?

The microcontroller in Roomba is more like the microcontrollers presented in the previous chapter: tiny, cheap, and works great for the limited functionality required of it. Putting a real OS on the microcontroller isn't feasible. Even if it was, iRobot hasn't released the specifications for firmware upgrade nor has the hacker community bothered to reverse-engineer how to do it. Instead, you can add one of these OS-capable embedded boards to the robot, laying it on top, and communicating to the robot through its ROI protocol. So, it is Linux *on* Roomba, just not Linux *in* Roomba.

The small systems that run a stripped down version of Linux often have more in common with tiny microcontrollers like the Basic Stamp than a desktop computer. In the computer/microcontroller comparison of Figure 13-1, they're definitely on the microcontroller side. Unlike many operating systems, Linux is highly scalable. It's used on terabyte-sized supercomputers and yet can be shrunk down to a few megabytes to run on a microcontroller. These Linux-capable controllers are much simpler than a full Linux server or desktop system and you don't have to be a Linux expert to make them work. You do need to have familiarity with the command line, as in the previous chapters, but you don't need to be a professional Linux system administrator. If you're new to Linux, this is a great way learn the basics. All the techniques used below are applicable to larger Linux systems

Single Board Computers

There is a vast array of different types of embedded boards. Some of these systems are called single board computers (SBCs) with the exemplar being the PC/104 standard. The PC/104 standard defines a rugged 3.5″ × 3.7″ board with edge connectors that have all the I/O functionality of an IBM/PC-compatible system. There are many different makers of PC/104 boards. CPU speeds on these systems range from 100 to 500 MHz and memory ranges from 1 to 256 MB. On them you can run DOS (Windows, or Linux on some) and use standard PC peripherals. An interesting aspect of PC/104 is that the standard ISA bus is present on the edge connectors and you add peripherals by stacking components like building blocks. The PC/104 standard is used extensively in industrial environments, often in specialized data acquisition roles. The industrial users of PC/104 aren't usually concerned about minimizing power consumption, cost, or space, so a PC/104 system with Wi-Fi will often draw upwards of 1000 mA of current, cost more than $300 for a single system, and end up taking up a 3.5″ × 3.7″ × 3″ volume. PC/104 is interesting, small, and low-power compared to a full *x*86-compatible PC, but there are smaller and lower power SBCs that are even more interesting.

The Gumstix system (`http://gumstix.org/`) is perhaps the smallest and lowest power SBC available. It's not PC-compatible, preferring to throw out that somewhat bulky design for a more streamlined one. The core of this new architecture is the Intel XScale processor, a RISC-based architecture called ARM that Linux has been ported to and works well on. The XScale runs at up to 400 MHz and has 64 MB of RAM, with several useful built-in peripherals like serial and Ethernet. And it's all in the size of a stick of gum: 0.8″ × 3.1″ × 0.25″. Similar to PC/104, you add additional peripherals by stacking boards on its connectors. One of the most popular board is a Compact Flash (CF) board that a Wi-Fi card can be plugged into. A basic Wi-Fi Gumstix system would draw about 400 mA and run about $220. Such a system is a natural for adding to Roomba, and in Chapter 16 you can see an example of Gumstix-controlled

Roomba. The Gumstix board has two main drawbacks, however: it cannot function as a USB host, so you can't add fun USB devices, and it's a bit expensive for casual hacking.

Fortunately, a few other consumer electronics companies have shown the same foresight that iRobot has by creating open systems that are amenable to hacking. The most famous of these is Linksys and their WRT54G series of wireless routers, which you'll be using next.

Wireless Routers as Hacker Toys

Most wireless routers in the home today use a CPU and board architecture created by Broadcom, an innovative chipmaker who also supplies the video processor for the iPod and the Wi-Fi chips in the Nintendo DS. The Broadcom chip used in these routers is a specialized version of the MIPS32 processor, a RISC-architecture chip on which Linux also works well. The insides of one of these routers look not much different from other SBCs like the Gumstix except they have an added chip or two for the Wi-Fi radio.

The best part about these specialized wireless router SBCs is that they can be purchased at any consumer electronics store for around $50. For the cost of a few weeks' worth of Starbucks, you can get a tiny Wi-Fi computer that runs a modern operating system and that you can reprogram yourself.

Firmware Upgrade or Replacement

All SBCs need some way of loading software into them. The three most common ways are:

- A bootloader (like in Arduino) sits on a serial or Ethernet port awaiting code.

- A replaceable memory card is programmed and inserted into the SBC.

- A special program is invoked from inside an old version of the software.

The last option is what is used normally with wireless routers, and you've likely seen how to do this from the Upgrade Firmware section of your router's web interface. Almost all SBCs store their code in non-volatile flash memory so that it stays around if you unplug the box. There is the possibility that if the user invokes Upgrade Firmware and then removes the router's power, the box will be only partially programmed and thus totally non-functional. This is called *bricking* your router, and it's a danger you face any time firmware upgrades are done.

To survive a bricking error, Linksys (and a few other vendors) provides a failsafe that isn't affected by bad or partial firmware. There is a special bootloader in ROM that knows enough to get on your LAN (either via a static IP address or DHCP) and then receive a new firmware through Trivial File Transfer Protocol (TFTP). TFTP is a rudimentary method of transferring files with no error-correction like FTP. On an idle Ethernet LAN, transmission errors are virtually non-existent so the lack of error correction is not an issue. After the bootloader has received the new firmware via TFTP, it runs a checksum on it to verify it's error-free, and then programs the flash ROM and reboots.

Note   This bootloader is called `boot_wait` by the router hacking community, after the variable used to turn it on. Having `boot_wait` means hackers don't need to fear inadvertently screwing up their routers, which gives them (and you) the freedom to experiment.

Linksys WRT54G

If you read Slashdot.org regularly you probably recall some hubbub a few years ago about Linksys and Linux. Linksys used Linux as the OS for their wireless routers but didn't abide by the GNU GPL license that covers Linux usage. Namely, any company using GPL software in a product is required to provide the source code to that software and any changes they made to it. The license allows proprietary drivers (like those for the Broadcom wireless chips) to remain proprietary for trade-secret reasons, but anything that's part of Linux needs to be released. By having that clause in the license, Linux is always being improved. No doubt the Linksys or Broadcom engineers found some bugs in Linux with respect to their CPU and fixed them. If they release those bug fixes in the form of modified source code, others will benefit. The GPL is about enabling the entire community of software developers to advance as quickly as possible.

Figure 14-1 shows a typical Linksys WRT54G. This little box contains an Ethernet port, a 4-port Ethernet switch, a Wi-Fi interface, and typically 4 MB of flash ROM and 16 MB of RAM. It draws less than 500 mA during normal operation, putting it within the ballpark of the other SBCs considered in this chapter.

FIGURE 14-1: Linksys WRT54G, the box that started it all

Linksys is always slightly changing the internals of the box. The current version 5 of the WRT54G has only 2 MB of flash. This version is too small to run a usable Linux system, and so Linksys has switched to VxWorks. VxWorks is a real-time operating system most famous for being the OS on several Mars probes. Unfortunately, it seems less than ideal for a wireless router and many have found this new v5 version to be buggy and crash prone. Linksys still makes several other Linux-capable boxes, however, and you can find older model WRT54Gs to use quite cheaply. The new standard for hacking is the WRT54GL (L for Linux, really), which is still perhaps the easiest router to get running with your own Linux.

OpenWrt

OpenWrt (`http://openwrt.org/`) is one of many Linux distributions for wireless routers. (Others include DD-WRT, HyperWRT, and Sveasoft.) It is named in honor of the WRT54G even though OpenWrt runs on several dozen different models of network hardware. Started in 2004, it deviated from the other hacker versions of Linux for the WRT54G by working from the bottom up to get a working system instead of making slight perturbations in the source code Linksys released.

OpenWrt Software

The most noticeable difference between OpenWrt and the other firmware releases is its modularity. OpenWrt is more like a distribution, like RedHat or Ubuntu. It has a network-based packing system with more than 100 official software packages and even more unofficial ones created by hackers worldwide. The OpenWrt developers provide extensive documentation on building your own code and packaging it. If you can write a simple C program, you can write native code for OpenWrt. If you prefer scripting languages to C, both PHP and Perl have been ported to OpenWrt. And of course, it comes with a command-line shell for any shell programming that's needed.

Besides the many standard network applications like an HTTP and FTP server, OpenWrt can also handle VoIP calls or act like a phone PBX with Asterix. It can be a VPN endpoint with OpenVPN or run a BitTorrent client. Virtually any non-GUI application that already runs on Linux will run on OpenWrt. New packages are being created all the time. Anyone can submit packages to the OpenWrt community for inclusion into the official distribution.

Note Be sure to visit the OpenWrt Documentation site at `http://wiki.openwrt.org/` for more in-depth documentation about OpenWrt in general. To search both the official and the many third-party package repositories, visit `http://www.ipkg.be/`.

OpenWrt Hardware

The OpenWrt community maintains a database of devices that work or don't work with OpenWrt at `http://wiki.openwrt.org/TableOfHardware`. There are at least 50 different network devices that work with OpenWrt. Besides Broadcom chips, OpenWrt has been ported to network processors made by Aetheros and Texas Instruments. Broadcom is the most supported and most familiar, however.

The peripherals supported by OpenWrt are constrained by the type of hardware it's running on. Obviously if the wireless router has no USB interface, it cannot support USB devices. Most peripherals that both work on Linux and have some way to be hooked up to a router work in OpenWrt. These include devices like IDE drives, flash drives, disk drives, serial adapters, webcams, and printers.

How OpenWrt Differs from Regular Linux

Unlike the Linux you may be familiar with, OpenWrt is pared down severely in order to fit in the 4 MB flash chip. The most striking difference is the lack of permanent disk space. The disk is really the flash chip, and the base OpenWrt system fills most of that up. OpenWrt also creates a RAM disk with the available RAM and that offers a good temporary storage location. Many OpenWrt setups operate entirely in a read-only file system mode where the flash disk cannot be written to. This is more secure, but makes for a miserable hacking experience.

To store changeable configuration information like network setup and Wi-Fi settings, a special 64 KB section of the flash chip is reserved for what is known as NVRAM (for non-volatile RAM, even though it's not RAM any longer). OpenWrt, the bootloader, and the stock vendor firmware all use NVRAM. This means that firmware upgrades don't erase your network configuration.

Note   When you have OpenWrt up and running, type **nvram show** on the command line to see all the nvram variables.

To provide the command line tools you expect in a Linux system, OpenWrt uses BusyBox (www.busybox.net/), a single program that combines most of the tiny utilities everyone uses. So if you were to look in the /bin directory of an OpenWrt machine, you'd see the following:

```
lrwxrwxrwx    1 root      root              7 Mar 27 00:06 ash ->
busybox
-rwxr-xr-x    1 root      root         551416 Mar 27 00:06 busybox
lrwxrwxrwx    1 root      root              7 Mar 27 00:06 cat ->
busybox
lrwxrwxrwx    1 root      root              7 Mar 27 00:06 chmod ->
busybox
lrwxrwxrwx    1 root      root              7 Mar 27 00:06 cp ->
busybox
lrwxrwxrwx    1 root      root              7 Mar 27 00:06 dd ->
busybox
lrwxrwxrwx    1 root      root              7 Mar 27 00:06 df ->
busybox
[...]
```

Using BusyBox saves an enormous amount of space. Another big space saver is the use of uCLib (http://uclibc.org/), a C library that is much smaller than the GNU C Library that is standard with Linux. The standard C library contains hundreds of useful functions like printf(),malloc(),floating point arithmetic, and so on. Every program in Linux uses the C library, so making it as small as possible is critical.

Most aspects of OpenWrt are inspected and refactored in a similar way. For example, instead of the standard OpenSSH to provide an SSH server, OpenWrt uses Dropbear, a complete SSH server and client in only 220 KB.

Parts and Tools

The parts for this project are:

- Asus WL-HDD WLAN Hard drive box, Newegg part number WL-HDD2.5
- Generic USB-to-serial adapter (not Keyspan), Newegg part number SBT-USC1K or similar
- Roomba serial tether (or Roo232 from RoombaDevTools.com)
- 7805 +5 VDC voltage regulator IC, Jameco part number 51262
- Two 1 µF polarized electrolytic capacitors, Jameco part number 94160PS
- General-purpose circuit board, Radio Shack part number 276-150
- TO-220 heat sink for voltage regulator, Jameco part number 326617
- Battery holder for 6 AA batteries, Jameco part number 216223
- Six 2700 mAh NiMH AA batteries, SparkFun part number Batt-NiMH
- Battery charger, 8-bay for NiMH/NiCd batteries, Spark Fun part number Batt-8Charger
- Power plug, female, 3.5 mm × 1.3 mm, Jameco part number 237227CM

Choosing the Right Wireless Router for Roomba

Adding an OpenWrt-capable box to Roomba can be made easy or difficult, but it'll either be easy in software and difficult in hardware or vice versa. Almost every OpenWrt router contains a UART (serial port) that is immediately visible and usable from Linux (as /dev/ttyS0). This port isn't brought out to the back panel though, so you'd need to open the router, find the serial ports, and build a 3.3V-to-5V signal converter. It's not too difficult if you're handy with a soldering iron, but it definitely voids the warranty. For an example of bringing out the serial ports to RS-232 jacks on a WRT54GS, see www.rwhitby.net/wrt54gs/serial.html. The cool thing about hooking up the serial port is that you can watch boot messages from the router and log into OpenWrt since it treats ttyS0 as the console.

The other common way of adding a serial port is to get a wireless router with a USB port and add a USB-to-serial adapter. This avoids opening up the router and the hardware becomes as easy as plugging in a USB cable. But this method can be difficult in software due to driver issues. Not all wireless routers have USB ports; usually just the higher-end or more portable ones do. Some of the routers with USB that people have reported success with are:

- Linksys WRTSL54GS
- Asus WL-500G, WL-500GD
- Asus WL-700E
- Asus WL-HDD

Asus is a bit over-represented in that list. Recently they seem to have taken the torch from Linksys in being a more hacker-friendly company. The Linksys WRTSL54GS is a great box and will be used in the next chapter. For this project, the Asus WL-HDD will be used.

Asus WL-HDD

The WL-HDD is a really interesting device (see Figure 14-2). On the one hand, it is a Wi-Fi access point. On the other hand, it's a hard drive box for any 2.5″ IDE hard drive. Instead of connecting to the hard drive via USB or Firewire, you connect to it via Ethernet or Wi-Fi. And others can connect to it, too. In a pinch it works as a network hub and file server for a small group of people. It also has a USB host port for attaching a USB thumb flash drive. It's as big as a paperback novel and draws less than 500 mA of current. To run all these peripherals, the WL-HDD has a 125 MHz Broadcom MIPS CPU, 16 MB of RAM, and 4 MB of flash ROM for its firmware. It costs about $80 and is pretty much the perfect hacker toy.

FIGURE 14-2: Asus WL-HDD, about as big as a 3.5″ hard drive

The WL-HDD is made to be taken apart by the user. Pull on the back panel while holding the case and the entire logic board slides out, ready for you to add a 2.5″ IDE hard drive (see Figure 14-3). This project doesn't require you to install a hard drive, and if you do, it will almost double the power usage when the drive is active.

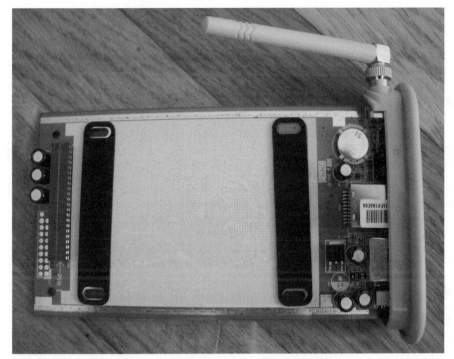

FIGURE 14-3: Asus WL-HDD insides, ready to take a 2.5″ IDE hard drive

If you install OpenWrt, you can even plug a printer into the USB port and make it a print server.

Note   For lots of interesting hacker info on the Asus WL line of routers, visit http://wl500g.info/ and http://wiki.openwrt.org/OpenWrtDocs/Hardware/Asus/WL-HDD.

Installing OpenWrt

Thanks to boot_wait, installing new firmware is a snap and entirely reversible. It's really just a matter of upgrading the firmware on the wireless router. This is something these devices are designed to do and is entirely reversible. To make the process go as smoothly as possible, you need to do a bit of preparation and learn a bit about the different varieties of OpenWrt.

Note Installing firmware technically voids the warranty of your wireless router. But since you can reflash the router back to the factory firmware, there's no way to tell if it ever had OpenWrt on it. Be aware that once you start using OpenWrt you may get no warranty service from the vendor.

Preparing Your Environment

It doesn't matter which type of computer you use, but it needs to have a TFTP client on it. All Unix-based OSes like Linux and Mac OS X have TFTP already installed as just `tftp`. Windows 2000/XP ships with `TFTP.EXE`, which should also work. On Windows, you should install the free Cygwin (`http://cygwin.com/`) environment to make your system as Linux-like as possible since almost all OpenWrt development takes place on Linux.

Next, prepare a totally isolated network that only has the wireless router you're installing OpenWrt on and your computer. A single Ethernet cable should be sufficient to create this isolated network, as the WL-HDD has an auto-switching Ethernet port. If that doesn't work, find an old hub or switch and plug only your computer and the OpenWrt box into it.

You should also review how to change IP addresses on your operating system. If your OS supports the concept of network Locations (Mac OS X and Ubuntu Linux do), it's useful to create two new locations:

- A location with a static IP address in the `192.168.1/24` network, say `192.168.1.111`, with netmask `255.255.255.0`. The default router and other settings are immaterial.

- A static IP address in your home network. If your home network is `10.10.0/24`, create an address `10.10.0.111` with a netmask `255.255.255.0`. Copy all the other settings from your normal setup.

Also, pick out a second static IP address on your network that you'll assign to the WL-HDD when it's ready.

In the following examples, the home network is `192.168.0/24` with a default gateway and DNS server at `192.168.0.1`. The WL-HDD is assigned the IP address `192.168.0.100`. The wireless network everything connects to has an ESSID of todbot with no encryption.

Note Your current network settings can be found in the Network Preferences control panel of your operating system. Be sure to write them down before changing anything. If your OS supports Locations, use them to switch between network configurations.

Getting the Right Firmware

When you first go to the OpenWrt download site at `http://downloads.openwrt.org/`, it's a little daunting. There's no single link to download and just install. OpenWrt works on many platforms and is in constant development. At the time of this writing, the most recent version of OpenWrt is White Russian RC5. By the time you read this, the next major version, Kamikaze, might be out. White Russian RC5 is available at `http://downloads.openwrt.org/whiterussian/rc5/`.

Within the White Russian RC5 release are three main variants: bin, micro, and pptp. The only difference between them is the standard packages that come pre-installed. The bin variety is the standard release, whereas micro is the smallest possible release, and pptp is just like the standard but adds PPTP protocol handling ability. PPTP is a method for creating virtual private networks between two networks. Many people use PPTP to securely join two network that are physically separated. OpenWrt includes it for this reason but it's not used here because these projects are not meant to replace your main network router. The micro version lacks several tools needed, so the bin variant of OpenWrt is used.

In the standard bin release you'll find firmware files for many different wireless routers. The ones you want for the WL-HDD are labelled openwrt-brcm-2.4-<type>.trx, where <type> is the type and size of the flash file system. For the WL-HDD, which has only 4 MB of flash, you need either the jffs2-4MB or squashfs versions. The SquashFS variant creates a read-only file system and saves space but is harder to use. For this project, I use the 4 MB JFFS firmware.

Thus the firmware you want for the WL-HDD is openwrt-brcm-2.4-jffs2-4MB.trx.

Download the file to your PC:

```
% wget http://downloads.openwrt.org/whiterussian/rc5/bin/↵
openwrt-brcm-2.4-jffs2-4MB.trx
```

Note The brcm variant is for any generic Broadcom-based CPU. All supported Asus routers use this type.

Installing the Firmware

You could install the OpenWrt firmware via the stock Asus web interface, but the TFTP method is often easier. Plus it's good to know how to do it in case something goes catastrophically wrong. Firmware installation on Asus routers is a little different than the normal method with Linksys routers. Instead of a straight TFTP transfer, the WL-HDD must be sent a special TFTP command to enable it to accept new firmware. The details of that command are described at http://wiki.openwrt.org/OpenWrtDocs/Hardware/Asus/Flashing. On that page is a link to flash.sh, a script to encapsulate those details. It's written for the WL-500g but can be easily modified for the WL-HDD (by changing all occurrences of 192.168.1.1 to 192.168.1.220). The Roombahacking.com site has this modified version, called flash-wl-hdd.sh, available for download, and that's what is used in the steps below.

To install OpenWrt on the WL-HDD:

1. Disable all network interfaces on your PC except the one Ethernet port that is to be connected to the WL-HDD.

2. Configure your PC to be on the 192.168.1/24 network.

3. Plug the Ethernet cable between your PC and the WL-HDD.

4. Press and hold the WL-HDD reset button. While doing so, plug in the power.

5. Wait for the yellow PWR LED to blink slowly (about once per second). This indicates the WL-HDD is in the bootloader.

6. Test if you can ping the WL-HDD:

```
% ping 192.168.1.220
PING 192.168.1.220 (192.168.1.220) 56(84) bytes of data.
64 bytes from 192.168.1.220: icmp_seq=1 ttl=64 time=0.45 ms
64 bytes from 192.168.1.220: icmp_seq=2 ttl=64 time=0.69 ms
64 bytes from 192.168.1.220: icmp_seq=3 ttl=64 time=1.21 ms
64 bytes from 192.168.1.220: icmp_seq=4 ttl=64 time=0.68 ms
^C
```

7. On your computer, run the `flash-wl-hdd.sh` script:

```
% ./flash-wl-hdd.sh openwrt-brcm-2.4-jffs2-4MB.trx asus
Confirming IP address setting...
tftp> Flashing 192.168.1.220 using openwrt-brcm-2.4-jffs2-
4MB.trx...
tftp> Sent 2162688 bytes in 2.3 seconds
tftp> Please wait until led stops flashing.
```

8. Wait for it to complete. It's done when the PWR LED is on with no blinking and the green WLAN LED is flashing.

9. On your computer, go to the web interface at `http://192.168.1.220/` to verify it's functioning. You should see something like Figure 14-4 (in the following section).

10. Don't try to make any changes yet. Reboot once more. Remove power from the WL-HDD, wait 10 seconds, and plug the power back in. This second reboot allows the JFFS file system to decompress and become fully usable.

11. When the WLAN LED lights again, you can log back into the web interface and begin configuring.

Note The default IP address of the WL-HDD is `192.168.1.220` and the default IP address of OpenWrt is `192.168.1.1`. If you can't ping the WL-HDD at one address, always try the other.

Tip If you're having network problems, make sure you can ping the various parts of your network. See the "Debugging Network Devices" sidebar in Chapter 11 for details.

Basic OpenWrt Configuration via the Web Console

Congratulations, you now have your own tiny Linux box to do with as you will. The default configuration is for a wireless router, but for Roomba purposes, it's better to configure it as a wireless client of your existing wireless network (like a laptop). With your computer and the WL-HDD still on the isolated network (with the Ethernet cable), do the following:

1. Log in to the OpenWrt web interface by going to `http://192.168.1.220/`. Figure 14-4 is what you should see. Besides giving you versioning information, the top right of the home page gives you useful status information like uptime and system load. OpenWrt is more stable than the vendor-provided firmware (particularly Linksys), and it's fun to watch your uptime grow. At the bottom of the page, you'll see the MAC address, which is useful when diagnosing Ethernet and ARP cache problems.

2. Set an administrative password. When you click any of the configuration tabs for the first time, OpenWrt won't let you do anything until you set a password, as in Figure 14-5. This password is for both the web interface and SSH. Click Set when you're done and OpenWrt will let you continue. After you set the password, you'll be presented with a login window. The username is root and the password is the one you just saved.

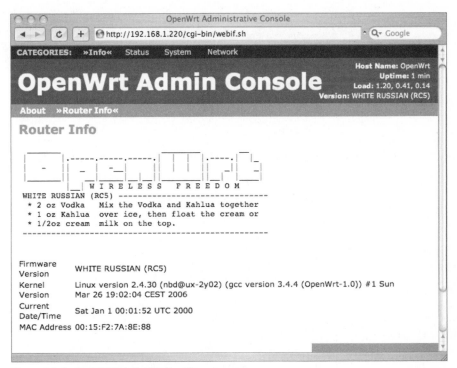

FIGURE 14-4: OpenWrt web interface home page

FIGURE 14-5: Setting a password in OpenWrt

3. Configure the IP address.

- Click the Network category and then the LAN subcategory.

- Input the static IP address you chose for the WL-HDD.

- Input a netmask of 255.255.255.0, the standard for most all networks.

- Also input your home network's default gateway and DNS server settings. Figure 14-6 shows the configuration settings for a WL-HDD with IP address 192.168.0.100 using a router and DNS server on 192.168.0.1.

Note   OpenWrt will always pre-fill the DNS server text box with 192.168.1.1 in an attempt to help you out. If that's correct for you, great. Otherwise, just ignore it.

- Click Save Changes but do *not* click Apply Changes yet.

Note One of the nice things about OpenWrt compared to vendor-supplied is that you can stack up changes to be made and then apply them all at once. Another nice thing is that applying changes doesn't require a reboot.

FIGURE 14-6: Changing the IP address in OpenWrt

4. Disable the WAN port.

   ▪ Select the WAN sub-category as shown in Figure 14-7.

   ▪ The WL-HDD only has one Ethernet port, so the distinction between LAN and WAN is a bit meaningless here. To avoid confusion just disable it entirely.

   ▪ Click Save Changes when done.

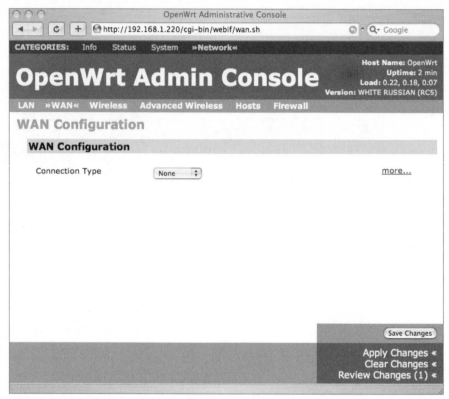

FIGURE 14-7: Disabling the WAN port in OpenWrt

5. Enable Wireless client mode.

- Select the Wireless sub-category. Second to the IP address, this is the most important configuration page. Normally you configure OpenWrt to act as an access point. Instead it will be a client on your existing wireless network like any other client.

- In the ESSID text box, type the name of your wireless network.

- In the Mode pop-up menu, change Access Point to Client (Bridge).

- Leave Channel at Auto and leave ESSID Broadcast at Hide.

- If you use encryption on your network, adjust the encryption settings appropriately. Figure 14-8 shows the settings for connecting to an unencrypted Wi-Fi network named todbot.

- Press Save Changes when done.

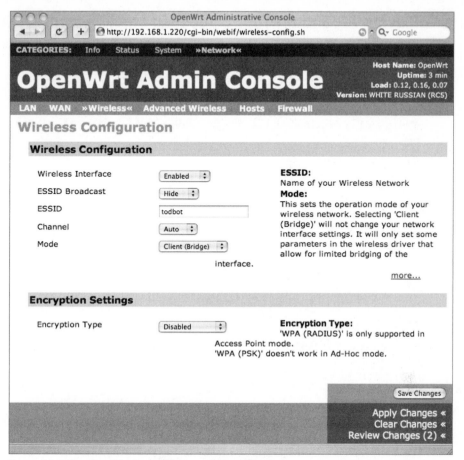

FIGURE 14-8: Configuring OpenWrt as a wireless client

6. Apply changes.

   ■ At this point you have all the changes queued up and you can review them with Review Changes.

   ■ Or, just press Apply Changes and you'll see OpenWrt reconfiguring itself.

   Then, because you changed its IP address, you won't be able to contact it. It will have joined your wireless network, so you can connect to it wirelessly now.

7. Change your PC's network configuration to be back on your wireless net. Use your browser to go to the new IP address of the WL-HDD. You should again see your router's home status page like in Figure 14-4.

The WL-HDD is now fully configured network-wise and it's time for some fun.

Playing Around on Your Embedded Linux System

Just like any good Linux system, OpenWrt has an SSH server. You can log into it like any other Linux box:

```
% ssh root@192.168.0.100
root@192.168.0.100's password:

BusyBox v1.00 (2006.03.27-00:00+0000) Built-in shell (ash)
Enter 'help' for a list of built-in commands.
```

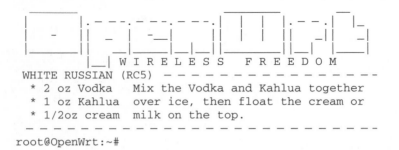

```
WHITE RUSSIAN (RC5) — — — — — — — — — — — — — —
  * 2 oz Vodka   Mix the Vodka and Kahlua together
  * 1 oz Kahlua  over ice, then float the cream or
  * 1/2oz cream  milk on the top.
— — — — — — — — — — — — — — — — — — — — — — — —
root@OpenWrt:~#
```

Of course, unlike a normal Linux system, the disk space situation is very different:

```
root@OpenWrt:~# df -h
Filesystem              Size      Used Available Use% Mounted on
/dev/root               3.2M      1.8M      1.4M  57% /
none                    7.0M     20.0k      7.0M   0% /tmp
```

Notice you only have 3.2 MB of disk total, and over half of it is already in use. But otherwise, it's very much like any other command line Linux system. You can ping Google:

```
root@OpenWrt:~# ping google.com
PING google.com (64.233.167.99): 56 data bytes
64 bytes from 64.233.167.99: icmp_seq=0 ttl=246 time=82.1 ms
64 bytes from 64.233.167.99: icmp_seq=1 ttl=246 time=84.7 ms
64 bytes from 64.233.167.99: icmp_seq=2 ttl=246 time=83.5 ms
```

Inspecting the Running System

As with any Linux system, you have access to a variety of command line tools to inspect the running system. You can run top to see what processes are running.

```
Mem: 8116K used, 6232K free, 0K shrd, 0K buff, 3140K cached
Load average: 0.00, 0.00, 0.00    (State: S=sleeping R=running,↵
W=waiting)
```

PID	USER	STATUS	RSS	PPID	%CPU	%MEM	COMMAND
2201	root	R	388	2196	0.9	2.7	top
2195	root	S	600	296	0.5	4.1	dropbear
2196	root	S	436	2195	0.0	3.0	ash
296	root	ˉS	400	1	0.0	2.7	dropbear
299	root	S	380	1	0.0	2.6	httpd
291	nobody	S	364	1	0.0	2.5	dnsmasq

```
   1 root      S       360     0  0.0  2.5 init
  62 root      S       344     1  0.0  2.3 syslogd
 309 root      S       344     1  0.0  2.3 crond
2149 root      S       332     1  0.0  2.3 wifi
  64 root      S       304     1  0.0  2.1 klogd
 301 root      S       260     1  0.0  1.8 telnetd
```

And you can run other common Linux programs like `netstat`, `ifconfig`, and `ps`. If you're familiar with Linux, you'll find most of the low-level administration and inspection tasks available to you. For example, the entire /proc virtual file system exists to give you visibility into the running Linux kernel.

Even writing little shell scripts on the command line is possible:

```
root@OpenWrt:~# while [ 1 ] ; do
> echo "OpenWrt is the best!"
> sleep 1
> done
OpenWrt is the best!
OpenWrt is the best!
OpenWrt is the best!
```

The ipkg Packaging System

You've no doubt noticed the clock is wrong. The easiest way to fix that is to get an NTP client from the `ipkg` repository. Ipkg is a code packaging and installation technology, a package repository, and a command line program. It takes away a lot of the drudgery of installing new software. Listing 14-1 shows a typical interaction with the `ipkg` command. If you're familiar with other network-based packaging systems like `ports`, `yum`, or `apt-get`, using `ipkg` will be immediately familiar to you.

Listing 14-1: Using ipkg to Find and Install ntpclient

```
root@OpenWrt:~# ipkg update
Downloading
http://downloads.openwrt.org/whiterussian/packages/Packages
Updated list of available packages in /usr/lib/ipkg/lists/↵
whiterussian
Downloading http://downloads.openwrt.org/whiterussian/↵
packages/non-free/Packages
Updated list of available packages in /usr/lib/ipkg/lists/↵
non-free
Successfully terminated.
root@OpenWrt:~# ipkg list | grep ntp
ntpclient - 2003_194-2 - NTP client for setting system time
from NTP servers.
openntpd - 3.7p1-1 - OpenNTPD is a FREE, easy to use
implementation of NTP
```

Continued

Listing 14-1 *Continued*

```
root@OpenWrt:~# ipkg install ntpclient
Installing ntpclient (2003_194-2) to root...
Downloading
http://downloads.openwrt.org/whiterussian/packages/ntpclient_↵
2003_194-2_mipsel.ipk
Configuring ntpclient
ntSuccessfully terminated.
root@OpenWrt:~# date
Sat Jan  1 00:29:08 UTC 2000
root@OpenWrt:~# ntpclient -c 1 -s -h pool.ntp.org
38914 01341.848   19061.0   2251.5    9917.4   50613.4           0
root@OpenWrt:~# date
Mon Jul 17 23:21:59 UTC 2006
```

Reverting to Factory Firmware

In the event you run into some problems with OpenWrt or if you want to sell the WL-HDD, you can always re-flash with the factory firmware. The steps are almost exactly the same as flashing OpenWrt:

1. Go to `http://support.asus.com/` and find the latest firmware file for the WL-HDD. You will download a ZIP file and inside it will be a file named something like WLHDD_1.2.3.9_en.trx. Notice how the .trx extension is the same as the OpenWrt firmware you flashed.

2. Use the method above with `flash-wd-hdd.sh` to reflash the WL-HDD. There is a utility program for Windows on the ASUS site that also supposedly will help you reflash. Also, inside of OpenWrt, there is a Firmware Upgrade page under the System category.

There have been no reported cases of irreparably bricking an ASUS wireless router by using OpenWrt, so you shouldn't be reluctant to try it.

Controlling Roomba in OpenWrt

Now that you have a working tiny Linux system, it's time to figure out how to get it to talk to the Roomba. As mentioned earlier, most of the chips OpenWrt supports have one or two built-in serial ports (UARTs) that you could solder to if you opened up the box. The WL-HDD doesn't appear to have one of those chips. That's okay, because not only does the WL-HDD have a USB port, but OpenWrt has several USB serial drivers.

USB Serial Port Drivers

The Linux USB serial device drivers that have been packaged for OpenWrt are: `ftdi_sio`, `pl2303`, `belkin_sa`, and `mct_u232`. These four drivers enable you to communicate with 90 percent of the USB-to-serial adapters out there and all of the generic ones.

Figure 14-9 shows a common generic USB-to-serial adapter. The actual chip for the adapter is embedded in the hood of the RS-232 connector. These are available from various online retailers like NewEgg.com for around $9. There's no indication as to which chip one uses; you have to plug it in to find out. But for only $9 it's a pretty cheap experiment, and if it isn't supported you can probably use it on a fully-fledged PC.

The Keyspan adapter you've used in the previous chapters doesn't use any of these drivers. There does exist an open-source Linux driver for the Keyspan adapter. It just hasn't been packaged for OpenWrt yet. There's no reason to think that it wouldn't work. Most users of USB serial on OpenWrt are connecting to GPS devices and PDAs, both of which tend to use the same USB-to-serial chips as the generic adapters.

The RooStick uses the Silicon Labas CP2103 chip, which appears to work with the standard Linux `cp2101` driver. Unfortunately that's another driver that hasn't been packaged for use with OpenWrt yet. There's no reason to think it couldn't be and it may be part of the next OpenWrt distribution.

On the positive side, the Arduino board uses an FTDI chip that is supported by the `ftdi_sio` driver. This means you could program an Arduino board to do something based on serial commands and then plug the board into the WL-HDD. This is very powerful. It's so easy to add sensors and actuators to Arduino that Arduino could become a sort of I/O co-processor for the WL-HDD.

 Note There have been sporadic reports of problems with combinations of some WL-HDDs with certain USB-to-serial adapters. If you have problems with one serial adapter, try another. If you think the problem is due to the WL-HDD, try the Linksys WRTSL54GS discussed in Chapter 15. It's almost twice as expensive but has a more advanced USB interface with fewer reported problems. This is the disadvantage of hacking consumer electronics: there's no guarantee the vendor will keep the device the same.

Installing USB Serial Port Drivers in OpenWrt

Thanks to `ipkg`, installing new kernel modules is a snap. Kernel modules are packaged up just like normal programs. The two kernel module packages needed to get serial working are:

- `kmod-usb-ohci`: Contains the drivers `usbcore` and `usb-ohci` needed to talk to the USB interface.

- `kmod-usb-serial`: Contains the general driver `usbserial` and the device-specific drivers `ftdi_sio`, `pl2303`, `belkin_sa`, and `mct_u232`.

FIGURE 14-9: Generic USB-to-serial adapter, uses PL2303 chip

Listing 14-2 shows how to use `ipkg` to install the drivers and `insmod` to load the drivers. To determine if these drivers loaded successfully, the best way is to check the output of `dmesg` and look for appropriate status or error messages. The last part of Listing 14-2 shows what you should expect to see. The most important line is the last one, which tells you which device file the driver has created for you to use. In the example, the device is `/dev/usb/tts/0`. OpenWrt uses the new devFS naming scheme for device files, instead of the more familiar `/dev/ttyUSB0`.

Listing 14-2: Installing Serial Port Drivers in OpenWrt

```
root@OpenWrt:~# ipkg install kmod-usb-ohci kmod-usb-serial
Installing kmod-usb-ohci (2.4.30-brcm-3) to root...
Downloading
http://downloads.openwrt.org/whiterussian/packages/↵
kmod-usb-ohci_2.4.30-brcm-3_mipsel.ipk
Installing kmod-usb-serial (2.4.30-brcm-3) to root...
Downloading
http://downloads.openwrt.org/whiterussian/packages/↵
```

Listing 14-2 *Continued*

```
kmod-usb-serial_2.4.30-brcm-3_mipsel.ipk
Configuring kmod-usb-ohci
Configuring kmod-usb-serial
Successfully terminated.

root@OpenWrt:~# insmod usbcore
Using /lib/modules/2.4.30/usbcore.o
root@OpenWrt:~# insmod usb-ohci
Using /lib/modules/2.4.30/usb-ohci.o
root@OpenWrt:~# insmod usbserial
Using /lib/modules/2.4.30/usbserial.o
root@OpenWrt:~# insmod pl2303
Using /lib/modules/2.4.30/pl2303.o
root@OpenWrt:~# dmesg
[...]
usb.c: registered new driver usbdevfs
usb.c: registered new driver hub
PCI: Setting latency timer of device 00:04.0 to 64
usb-ohci.c: USB OHCI at membase 0xb8004000, IRQ 2
usb-ohci.c: usb-00:04.0, PCI device 14e4:4715
usb.c: new USB bus registered, assigned bus number 1
hub.c: USB hub found
hub.c: 2 ports detected
usb.c: registered new driver serial
usbserial.c: USB Serial support registered for Generic
usbserial.c: USB Serial Driver core v1.4
usbserial.c: USB Serial support registered for PL-2303
hub.c: new USB device 01:02.0-1, assigned address 2
pl2303.c: Prolific PL2303 USB to serial adaptor driver v0.11
usbserial.c: pl2303 converter detected
usbserial.c: PL-2303 converter now attached to ttyUSB0 (or
usb/tts/0 for devfs)
```

Making the Drivers Load on Reboot

Although the drivers are loaded now, they won't reload on reboot. To have them load automatically each time the WL-HDD is rebooted, do the following to add the module names to the /etc/modules file:

```
root@OpenWrt:~# echo "usbcore"   >> /etc/modules
root@OpenWrt:~# echo "usb-ohci"  >> /etc/modules
root@OpenWrt:~# echo "usbserial" >> /etc/modules
root@OpenWrt:~# echo "pl2303"    >> /etc/modules
```

The order of the above lines is important. The usbcore driver must load before usb-ohci, and so on.

Debugging USB Devices

If you ever have problems with USB devices on Linux (any Linux, not just OpenWrt), you have options to diagnose them. The first is to look at all the dmesg output you can, as shown in Listing 14-2. The next is to see which kernel modules are currently loaded:

```
root@OpenWrt:~# lsmod
Module                  Size    Used by      Tainted: P
ftdi_sio                21848   0 (unused)
usb-ohci                19204   0 (unused)
pl2303                  12552   0
usbserial               23868   0 [ftdi_sio pl2303]
switch-core             4896    0
wlcompat                14896   0 (unused)
usbcore                 74792   1 [ftdi_sio usb-ohci pl2303
usbserial]
wl                      423640  0 (unused)
diag                    3320    0 (unused)
```

If you see something missing (or something that shouldn't be there), use insmod and rmmod to fix things up. To see which kernel modules are available, look in the /lib/modules/2.4.30/ directory. Each file is a different kernel module.

The /proc virtual file system contains huge amounts of interesting data about the running kernel. Use cat /proc/cpuinfo to get details on the processor or cat /proc/pci to see what exists on the internal PCI bus. Similarly for USB, you can see what's connected on the USB bus by doing as in Listing 14-3. If you understand USB, it gives you a lot of good information. It's a bit verbose, however, and if all you really want to know is if something is plugged in, install lsusb and run it.

Note The dmesg program spews many apparent errors. Usually you can ignore these. OpenWrt is still a work in progress, and some parts of the system are a little too chatty for users who aren't developers. As you become more familiar with OpenWrt, you'll get to learn which messages are true errors and which are not. As OpenWrt approaches a 1.0 release, the volume of dmesg output has been going down.

Listing 14-3: Inspecting USB Devices via /proc and lsusb

```
root@OpenWrt:~# cat /proc/bus/usb/devices
T:  Bus=01 Lev=00 Prnt=00 Port=00 Cnt=00 Dev#=  1 Spd=12    MxCh=
2
B:  Alloc=  0/900 us ( 0%), #Int=  0, #Iso=  0
D:  Ver= 1.10 Cls=09(hub  ) Sub=00 Prot=00 MxPS= 8 #Cfgs=  1
P:  Vendor=0000 ProdID=0000 Rev= 0.00
S:  Product=USB OHCI Root Hub
S:  SerialNumber=b8004000
C:* #Ifs= 1 Cfg#= 1 Atr=40 MxPwr=  0mA
I:  If#= 0 Alt= 0 #EPs= 1 Cls=09(hub  ) Sub=00 Prot=00
Driver=hub
```

Listing 14-3 *Continued*

```
E:   Ad=81(I) Atr=03(Int.) MxPS=   2 Ivl=255ms
T:   Bus=01 Lev=01 Prnt=01 Port=00 Cnt=01 Dev#=   2 Spd=12   MxCh=
0
D:   Ver= 1.10 Cls=00(>ifc ) Sub=00 Prot=00 MxPS=64 #Cfgs=   1
P:   Vendor=0557 ProdID=2008 Rev= 3.00
S:   Manufacturer=Prolific Technology Inc.
S:   Product=USB-Serial Controller
C:* #Ifs= 1 Cfg#= 1 Atr=a0 MxPwr=100mA
I:   If#= 0 Alt= 0 #EPs= 3 Cls=ff(vend.) Sub=00 Prot=00
Driver=serial
E:   Ad=81(I) Atr=03(Int.) MxPS=  10 Ivl=1ms
E:   Ad=02(O) Atr=02(Bulk) MxPS=  64 Ivl=0ms
E:   Ad=83(I) Atr=02(Bulk) MxPS=  64 Ivl=0ms
root@OpenWrt:~# ipkg install lsusb
[...]
root@OpenWrt:~# lsusb
Bus 001 Device 001: ID 0000:0000
Bus 001 Device 004: ID 0557:2008 ATEN International Co., Ltd
UC-232A ↵
Serial Port [pl2303]
```

Scripting Language Control

Now you have an embedded Linux system with a functioning serial port. It is time to hook it up to Roomba. Since the USB-to-serial adapter has the same RS-232 connector as the serial tether of Chapter 3, just connect it to the adapter and Roomba and you're good to go.

There have been efforts to get Java working in OpenWrt via the SableVM (http://sablevm.org/) or JamVM (http://jamvm.sourceforge.net/), but getting a working Java VM with the system classes in the remaining space of the WL-HDD would be difficult. There are many alternatives to Java, however, and it would be interesting to try some of them.

Shell Script Control

Linux supports many ways of dealing with serial ports. Unix-like operating systems treat everything as files, even devices such as serial ports. This level of abstraction enables you to deal with almost any device in any language, all just by opening files. Need to format a disk? Open /dev/hdd. Need to play an audio file? Send your WAV file to /dev/audio. You can even control Roomba from the command line:

```
% printf "\x89\x00\xc8\x80\x00" > /dev/usb/tts/0   # DRIVE
straight at 200 mm/s
```

(You use \x89 to send bytes using hexadecimal codes.)

Treating devices as files is great for sending and receiving data, but in order to configure the device something else is needed. For serial devices, you need to set the baud rate and other parameters, and the standard way of setting those parameters is with the stty program. Thus, the full set of command line commands to make a Roomba drive straight is:

```
% stty -F /dev/usb/tts/0 57600 raw -parenb -parodd cs8 -hupcl -
cstopb clocal
% printf "\x80" > /dev/usb/tts/0                      # ROI START
% printf "\x82" > /dev/usb/tts/0                      # ROI CONTROL
% printf "\x89\x00\xc8\x80\x00" > /dev/usb/tts/0      # ROI DRIVE
```

The stty gobbledygook sets the serial port to 57600 8N1 with no input or output data manipulation ("raw" mode). The stty program is normally used to help control serial terminals connecting to the computer, which is why there's so much configuration to undo all that.

If you wanted, you could put those four commands in a file, call it roomba_forward.sh, and run it whenever you wanted. That's the beauty of shell script programming: You can find a set of typed commands that perform a task you want to repeat and put them in a file so you only have to remember one thing instead of four.

Note The stty program has slightly different syntax on different versions of Unix. For example, on Mac OS X uses the syntax stty -f <port> instead of stty -F <port>.

Perl Script Control

But writing shell programs of any complexity is frustrating; its syntax is strange compared to Java or C. While Java may be out of reach for now, a version of Perl called microPerl is part of the standard OpenWrt distribution. It gives you most all of Perl's capability but in a few hundred kilobytes. With microPerl you can write nicely structured code but still be able to easily modify it on the WL-HDD itself.

Listing 14-4 shows a microPerl script called roombacmd.mpl that enables you to control Roomba from the command line. It is a microPerl program, but it's also a normal Perl program. Swap the two lines at the top and it will run on any system with Perl (and stty).

The roomba_init() function both executes the appropriate stty command as well as puts the Roomba into SAFE mode so it can receive DRIVE instructions. The rest of the functions are just implementations of various DRIVE functions you've seen before.

Listing 14-4: (micro)Perl Script roombacmd.mpl

```
#!/usr/bin/microperl
##!/usr/bin/perl
# in case we have stty in the current directory
$ENV{'PATH'}="$ENV{PATH}:.";

sub usage() {
```

Listing 14-4 *Continued*

```
    printf "Usage: $0 {serialport} ".
           "{init|forward|backward|spinleft|spinright|stop}\n";
    exit(1);
}
sub roomba_init() {
    # this style stty is for linux
    system("stty -F $PORT 57600 raw -parenb -parodd cs8 -hupcl
-cstopb clocal");
    printf "\x80" > $PORT;    sleep 0.2;
    printf "\x82" > $PORT;    sleep 0.2;
}
sub roomba_forward() {
    $vel="\x00\xc8";
    $rad="\x80\x00";
    printf "\x89$vel$rad" > $PORT;
}
sub roomba_backward() {
    $vel="\xff\x38";
    $rad="\x80\x00";
    printf "\x89$vel$rad" > $PORT;
}
sub roomba_spinleft() {
    $vel="\x00\xc8";
    $rad="\x00\x01";
    printf "\x89$vel$rad" > $PORT;
}
sub roomba_spinright() {
    $vel="\x00\xc8";
    $rad="\xff\xff";
    printf "\x89$vel$rad" > $PORT;
}
sub roomba_stop() {
    $vel="\x00\x00";
    $rad="\x00\x00";
    printf "\x89$vel$rad" > $PORT;
}

# If not enough arguments were passed, return
usage() if( @ARGV < 2 );

$PORT = $ARGV[0];
$CMD  = $ARGV[1];

if( $CMD eq 'init' ) {
    roomba_init();
}
elsif( $CMD eq 'forward' ) {
    roomba_forward();
}
```

Continued

Listing 14-4 Continued

```
elsif( $CMD eq 'backward' ) {
    roomba_backward();
}
elsif( $CMD eq 'spinleft' ) {
    roomba_spinleft();
}
elsif( $CMD eq 'spinright' ) {
    roomba_spinright();
}
elsif( $CMD eq 'stop' ) {
    roomba_stop();
}
else {
    usage();
}
```

In order to use the `roombacmd.mpl` script, you must first install `microperl` and get `stty` (if it's not already on your system). Listing 14-5 shows the commands needed to do that, as well as download the `roombacmd.mpl` script and install everything in a non-volatile location.

Listing 14-5: Installing and Using roombacmd.mpl

```
root@OpenWrt:~# ipkg install microperl
root@OpenWrt:~# ipkg install libgcc
root@OpenWrt:~# wget
http://roombahacking.com/software/roombacmd/roombacmd.mpl
root@OpenWrt:~# wget
http://roombahacking.com/software/roombacmd/stty.tar.gz
root@OpenWrt:~# tar xvzf stty.tar.gz
root@OpenWrt:~# chmod +x ./stty
root@OpenWrt:~# chmod +x ./roombacmd.mpl
root@OpenWrt:~# mv stty roombacmd.mpl /usr/bin
```

With everything in place you can now control Roomba from the command line of the WL-HDD. Listing 14-6 shows a simple Logo-like example of drawing a square with the Roomba.

Listing 14-6: Controlling Roomba from the Command Line

```
root@OpenWrt:~# # draw a square
root@OpenWrt:~# ./roombacmd.mpl /dev/usb/tts/0 init
root@OpenWrt:~# ./roombacmd.mpl /dev/usb/tts/0 forward    ;
sleep 5
```

Listing 14-6 *Continued*

```
root@OpenWrt:~# ./roombacmd.mpl /dev/usb/tts/0 spinright ; ↵
sleep 1
root@OpenWrt:~# ./roombacmd.mpl /dev/usb/tts/0 forward   ; ↵
sleep 5
root@OpenWrt:~# ./roombacmd.mpl /dev/usb/tts/0 spinright ; ↵
sleep 1
root@OpenWrt:~# ./roombacmd.mpl /dev/usb/tts/0 forward   ; ↵
sleep 5
root@OpenWrt:~# ./roombacmd.mpl /dev/usb/tts/0 spinright ; ↵
sleep 1
root@OpenWrt:~# ./roombacmd.mpl /dev/usb/tts/0 forward   ; ↵
sleep 5
root@OpenWrt:~# ./roombacmd.mpl /dev/usb/tts/0 stop
```

Making It All Truly Wireless

Until now you've been powering the WL-HDD from its AC adapter. That's fine for debugging, but the point is to make the WL-HDD completely stand-alone and mobile on top of the Roomba. So the next step is to figure out how to power the WL-HDD.

Asus products usually ship with a high-quality switching power supply that outputs a regulated +5 VDC instead of the heavy bulky wall warts shipped with most consumer electronics and which put out a noisy and only approximate +12 VDC. The advantage of the bulky wall wart to the hacker is that there must be voltage regulator inside the device and the device can take a wide range of input power. This is not so for the WL-HDD, so you have to construct an external voltage regulator to power it off the Roomba voltage of +16V.

Battery Pack Instead of Roomba

The WL-HDD is a pretty useful device by itself. Instead of tying it to Roomba power, you could instead make a battery pack and then you could stick the WL-HDD up to whatever you wanted. Imagine making your own network access point and file server at a café without needing any wires at all. Instead of using the +16V supply from Roomba, you can create a supply using just enough batteries to make a voltage regulator work. The 7805 needs at least 6V in order to regulate down to 5V. The nominal voltage across AA cells is between 1.2 and 1.5V. Six cells makes 7.29V, so that's about optimal. Any more and you'd just be causing the 7805 to work harder regulating a higher voltage down.

Figure 14-10 shows a battery pack circuit for the WL-HDD. SparkFun sells some 2700 mAh NiMH AA cells for $2 apiece. This is an amazing amount of capacity in an AA battery. Normal NiCd rechargable AA cells only put out at most 1000 mAh, so this is about a three-times improvement in capacity. SparkFun also sells a good but inexpensive charger for the batteries.

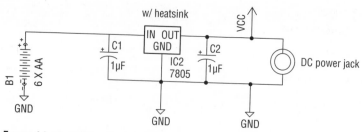

FIGURE 14-10: Battery power supply circuit for WL-HDD

The WL-HDD draws enough current that a heat sink should be used on the 7805 voltage regulator. Figure 14-11 shows a close-up of how one can attach the heat sink with a small screw and nut. You generally don't need heat sink compound if the power draw is lower than about 500 mA.

FIGURE 14-11: Attaching a heatsink to the 7805 voltage regulator

Building and Testing the Battery Pack

Using a spare piece of circuit board, a six-cell battery holder, and a power plug, construct the circuit and place it in an enclosure, like Figure 14-12. Continuing the trend of using enclosures purchasable at the grocery store, this enclosure is from a soap travel holder. The top of the holder isn't shown. The power plug in the example is taken from an old DC wall wart, but Jameco sells a matching plug that'll work with the WL-HDD. The most important thing about wiring up the plug is that the center conductor is positive and the outer conductor is ground.

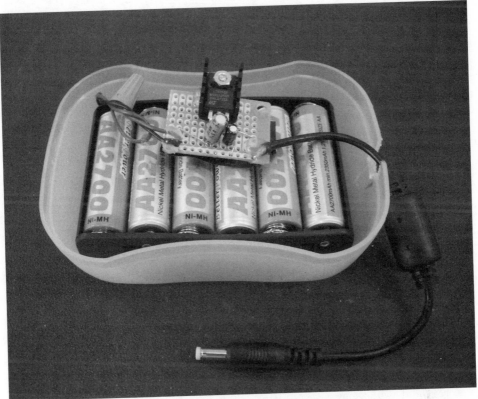

FIGURE 14-12: The power supply constructed and in its enclosure

When you've wired up the circuit, insert the batteries and check the voltages. You should be getting +5 VDC on the center pin of the power jack. When you're satisfied, plug it into the WL-HDD and watch it boot up. If the PWR LED light doesn't immediately come on, pull out the plug and check over all your wiring. Otherwise, you're up and running just like before, but now you're completely wireless.

In Figure 14-13, the lead from the battery is split and run through an ammeter (a multi-meter configured to measure current) to check how much current the WL-HDD actually draws from the batteries. It draws about 450 mA, with slight fluctuations based on CPU usage. With the 2700 mAh batteries, you should get about 2700 / 450 = 6 hours from a fully-charged battery pack. That's about three times longer than Roomba lasts on its battery pack.

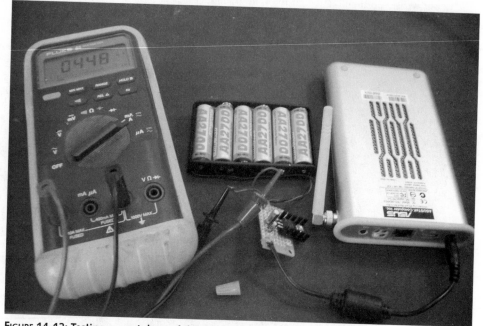

FIGURE 14-13: Testing current draw of the WL-HDD

WL-HDD on the Go

If you would like to use the WL-HDD without Roomba, you can easily lash together the battery pack to the WL-HDD and take it anywhere you want. Figure 14-14 shows how you could do it with a long piece of Velcro. The resulting bundle fits pretty easily in a book bag. A better enclosure could make an even smaller bundle. Insert the power plug to start up and pull it out to power down. Unlike most OSes, OpenWrt is very tolerant of sudden power removal.

FIGURE 14-14: Portable battery-powered Wi-Fi hard drive ready to go

To set the WL-HDD up for use with Roomba, gather up the battery pack, USB-to-serial adapter, and Roomba serial tether and place them all in a slightly larger enclosure. Figure 14-15 shows a plastic sandwich holder (grocery store again) holding all three, with their cables sticking out of a drilled hole in the side to permit them to easily plug in. This rather bulky assemblage can then be placed on top of Roomba and secured with a bit of Velcro or double-sided tape as in Figure 14-16. Carrying around the 15 feet of cable in the serial tether is rather silly, as well as the conversion to RS-232 signal levels and back again. A more svelte design would bypass the length and levels of the example shown here. A RooStick is almost perfect, if only the drivers existed. The cellphone sync cable hack mentioned in the next chapter is also possible and would be quite tiny.

Note   For another take on the WL-HDD battery pack, with an even better enclosure, see www
.aximsite.com/boards/showpost.php?p=1105083.

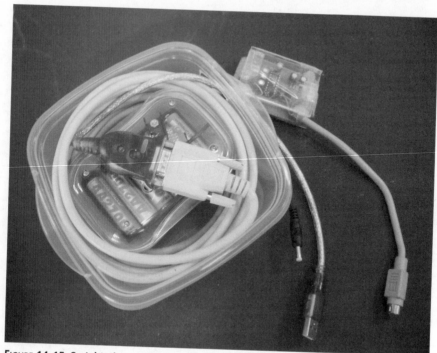

FIGURE 14-15: Serial tether, USB-to-serial adapter, and battery pack, sealed for freshness

FIGURE 14-16: Wireless Linux on a Roomba

Summary

You now have a wireless embedded Linux system that controls Roomba, all for about $100. Having a true general-purpose preemptive multitasking operating system enables you to experiment with high-level robotic cognition at the same level as academic researchers. Putting Linux on Roomba completely fulfills the robot's promise of being an inexpensive robotics platform. Not only can you create simple microcontroller programs, but now you're able to write complex logic based on many types of stimuli (even those from out on the Net).

Not only complex action is possible, but increased data storage, too. You could make Roomba log every action it makes every tenth of a second for an entire cleaning cycle. A few megabytes of data is out of the question with a microcontroller, but for a Linux Roomba with a 1 GB flash drive, it's no big deal.

With a real OS comes the ability to add increasingly complex input devices. Microcontroller boards like Arduino are good for a few sensors; with OpenWrt and a few serial adapters you could control a whole handful of Arduinos chock full of sensors. A USB interface opens a whole unexplored space of the many USB devices out there, all waiting to be used for robotics.

RoombaCam: Adding Eyes to Roomba

in this chapter

☑ Add a Linksys WRTSL54GS to Roomba

☑ Add a camera to Roomba

☑ Install and use webcam drivers in OpenWrt

☑ Add Flash memory storage

☑ Control Roomba from C

☑ Create a web-based command center

☑ Build a small USB-to-serial dongle

Adding vision to a robot is no small feat. You need sufficient processing power to handle the higher data rate of video, sufficient storage space to store intermediary images for processing, and an interface to hook a camera to. This is all just to get video into the robot. It says nothing of the kind of processing needed to parse the incoming video and recognize even simple things like color or motion, let alone people or places.

Vision systems in robotics and artificial intelligence are a huge area of research. Several applications have been created (many free research projects that run on Linux), each tackling one subset of the vision problem like motion detection or spot following. These tools all require a prepared environment with consistent environmental parameters, or they must be tuned to your particular environment.

Rather than attempt to show one of these systems in use, this chapter adds vision to a Roomba robot to create a remote telemetry vehicle, much like a Mars rover. The on-board camera becomes your view into the world as seen by the robot. Unlike a simple video camera, however, the camera data is run through the embedded Linux system on the robot, allowing processing and acting on vision data.

To add a camera, a more capable embedded Linux system than the one used in Chapter 14 is needed. The Asus WL-HDD from that chapter is a great box but is a bit underpowered for handling video. Its USB subsystem also seems to have problems when communicating with multiple devices simultaneously. There are several upgrades possible and the Linksys WRTSL54GS is used here.

The small systems that run stripped-down versions of Linux have more in common with the tiny microcontrollers like the Basic Stamp of Chapter 13 than a desktop computer. In the computer/microcontroller comparison of Figure 13-1, these Linux-capable devices are definitely on the microcontroller side. They are much simpler than a full Linux system, and you don't have to be a Linux expert to make them work. You do need to have familiarity with the command line, as in the previous chapters, but you don't need to be a professional Linux system administrator. If you're new to Linux, this is a great way learn the basics. And all the techniques used in this chapter are applicable to larger Linux systems.

This chapter uses many of the same techniques of installing the OpenWrt embedded Linux from Chapter 14. Installing and configuring OpenWrt is very similar on the various routers it supports. Refer to Chapter 14 for details on getting OpenWrt running for the first time.

Parts and Tools

The parts for this project are:

- Linksys WRTSL54GS wireless router, Newegg N82E16833124057
- Creative Instant Webcam, or equivalent, Newegg N82E16830106120
- USB flash drive, any capacity, Newegg N82E16820211027
- USB hub, 4-port travel size, Newegg N82E16817201604
- Generic USB-to-serial adapter, or USB dongle (see sidebar below)
- Power plug, standard barrel connector, Radio Shack part number 274-1569
- 9V battery snap connector, Radio Shack part number 270-325
- 8-cell AA battery pack, Radio Shack part number 270-407
- Eight AA cells, 2700 mAh NiMH, SparkFun part number Batt-NiMH
- 1N4004 diode, Radio Shack part number 276-1103

It turns out all the component parts for this project are available from Radio Shack. The AA cells should be rechargeable, and preferably the nice 2700 mAh NiMH batteries from SparkFun should be used.

Upgrading the Brain

Even though an entirely different wireless router from a different vendor is being used, the underlying chipset is essentially the same. Thus any binaries you have from the previous project will work in this one, too. And since you're using OpenWrt, much of the configuration is the same as in the previous project.

Linksys WRTSL54GS

The WRTSL54GS (or SL because the name is so ungainly) is composed of a 266 MHz CPU, 8 MB flash and 32 MB RAM. So it's essentially double everything the Asus WL-HDD is, except the price. The SL can be had for a little over $100, whereas the WL-HDD can be found for around $70. The SL is a little bigger. It's shown in Figure 15-1, and you can see the USB port that makes it so interesting compared to other wireless routers.

Internally the SL has two hardware serial ports that are recognized by OpenWrt, but they operate at 3.3V, so you'll need a voltage converter if you plan on using them with Roomba or any other 5V-based system. The SL can be taken apart easily by removing the four rubber feet on the bottom of the unit to reveal four Phillips screws. Of course, like installing OpenWrt, opening the box voids the warranty.

FIGURE 15-1: Linksys WRTSL54GS; note the USB port

Installing OpenWrt

Firmware installation on the WRTSL54GS is almost exactly the same as on the WL-HDD. Both boxes have the `boot_wait` bootloader that enables you to install (through TFTP) a new firmware image to it during the first few seconds of it powering on.

Unlike the WL-HDD, which is a very standard Broadcom-based device and thus uses the generic bcrm firmware, the SL needs a firmware specifically built for it. You'll still want the JFFS file system used previously because it enables you to save programs to the flash disk. Following the naming scheme described in the previous chapter, the firmware you want is:

```
openwrt-wrtsl54gs-jffs2.bin
```

And the full URL for the correct firmware is:

```
http://downloads.openwrt.org/whiterussian/rc5/bin/
openwrt-wrtsl54gs-jffs2.bin
```

To install the firmware, follow the same preparation and setup procedure from the previous chapter. You can use either `flash.sh` from Chapter 14 or the Linksys web interface as in Figure 15-2. The SL by default comes up on `192.168.1.1` like OpenWrt so you can browse directly to it. After the firmware is programmed, the router will reboot, the browser page will automatically refresh, and you'll be presented with the familiar OpenWrt home page.

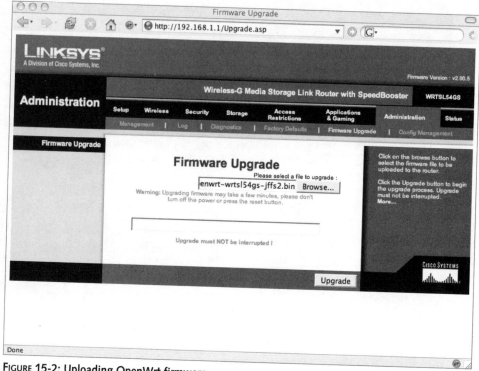

FIGURE 15-2: Uploading OpenWrt firmware

When you have OpenWrt running, go through the same steps to get it configured as a wireless client and install the serial drivers. The configuration is the same; it's one of the benefits of using OpenWrt. The USB driver is different, so instead use the `kmod-usb-ohci` package and run `insmod usb-ohci`. Otherwise, the only difference is when typing `df -h` or running `top`; you'll notice you have more disk space and memory.

If OpenWrt is updated from White Russian RC5, you should try the updated version. It will have more features and work better. If it's not out yet, but you want to get a taste for what's in store, seek out how to build Kamikaze from the OpenWrt wiki.

Adding a Battery Pack or Roomba Power

Externally the WRTSL54GS has a 12 VDC input using a standard barrel power connector and a cheap 12 VDC wall wart. Those wall warts put out noisy power, so there must be something like a 7805 voltage regulator inside the SL. When you open it, you find that instead of a simple linear voltage regulator like the 7805, there's an MP1410ES DC-DC switching converter IC. This chip takes a wide range of DC input between 5 and 20 V and outputs a clean 3.3V. This means you can forgo building an external voltage regulator and instead just build a cable going from the battery pack to a barrel connector.

Figure 15-3 shows the simple schematic and Figure 15-4 shows the complete power adapter. There are two things to note. First, a 9V connector is used because the 8-cell AA battery holder used has a 9V-style snap connector on it (8 × 1.2V = 9.6V, close to 9V). Second, a power diode is inserted in series with the positive power lead. Those 9V snap connectors are easy to plug and unplug, but they're also easy to temporarily mix up and apply a reverse voltage to your gadget. By adding the $0.20 diode, you can avoid blowing up your $100 router. The diode does end up making the output voltage about a volt lower, but the SL doesn't seem to mind.

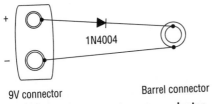

9V connector   Barrel connector

FIGURE 15-3: Schematic for power adapter

In Figure 15-4 it's hard to see where the diode is installed. It's within the heat-shrink tubing on the red lead (the one that is above and to the right). Figure 15-5 shows the diode being soldered in between the two power leads. Make sure the white bar of the diode points away from the 9V connector when soldering.

Figure 15-6 shows the power adapter and battery pack in use. The WRTSL54GS is totally portable and pretty easy to carry around. A portable wireless router not connected to the Internet may not seem all that useful until you hear about mesh networking. A mesh network is a group of routers communicating in a manner that enables them all to provide Internet connectivity when only a few of them at the edge of the group have Internet access. A portable and quickly deployed mesh network could be the critically needed infrastructure during emergencies that wipe out normal cell and telephone service.

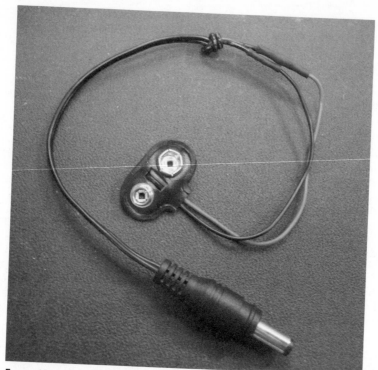

FIGURE 15-4: The finished power adapter

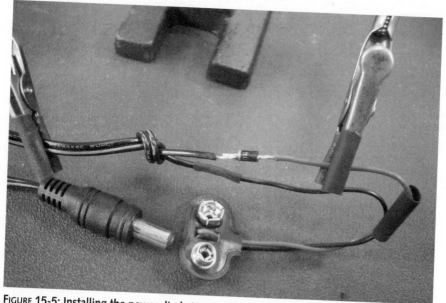

FIGURE 15-5: Installing the power diode to prevent reverse polarity

FIGURE 15-6: Power cable and battery pack in use

Adding a Camera

There are literally hundreds of different models of USB webcams that have been sold over the years. Unfortunately none of them seem to use the USB video device class, so each webcam must have a separate driver instead of them all using a general driver. In practice, there are only a handful companies that make webcam chipsets, turning the problem into writing a dozen drivers instead of hundreds. In Windows this isn't as relevant because of the vendor-supplied drivers (and you hope the vendor continues to release upgrades when the OS is updated), but in Linux these drivers are written by the community. Because it takes time to figure out how to talk to a webcam without any documentation and because the webcam maker can change the internals of the camera without changing the model name, finding a working webcam under Linux is often hit-or-miss.

This problem is compounded in OpenWrt because embedded systems have their own set of restrictions. Of the several webcam drivers in Linux, the driver for the SPCA5XX chips seems to have the best support under embedded Linux. This driver, called spca5xx, works well in more than 100 webcams, with new ones being added regularly. The author, Michel Xhaard, has even created an embedded version of the driver, called spca5xx_lite. This small driver is only for those webcams that support JPEG streaming, a feature where the camera does the JPEG compression and the driver needs to do very little to make an immediately viewable image. Currently, 83 webcams support this mode and thus are compatible with spca5xx_lite. This is the driver that is used here.

If you're lucky, you have an old webcam sitting around that works with spca5xx. If not, there are many to choose from. One of the cheapest is the Creative Instant webcam (see Figure 15-7), available for about $25.

Note

Visit http://mxhaard.free.fr/spca5xx.html for a complete list of cameras and their level of support. See http://mxhaard.free.fr/embedded.html for information about spca5xx_lite.

Note

For great tutorials on doing interesting things with OpenWrt, including a great tutorial on getting spca5xx cameras working, see Macsat at www.macsat.com/. Macsat's tutorial was indispensable for this section and his compilations of the spca5xx drivers and tools that are used here. The http://roombahacking.com/ site mirrors them as a convenience.

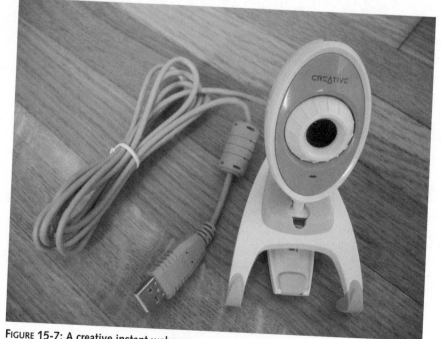

Figure 15-7: A creative instant webcam

Installing Drivers

Similar to the USB serial adapter installation, two drivers must be downloaded and installed. The first is the generic video device driver, called `videodev`. The second is `spca5xx_lite`. The steps to get and install the drivers are:

```
root@OpenWrt:~# ipkg install videodev
root@OpenWrt:~# wget
http://roombahacking.com/software/openwrt/spca5xx_lite.o.gz
root@OpenWrt:~# gunzip spca5xx_lite.o.gz
root@OpenWrt:~# mv spca5xx_lite.o /lib/modules/`uname -r`
root@OpenWrt:~# insmod videodev
root@OpenWrt:~# insmod spca5xx_lite
root@OpenWrt:~# echo "videodev"     >> /etc/modules
root@OpenWrt:~# echo "spca5xx_lite" >> /etc/modules
```

As before, the `insmod` commands load the drivers immediately. Adding the driver names to `/etc/modules` causes the drivers to be loaded on reboot. With the drivers installed, plug in the webcam and watch the output of `dmesg`. You should see the driver recognize the camera. For the camera shown in this example, you see:

```
root@OpenWrt:~# dmesg
[...]
usb.c: registered new driver spca5xx
spca_core.c: USB SPCA5XX camera found. Type Creative Instant P0620
spca_core.c: spca5xx driver 00.57.06LE registered
```

Taking Pictures

The driver creates a standard Video For Linux (v4l) device, so you could use any v4l program. But the easiest way to start capturing images is to get the `spcacat` program, from the author of the `spca5xx` driver. The program expects devices to be in `/dev`, not in a subdirectory, so a symlink (also known as a symbolic link, a file that points to another file) needs to be created.

```
root@OpenWrt:~# wget
http://roombahacking.com/software/openwrt/spcacat.gz
root@OpenWrt:~# gunzip spcacat.gz
root@OpenWrt:~# chmod +x spcacat
root@OpenWrt:~# mv spcacat /usr/bin
root@OpenWrt:~# ln -s /dev/v4l/video0 /dev/video0
root@OpenWrt:~# spcacat -d /dev/video0 -g -f jpg -p 1000 -o
```

You can run `spcacat` with no arguments to see what its options are. The notable ones being used here are:

- `-f jpg`: Data format is JPEG stream from camera.

- `-p 1000`: Pause 1000 milliseconds between images.

- `-o`: Overwrite image. Output image is `SpcaPict.tif` and is overwritten each time.

The `spcacat` program spews a lot of information as it interrogates the camera to figure out what it's capable of. It then prints a line each time it takes a picture. This is a lot of text and after a while you'll get tired of it. If so, create a little script called `camstart` and have it contain:

```
#!/bin/sh
ln -fs /dev/v4l/video0 /dev/video0
spcacat -d /dev/video0 -g -f jpg -p 1000 -o > /dev/null
```

Move your new `camstart` program to `/usr/bin` and use it instead:

```
root@OpenWrt:~# vi camstart              # write the above and save
root@OpenWrt:~# chmod +x camstart
root@OpenWrt:~# mv camstart /usr/bin
root@OpenWrt:~# camstart &
```

Viewing Images

You now have a program writing a new JPEG image to the file `/tmp/SpcaPict.tif` once a second and doing it entirely in the background. To view that image, you could copy it to your PC, but it's easy to put it within the document root of the OpenWrt built-in web server. This server is located at `/www`, so the easiest thing to do is create a symbolic link from the image to somewhere under `/www`, like so:

```
root@OpenWrt:~# ln -s /tmp/SpcaPict.tif /www/SpcaPict.tif
```

Now you can point your web browser at your OpenWrt box and view webcam images. If your WRTSL54G has an IP address of `192.168.0.101`, then go to `http://192.168.0.101/SpcaPict.tif`. Press Reload on your browser to get an updated image.

Note

You may wonder why you don't just run the webcam program from inside `/www`. It will write there without this symlink business. The problem is that `/www` is part of the OpenWrt flash memory. Flash memory has a limited number of write cycles, a few ten thousand. For a USB disk this is okay because you can easily replace it, but it's almost impossible to replace the flash soldered down inside the WRT. If the webcam program were allowed to continuously write to the flash, the flash memory would be worn out within a day.

Pressing Reload all the time is tedious. With a little JavaScript you can have the browser refresh the image for you. Listing 15-1 shows a small dynamic HTML page to accomplish this, called `roombacam.html`. When you have this file, put it anywhere in the `/www` directory. You may need to adjust the symlink to `SpcaPict.tif` so that it's in the same directory as `roombacam.html`. Figure 15-8 shows the resulting web page. You can now easily watch the world from the perspective of your Roomba as it goes about its business.

Listing 15-1: roombacam.html: Auto-Refreshes the Webcam Image

```
<html>
<head>
<title> Roomba Camera </title>
<script language="JavaScript">
<!--
```

Listing 15-1 *Continued*

```
function refreshIt() {
    if (!document.images) return;
      document.images['SpcaPict'].src =
                          'SpcaPict.tif?' + Math.random();
      setTimeout('refreshIt()',2000); // call again in 2000
msec
 }
//-- >
</script>
</head>
<body onLoad="setTimeout('refreshIt()',5000)">
<center>
<h2> Roomba Camera </h2>
<img src="SpcaPict.tif" name="SpcaPict">
<br/>
Image refreshed every 2 seconds.
</center>
</body>
</html>
```

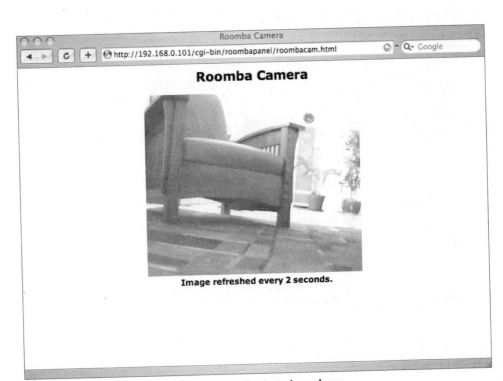

FIGURE 15-8: An auto-refreshing webpage for the Roomba webcam

Adding a Flash Drive

It would be nice to be able to save the images recorded by the Roomba. The router's flash memory shouldn't be used because of its limited lifespan and small space. The router's RAM is a little bigger but disappears when the power is removed. You could copy the images to another network device, but that requires the other device to be online constantly. A USB flash drive seems perfect. It is made for a lot of use and has huge amounts of space.

Figure 15-9 shows a typical USB thumb drive. You can get a 1 GB thumb drive for under $30 from several electronics retailers. Each JPEG image from the webcam takes up approximately 16 KB. On a 1 GB drive, you could store over 60,000 images.

FIGURE 15-9: Typical USB thumb drive, red to match this Roomba

Installing Drivers

Installing the storage drivers in OpenWrt is standard and nothing tricky. The kmod-usb-storage package contains several drivers that work in conjunction. Instead of trying to load them all manually, add them to /etc/modules and just reboot the router. It ends up being faster.

```
root@OpenWrt:~# ipkg install kmod-usb-storage
root@OpenWrt:~# ipkg install kmod-vfat
root@OpenWrt:~# echo "scsi_mod"   >> /etc/modules
root@OpenWrt:~# echo "sd_mod"     >> /etc/modules
root@OpenWrt:~# echo "sg"         >> /etc/modules
```

```
root@OpenWrt:~# echo "usb-storage" >> /etc/modules
root@OpenWrt:~# echo "fat"         >> /etc/modules
root@OpenWrt:~# echo "vfat"        >> /etc/modules
```

After rebooting, plug in the USB drive and watch dmesg. You should see something like Listing 15-2. The most important part is the Partition check section as it tells you the full path to the disk. Use it when mounting the disk.

Listing 15-2: dmesg Output for Detected USB Drive

```
root@OpenWrt:~# dmesg
[...]
hub.c: new USB device 01:02.0-1, assigned address 2
SCSI subsystem driver Revision: 1.00
Initializing USB Mass Storage driver...
usb.c: registered new driver usb-storage
scsi0 : SCSI emulation for USB Mass Storage devices
  Vendor: VBTM      Model: Store 'n' Go   Rev: 5.00
  Type:   Direct-Access                   ANSI SCSI
revision: 02
Attached scsi removable disk sda at scsi0, channel 0, id 0, lun
0
SCSI device sda: 2013184 512-byte hdwr sectors (1031 MB)
sda: Write Protect is off
Partition check:
 /dev/scsi/host0/bus0/target0/lun0: p1
WARNING: USB Mass Storage data integrity not assured
USB Mass Storage device found at 2
USB Mass Storage support registered.
```

Using It

OpenWrt should attempt to mount the disk at the path /mnt/disc0_1. This is mostly reliable but sometimes doesn't work. To mount it yourself, use the full path from the dmesg output and add the partition you want (usually part1 for the first partition). With the resulting device path, mount the disk and see how much space you have on it:

```
root@OpenWrt:~# mkdir /mydisk
root@OpenWrt:~# mount /dev/scsi/host0/bus0/target0/lun0/part1
/mydisk
root@OpenWrt:~# df -h
Filesystem                              Size    Used  Available Use% Mounted on
/dev/root                               7.1M    3.4M      3.7M  48% /
none                                    14.9M   20.0k    14.9M   0% /tmp
/dev/scsi/host0/bus0/target0/lun0/part1         982.7M   107.5M      ↵
875.2M  11% /mydisk
```

At this point, you can cd into the disk and re-run the spcacat command, but leave off the -o argument. Then it will create image after image with names like jpeg 07:21:2006-08:24:03-P0004.tif, which is the date, time, and image number of the capture.

For example, here's how you might capture 100 images at a one-second interval, saving them to the flash drive:

```
root@OpenWrt:~# cd /mydisk
root@OpenWrt:~# spcacat -d /dev/video0 -f jpg -g -p 1000 -N 100
```

When this is running, at any point you can just shut down the router, remove the USB drive, and stick it into your PC to get a visual record of what the router saw as a series of time stamped JPEGs. From there, you can make a time-lapse movie using QuickTime Pro or similar program.

Controlling Roomba from C

For controlling the Roomba robot, a compiled C program has a few advantages over the other methods presented so far:

- **Faster:** Although hard to see on modern PCs, interpreted languages like MicroPerl are noticeably slower on these embedded systems.

- **No need for stty:** Shelling out to an external program to set serial parameters just seems inelegant. In C, it's easy to do the equivalent without needing an external program.

- **Self-contained:** No execution environment is needed. Execution environments like the shell or Perl can add complexity. If the task is simple, use a simple implementation.

Also by programming in C, you have better control over the memory usage and timing of the code. This isn't critical for the simple Roomba controlling done here, but if you were analyzing a lot of Roomba sensor data, using space efficiently becomes important when your machine only has 64 MB of RAM.

The roombacmd program is divided into a set of basic Roomba commands, called roombalib.c, and the command-line parsing, called roombacmd.c. The code compiles and runs not only on OpenWrt Linux but on regular Linux and Mac OS X. It should also run in Windows with Cygwin.

Listing 15-3 shows the three primary functions in roombalib.c. The roomba_init_serialport() function uses standard Unix calls to accomplish what the stty line did in the Perl script. The roomba_drive() function looks very similar to the drive() method in RoombaComm you created in Chapter 5, and the roomba_read_sensors() function looks much like the same function in Arduino from Chapter 13.

The roombalib library defines a Roomba data structure that just contains the serial port name and file descriptor. Think of it as the data-only part of the RoombaComm object.

Listing 15-3: roombalib.c Basics

```c
#define COMMANDPAUSE_MILLIS 100
#define DEFAULT_VELOCITY 200
Roomba* roomba_init(char* portname)
{
    int fd = roomba_init_serialport(portname,B57600);
    if( fd == -1 ) return NULL;
    char cmd[1];

    cmd[0] = 128;      // START
    int n = write(fd, cmd, 1);
    if( n!=1 ) {
        perror("open_port: Unable to write to port ");
        return NULL;
    }
    roomba_delay(COMMANDPAUSE_MILLIS);

    cmd[0] = 130;      // CONTROL
    n = write(fd, cmd, 1);
    if( n!=1 ) {
        perror("open_port: Unable to write to port ");
        return NULL;
    }
    roomba_delay(COMMANDPAUSE_MILLIS);

    Roomba* roomba = calloc(1, sizeof(Roomba));
    roomba->fd = fd;
    roomba->velocity = DEFAULT_VELOCITY;
    return roomba;
}
void roomba_drive(Roomba* roomba, int velocity, int radius)
{
    char vhi = velocity >> 8;
    char vlo = velocity & 0xff;
    char rhi = radius   >> 8;
    char rlo = radius   & 0xff;
    if(roombadebug) fprintf(stderr,
                "roomba_drive: %.2hhx %.2hhx %.2hhx
%.2hhx\n",
```

Listing 15-3 *Continued*

```
                            vhi,vlo,rhi,rlo);
        char cmd[5] = {137, vhi,vlo, rhi,rlo};   // DRIVE
        int n = write(roomba->fd, cmd, 5);
        if( n!=5 )
            perror("roomba_drive: couldn't write to roomba");
}
int roomba_read_sensors(Roomba* roomba)
{
        char cmd[2] = {142, 0};     // SENSOR, get all sensor data
        int n = write(roomba->fd, cmd, 2);
        roomba_delay(COMMANDPAUSE_MILLIS);
        n = read(roomba->fd, roomba->sensor_bytes, 26);
        if( n!=26 ) {
            if(roombadebug) fprintf(stderr,
                "roomba_read_sensors: not enough read
(n=%d)\n",n);
            return -1;
        }
        return 0;
}
```

On Mac OS X and Linux desktop systems, you can compile the program by simply typing make in the directory, and the Makefile will build it for you:

```
demo% wget http://roombahacking.com/software/roombacmd/roombacmd-
1.0.tar.gz
demo% tar xzf roombacmd-1.0.tar.gz
demo% cd roombacmd
demo% make
gcc -Wall -I. roombacmd.c roombalib.c -o roombacmd
```

You can then run it with no arguments to see how to use it, as in Listing 15-4. This compiled version of roombacmd isn't usable in OpenWrt, but you can test it before making an OpenWrt version.

Listing 15-4: roombacmd Usage

```
demo% ./roombacmd
Usage: roombacmd -p <serialport> [OPTIONS]
Options:
  -h, -- help                    Print this help message
  -p, -- port=serialport         Serial port roomba is on
```

Listing 15-4 *Continued*

```
   -v, -- verbose=NUM          Be verbosive (use more for more
verbose)
   -f, -- forward              Go forward at current speed
   -b, -- backward             Go backward at current speed
   -l, -- spin-left            Spin left at current speed
   -r, -- spin-right           Spin right at current speed
   -s, -- stop                 Stop a moving Roomba
   -w, -- wait=millis          Wait some milliseconds
   -v, -- velocity=val         Set current speed to val (1 to
500)
   -S, -- sensors              Read Roomba sensors,display
nicely
   -R, -- sensors-raw          Read Roomba sensors,display in
hex
       -- debug                Print out boring details
Examples:
  roombacmd -p /dev/ttyS0 -v 250 -- forward
  roombacmd -p /dev/ttyS0 -- spin-left
  roombacmd -p /dev/ttyS0 -- sensors
  roombacmd -p /dev/ttyS0 -- stop
Notes:
- The '-p' port option must be first option and is required.
- All options/commands can be cascaded & are executed in order,
like:
     roombacmd -p /dev/ttyS0 -f -w 1000 -b -w 1000 -s
     to go forward for 1 sec, go back for 1 sec, then stop.
```

Building roombacmd for OpenWrt

To immediately try out a version of roombacmd for OpenWrt, a pre-compiled ipkg package has been made that you can fetch and install:

```
root@OpenWrt:~# wget http://roombahacking.com/software/openwrt/↵
roombacmd_1.0-1_mipsel.ipk
root@OpenWrt:~# ipkg install roombacmd_1.0-1_mipsel.ipk
```

You'll then have /usr/bin/roombacmd at your disposal on your OpenWrt system.

Building a version of roombacmd for OpenWrt is not as easy as building it for a desktop OS, because you have to cross-compile. That is, you have to produce an executable for one architecture on another. This is what you did when producing Arduino sketches, but Arduino hides all that from you. The language, GCC, is the same. But compiling embedded Linux programs is a bit more complex than compiling AVR programs. OpenWrt does make it quite easy with the company's SDK. The abbreviated steps are:

1. Get the OpenWrt SDK.

2. Create a directory with your source code and package control commands for `ipkg`.

3. Within that directory, compile using the OpenWrt makefiles.

4. Copy the resulting `ipkg` package to the router and install.

To save you some time, the `roomcmd` C code has been placed in such a directory set up to work with the OpenWrt SDK. Using it you can get started making modifications and improvements to suite your needs. To get the SDK and the bundled `roombacmd` source and to start cross-compiling, run these commands:

```
demo% wget http://downloads.openwrt.org/whiterussian/newest/↵
OpenWrt-SDK-Linux-i686-1.tar.bz2
demo% bzcat OpenWrt-SDK-Linux-i686-1.tar.bz2 | tar -xvf -
demo% wget http://roombahacking.com/software/roombacmd/roombacmd-
ipkg.tar.gz
demo% tar xvf roombacmd-ipkg.tar.gz
demo% mv roombacmd-ipkg OpenWrt-SDK-Linux-i686-1/package
demo% cd OpenWrt-SDK-Linux-i686-1
demo% make clean && make
```

The last line will spew a lot of text as it builds, and then when it's done you'll have the `roombacmd` ipkg file in `bin/packages`. Copy that file to your OpenWrt box and install it using `ipkg`.

The `roombacmd-ipkg` directory contains an `ipkg` package description and a Makefile that abides by the OpenWrt SDK's standards. It's like a meta-Makefile in that it contains descriptions on how an existing Makefile (the one for the regular Linux version of `roombacmd`) should be modified to compile for OpenWrt. OpenWrt has several recipes for how to do this, and `roombacmd-ipkg` uses one of them almost verbatim. If you're familiar with Makefiles, it's a little hairy but not too complex. If you have other C code you'd like to run on your OpenWrt system, you can use the preceding techniques to cross-compile them as well.

Note   Currently, you need a Linux system (Debian-based, like Ubuntu) to run the OpenWrt SDK. If you have VMware or VirtualPC, you can run virtual Linux.

Note   OpenWrt has a great how-to document on compiling for OpenWrt and building packages at `http://wiki.openwrt.org/BuildingPackagesHowTo`.

Controlling Roomba with CGI

OpenWrt ships with a rudimentary Common Gateway Interface (CGI) capability as part of its web server. With the `roombacmd` C program installed, it can be used in CGI programs. The most common language used for OpenWrt CGI programs is plain shell scripting. You can write CGI programs in C, but the cross-compiling gets to be a pain. If you've poked around the /www directory, you may have noticed the OpenWrt web interface is all written in shell.

Listing 15-5 shows `roombapanel.cgi`, a small CGI shell script used to control Roomba with `roombacmd`. The top part of it can be edited to fit the configuration of your system. The middle part is a simple HTML interface with buttons that point to the CGI and set the $QUERY_STRING environment variable to a valid `roombacmd` command. The last part does the work of commanding Roomba. If a button is pressed, the $cmd variable is set and that is used as the movement argument to `roombacmd`. Figure 15-10 shows what the CGI looks like when running. The PICPATH and SNAPPATH variables aren't used yet, but they will be.

Listing 15-5: roombapanel.cgi

```
#!/bin/sh
# edit this: serial port of the roomba
PORT="/dev/usb/tts/0"
# edit this: path to roombacmd
ROOMBACMD="/usr/bin/roombacmd"
# edit this: where the webcam is writing an image
PICPATH="/tmp/SpcaPic.tif"
# edit this: where archived ("snapshot") images should be
stored
SNAPPATH="/mydisk/archive/cam-`date -Is`"

me="$SCRIPT_NAME"
cmd="$QUERY_STRING"

cat <<EOF
Content-type: text/html

<html>
[... html interface to make buttons ...]
</html>
EOF

if [ "$cmd" ] ; then
    echo "cmd: $cmd"
    $ROOMBACMD -p $PORT -- $CMD
fi
```

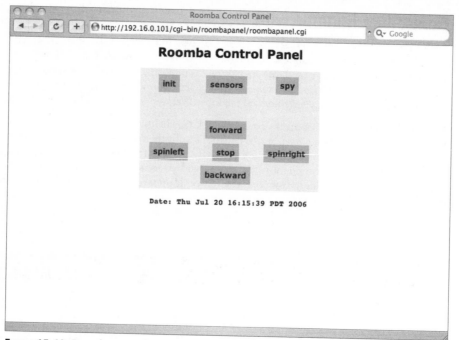

FIGURE 15-10: Roomba control panel CGI script in use

Putting It All Together

You now have all the components working and ready to assemble. Thankfully, there are no real interdependencies, so the isolated tests above are valid. The first thing to do is get all the hardware plugged in and working. Then you can finalize the software to control it all.

System Configuration

All the device drivers should be configured to load at boot. Listing 15-6 shows what the /etc/modules file should end up looking like. The order is important. The general driver (usb-serial, videodev, or so on) must be loaded before the device-specific driver (p12303, spca5xx_lite, or so on).

Listing 15-6: Final Contents of /etc/modules

```
root@OpenWrt:~# cat /etc/modules
wl
usbcore
usb-ohci
```

Listing 15-6 *Continued*

```
usbserial
pl2303
videodev
spca5xx_lite
scsi_mod
sd_mod
sg
usb-storage
fat
vfat
```

Make sure you save any programs you want saved to /usr/bin and not in /tmp. Otherwise, they'll disappear on reboot, since /tmp is a RAM disk. This includes roombacmd, camstart, stty, and any other scripts or commands you want.

The web and CGI files you create live in /www, which is already flash, so you won't have to worry about them disappearing.

Building a USB Serial Tether from a Phone Sync Cable

Using the RS-232 serial tether from Chapter 3 with a generic USB-to-serial adapter works but is very ungainly. The tether's long cable is hard to manage when the controlling computer is right on top of the Roomba. Building a tether with a shorter cable is definitely possible, but a really small solution is to use a USB data cable originally meant to sync mobile phones. Some mobile phones have logic-level serial ports on them instead of USB interfaces, and the sync cables contain small USB-to-serial adapters. These cables are available for about $20 from various places, including Radio Shack. Since this is a Roomba USB dongle, it's called the Roombongle.

Things you'll need to build the Roombongle:

- USB data cable for Nokia phones (AKA FutureDial Cable 22), Radio Shack part number 17-762

- Mini-DIN 8-pin cable, Jameco part number 10604

The construction is really straightforward:

1. Acquire the right USB sync cable. Figure 15-11 shows the packaging for the sync cable used here, and Figure 15-12 shows what the cable looks like. It's very important to get Cable 22, which has the bulge in the middle, and not one of the other cables. Other cables can be used, but you'll have to figure out the pinouts.

Continued

Building a USB Serial Tether from a Phone Sync Cable *Continued*

FIGURE 15-11: FutureDial Cable 22 USB sync cable for Nokia

FIGURE 15-12: The cable itself — make sure it has the bulge.

Building a USB Serial Tether from a Phone Sync Cable *Continued*

2. When you open up the bulge (see Figure 15-13), you'll see that it contains a Prolific PL-2303 USB-to-serial interface chip, just like the cheap USB-to-serial adapters. You already know that this chip is well supported on all the OSes (and the p12303 driver is part of any Linux, including OpenWrt). You can download the spec sheet for said chip and you'll see that it normally operates at 3.3V, but its inputs are 5V-tolerant and its 3.3V outputs are within the valid range for 5V logic. Thus, Roomba should understand it, and vice versa.

FIGURE 15-13: Inside the bulge, a PL-2303 USB-to-serial chip

3. Turn the board over and see the serial cable wires (see Figure 15-14, left side). For the Cable 22 sync cable, the wires are:

- **Black:** GND
- **White:** RXD, receive from PC (connect to TXD on Roomba)
- **Orange:** TXD, transmit from PC (connect to RXD on Roomba)

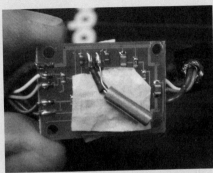

FIGURE 15-14: The wires on the left are the serial port; Black=GND, White=RXD, Orange=TXD

Continued

Building a USB Serial Tether from a Phone Sync Cable *Continued*

4. Note the position of those wires and unsolder the cable bundle from the board. Then take a Mini-DIN cable that you've figured out which wires are GND, TXD, and RXD for Roomba and solder them onto the board in the correct position. Figure 15-15 shows the finished task.

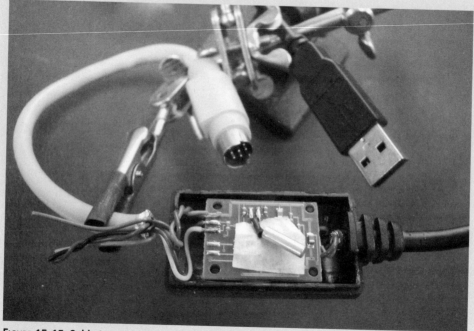

FIGURE 15-15: Soldering on a Mini-DIN 8-pin cable

5. Apply a little hot glue to the Mini-DIN cable where it meets the enclosure and snap on the other half of the enclosure. The unused wires from the Mini-DIN can be snipped off (just make sure they don't short together) or you can take the GND and Vpwr lines from and run them to an external connector. This will enable you to run projects off the robot's battery as well as using the USB dongle. Figure 15-16 shows the completed dongle with an external power connector. In this version, the 9V snap-on connector was used to mate with the WRTSL54GS power adapter made previously.

The knowledge to use the USB phone sync cables came from the NSLU2-Linux hackers at www.nslu2-linux.org/wiki/HowTo/AddASerialPort.

Building a USB Serial Tether from a Phone Sync Cable *Continued*

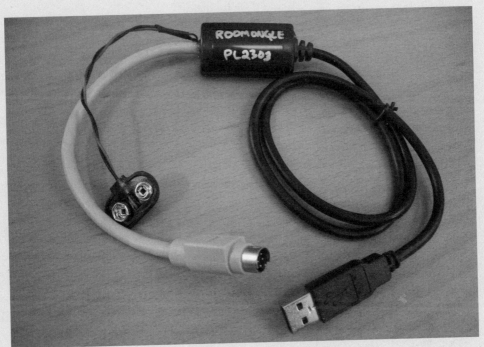

FIGURE 15-16: The finished dongle

The NSLU2 is a low-cost network storage device from Linksys, sort of like the Asus WL-HDD. The NSLU2-Linux hackers use the sync cables to connect to the internal serial ports on the NSLU2, which are very similar to the two internal serial ports on the WRTSL54GS. They've documented not just how to use Cable 22, but all the USB sync cables available from Radio Shack. If you have other OpenWrt-capable routers, you can use one of these sync cables to connect to these internal serial ports and have a serial console login to OpenWrt.

Plugging All the Devices Together

Thanks to a tiny USB travel hub, all the devices can fit in small box, as in Figure 15-17, showing a test fitting. The small box can hold the battery pack if you don't want to power everything off the robot. In Figure 15-17, the box was a small wooden gift box covered in black duct tape to give it extra strength and add a bit of grippiness. In one wall of the box, a slit was cut with a Dremel to allow the cables to pass through.

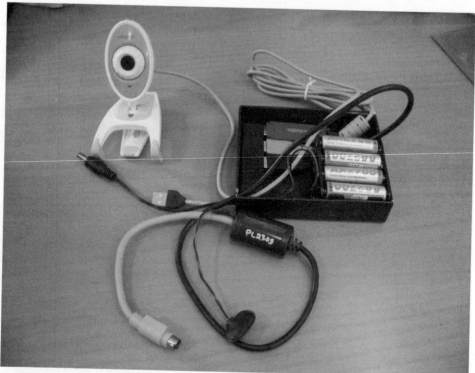

FIGURE 15-17: The peripherals plugged into a small USB travel hub

When plugged in to the SL, lsusb provides the following output:

```
root@OpenWrt:~# lsusb
Bus 003 Device 001: ID 0000:0000
Bus 002 Device 001: ID 0000:0000
Bus 002 Device 006: ID 058f:9254 Alcor Micro Corp. Hub
Bus 002 Device 007: ID 067b:2303 Prolific Technology, Inc. PL2303
Serial Port
Bus 002 Device 008: ID 041e:4034 Creative Technology, Ltd
Bus 002 Device 009: ID 08ec:0008 M-Systems Flash Disk Pioneers
Bus 001 Device 001: ID 0000:0000
```

Measuring Current Consumption

Power consumption is always an issue with portable devices. In this case it's worrisome because there are so many devices in the mix. It may be that the power budget is blown and none of this will actually last for any reasonable amount of time on a battery pack.

So the first thing is to test how much power just the WRTSL54GS draws, since it likely sucks the most power. Figure 15-18 shows the power consumption of the SL after it is booted and a file is being transferred to it. At 358 mA it draws less power than the WL-HDD, which has a CPU almost twice as slow. The SL is a newer design and doesn't have an IDE interface, which perhaps accounts for the difference. In any case, it's a pleasant surprise. At 400 mA current draw (being conservative) of just the SL, the 2700 mAh battery pack should last about six hours.

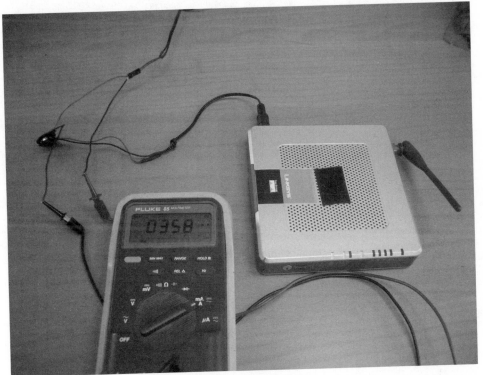

FIGURE 15-18: Measuring current draw of just WRTSL54GS

Figure 15-19 shows the current draw with the peripheral pack added. The webcam is taking pictures once per second and saving them to the flash drive. Doing that adds only about another 100 mA to the power consumption. This is great. This means that with the battery pack, the entire assembly can go for about five hours.

Alternatively, if you ran off the robot's 3000 mAh battery pack and don't run the vacuum motors, Roomba will draw about 500 mA, the SL and peripherals will draw another 500 mA, and you'll get a couple of hours of run time.

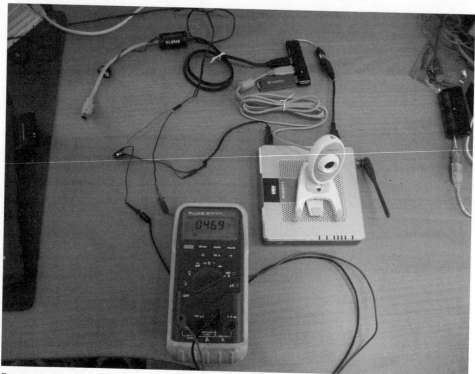

FIGURE 15-19: Measuring current of the webcam taking pictures to flash drive, with Internet access

Note If at some point you end up with a project that draws too much power for Roomba or a small battery pack, one solution is to use a small lead-acid battery. These have high charge ratings, are rechargeable, and can take high current demands.

Assembling the Roomba Command Center

You have the makings of a Roomba command center: a video panel showing live images from Roomba and a control panel to tell Roomba what to do. To combine these two browser windows into one, use an HTML frameset. Call it `roombapanel.html` and put it in the same directory as the other two pages:

```
<html>
<title> Roomba Command Center </title>
<frameset cols="50%,50%">
  <frame src="roombacam.html">
  <frame src="roombapanel.cgi">
</frameset>
</html>
```

The result is something like in Figure 15-20. However, there is a small change to the control panel. A new button called Snapshot has been added. As you drive around exploring the world from the robot's perspective, it would be nice to be able to save the current image. It would also be useful to save the Roomba sensor state at that point.

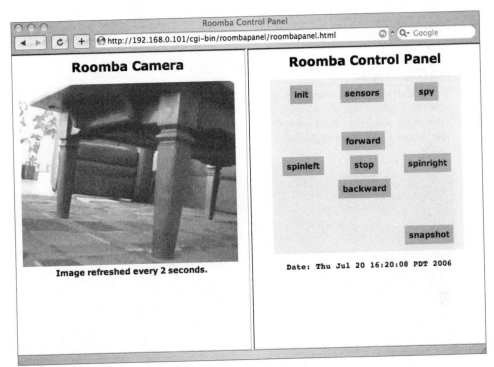

FIGURE 15-20: The finished command center

The snapshot command copies the current live webcam image to an archive location (the flash drive). It also dumps the current sensor state of Roomba to a text file. With both a webcam image and sensor telemetry data, you can reconstruct what Roomba was doing and remember your adventures. Listing 15-7 shows the changes needed to be made to roombapanel.cgi.

Listing 15-7: Changes to roombapanel.cgi to Add Snapshot Capability

```
if [ "$cmd" ] ; then
    echo "cmd: $cmd"
    if [ "$cmd" = "snapshot" ] ; then
        echo "saving image and sensors to $SNAPPATH!"
        cp $PICPATH "$SNAPPATH.tif"
        $ROOMBACMD -p $PORT -- sensors-raw > \
            "$SNAPPATH-sensors.txt"
```

Continued

Listing 15-7 *Continued*

```
    else
        $ROOMBACMD -p $PORT -- $CMD
    fi
fi
```

The Final Product

With the peripherals in a small pack, all that's left is to assemble it all and install it onto the Roomba. Figure 15-21 shows what the final Roomba Rover looks like. As before, small bits of Velcro hold down the router, peripheral box, and camera. If you want to use the Roomba buttons, you can situate the assembly on the dustbin.

FIGURE 15-21: Red Roomba Rover on the prowl

Viewing the world from the robot's perspective is an interesting experience. Between the different angle and the lower image quality and frame rate, exploring even mundane places seems like exploring a different world. It makes you wonder how alien the surface of Mars is, and how much is attributable to the difference of perspective.

A Final Fun Thing

Like most of the newer wireless routers, the WRTSL54GS has a button on its front. This button is available in OpenWrt by looking at /proc/sys/button. For example:

```
root@roombawrt:~# cat /proc/sys/button     # button not pressed
0
root@roombawrt:~# cat /proc/sys/button     # button pressed
1
```

It's easy to write a little program that monitors the button and takes a snapshot and archives it, like the snapshot command of the control panel. For example, Listing 15-8 shows just such a program that runs in the background. Now you can tell anyone to go and press the button, and you'll get a picture of them doing it. Alternatively, you could have it e-mail the picture and post it to Flickr. (There's a mini-sendmail ipkg package to help with this.)

Listing 15-8: Making the WRTSL54GS Button Take a Picture

```
#!/bin/sh
# edit this: where the webcam is writing an image
PICPATH="/tmp/SpcaPic.tif"
# edit this: where archived ("snapshot") images should be
stored
SNAPPATH="/mydisk/archive/cam-`date -Is`"
while [ true ]; do
    sleep 1
    if [ `cat /proc/sys/button` = "1" ]; then
        echo "Saving snapshot to $SNAPPATH"
        cp $PICPATH "$SNAPPATH.tif"
    fi
done &
```

Summary

You now have a vision-enabled, large storage capacity, Linux-running Roomba rover and a handful of common consumer electronics components. It's a great platform to start adding complex data acquisition algorithms and peripherals. You could easily add a cheap USB audio dongle and microphone so Roomba records what it hears as well as what it sees. Multiple cameras pose no problem as well, and you could add a backward-facing camera to supplement the front camera. Add four cameras and produce a real-time panorama of what your Roomba sees.

You can also build new C programs that run natively in OpenWrt to access all these peripherals. From the huge repositories of open-source software, you could find some interesting code and port it for use with OpenWrt. Most non-GUI Linux code will port easily. Find some of the free motion-tracking code and try it out on your Roomba. Program your Roomba to follow you throughout your house. Add USB audio and some powered speakers and you have a boom box that plays your favorite online radio station wherever you go.

Other Projects

Roomba is one of the most hacked consumer electronics devices in history, it seems. The projects in this book take one approach to Roomba hacking — home-built add-ons via the ROI port. However, there are many other ways to hack your Roomba. This chapter shows just a few options.

Although most Roomba projects are done by regular folk just for the fun of it, there are a few companies that have been formed to take those good ideas and become enablers for others. They've done the research and through trial and error built a gadget usable by anyone.

Roomba has been around for several years now, but only recently has it become more hackable because of the ROI port. The hard-won hacks that only apply to the original Roomba robot aren't covered here since in general they are quite difficult and aren't as applicable to the general Roomba-owning public.

In addition to the projects listed here, there are some good Roomba resources on the Net that regularly feature Roomba projects:

- `http://roombareview.com/`: The original Roomba fan site. The site contains Roomba and other domestic robot reviews, tips, and news. It also has a great forum with discussions on general Roomba and Scooba use and Roomba robotics.

- `http://roomba.pbwiki.com/`: A compilation of various Roomba hacks.

- `http://makezine.com/`: General hacker blog that occasionally features Roomba hacks. Part of Make magazine.

- `http://hackaday.com/`: General hacker blog that occasionally features Roomba hacks.

- `http://roombahacking.com/`: The site containing all the code for this book, including updates and news about other Roomba hacks.

Autonomous Roombas

The projects so far have shown you how to attach several different embedded computing systems to Roomba, but you're not limited to just those. Hackers across the globe are turning Roomba into a programmable robot using just about every conceivable embedded computer in every conceivable programming language. These designs are documented on the Net so you

in this chapter

- ☑ Control the mind of your Roomba

- ☑ Gumstix on a Roomba

- ☑ iPaq Roomba

- ☑ Explore Roomba APIs and applications

- ☑ Dress up your Roomba

- ☑ Brave the warranty-voiding hacks

can duplicate and improve upon their efforts. And if you don't want to build your own add-ons, there are companies that will sell you programmable ones. Already mentioned are the upcoming RooStamp and RooAVR from RoombaDevTools.com, allowing you to program Roomba with a Basic Stamp or an AVR chip, respectively.

Roomba Mind Control by Element Direct

`http://elementdirect.com/`

Another commercial microcontroller product that is very interesting is the Roomba Mind Control from Element Direct. The Mind Control consists of two components: a code stick that plugs into Roomba and a code stick programmer. The code stick is an extremely tiny AVR microcontroller housed in a cylindrical shell as big as the Mini-DIN connector of the Roomba ROI port. It is virtually invisible when plugged into Roomba. Figure 16-1 shows what it looks like.

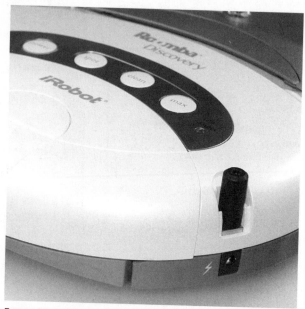

FIGURE 16-1: Roomba Mind Control microcontroller code stick plugged into the Roomba Discovery model

The programmer is a small board powered and controlled via USB. Figure 16-2 shows the programmer with a code stick plugged into it and being programmed. The currently Windows-only software used to program the code stick is derived from the same AVR GCC that is used in Arduino, so it's conceivable that Arduino or any other AVR GCC-based programming environment could work with the Mind Control.

FIGURE 16-2: Roomba Mind Control code stick being programmed

RoombaNet: Gumstix-Controlled Roomba

http://people.csail.mit.edu/bpadams/roomba/

The Gumstix boards were mentioned in Chapter 14 as a way to get an embedded Linux system onto the Roomba. Bryan Adams, a graduate student at MIT at the time, now working at iRobot, created what was perhaps the first Roomba ROI hack, which he called RoombaNet. It consists of a Gumstix board, a Compact Flash Wi-Fi card, and a custom adapter board. Figure 16-3 shows RoombaNet hooked up. It's quite small. Figure 16-4 shows a closeup of his board stack and you can see it's just a little over an inch tall.

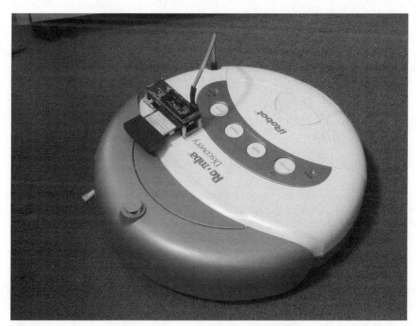

FIGURE 16-3: Tiny Wi-Fi Linux on a Roomba with RoombaNet

FIGURE 16-4: RoombaNet board closeup, barely over an inch tall

Bryan created a great web page about how he created RoombaNet, outlining the methods he used and pitfalls he encountered. As part of his PhD research, Bryan created evolved neural network programs that he then loaded onto his RoombaNet board. He had the Roomba wake up, undock from its charging base, run the neural net for a few minutes, and then return to its home base.

iPaq on a Roomba

    http://roomba.pbwiki.com/iPaq+Roomba+1

PDAs are small computers that run an embedded operating system and have a nice display and input device to boot. They seem perfect for Roomba hacking. Michael Menefee added iPaq Pocket PC PDAs to his Roomba models. On the hardware side, Mike created a powerful serial converter and power supply to convert the Roomba ROI signals to usable RS-232 and +5V DC. The iPaq connects directly to this, through its built-in serial port, and the power supply has additional connections and capacity for any additional peripherals.

For software, Michael installed Linux on the iPaq using Familiar (http://familiar.handhelds .org/). On top of that he ran Player (http://playerstage.sourceforge.net/), a robot device interface that provides a high-level and abstract view of a robot's actions. Researchers worldwide use Player as a common toolkit for experimenting with robotics. To allow Player to control the robot, Michael wrote a Roomba driver in C++. A small configuration file change to Player is all that's required to get it to use the Roomba driver.

Figures 16-5 and 16-6 show iPaq Roomba installed using two different iPaq models and two different Roomba models. Both have Wi-Fi and can run a web server or SSH server. They both run the Player server and thus can be controlled remotely. A third computer runs the Player Stage program for multi-robot interaction.

Mike has plans on adding a camera and other useful sensors. Thanks to Player and Linux these can be integrated without much extra code.

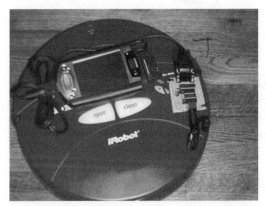

FIGURE 16-5: iPaq Roomba with an older iPaq on a red Roomba model

FIGURE 16-6: iPaq Roomba with a newer iPaq on a Roomba Discovery

Erdos

```
http://www.cs.hmc.edu/~dodds/erdos/
```

Erdos is an academic Roomba project created by Dr. Zachary Dodds and others at Harvey Mudd College. Recognizing the Roomba as a low-cost but durable robotics platform, Erdos is aimed at creating a teaching environment for robotics in the classroom. It has successfully been used in the introductory Computer Science classes at Harvey Mudd and other schools.

Erdos consists of a Bluetooth-controlled Roomba, a Python language Roomba library, and an overhead webcam-based vision system. The webcam provides the feedback needed to accurately determine the robot's position, and the Bluetooth connection allows state-of-the-art artificial intelligence algorithms be run from large computer systems.

The low-cost, open architecture, and ruggedness of Roomba make it a perfect classroom addition, and Dr. Dodds is promoting Erdos for use in classrooms worldwide.

Roomba Costumes and Personalities

With their jaunty tunes and non-threatening shape, Roomba robots are friendly little gadgets. Their owners become attached to them and even give them names. Is it any surprise that people want to dress up their Roombas?

RoomBud Costumes by myRoomBud

```
http://myroombud.com/
```

myRoomBud is a company that makes a line of Roomba costumes, called RoomBuds. Figure 16-7 shows some of the costumes available. They are extremely cute and leave Roomba totally functional. The wall sensor, side brush, and remote control sensor are all unobstructed. It does cover up the buttons on the top, but if you use the remote this isn't really an issue. Each RoomBud outfit has a cute name like Roobit the Frog or Mooba the Cow. The costumes transform Roomba from a vacuum cleaner to a pet (if it wasn't already one for you). In addition to the Roomba outfits, myRoomBud is working on a line of costumes for the iRobot Scooba, called ScooBuds.

The team of four young entrepreneurs who started myRoomBud are all Roomba enthusiasts who wanted Roomba outfits for themselves. After creating costumes for their own Roombas and for family members' Roombas, they decided they were onto something and started selling them online. There was definitely demand for Roomba fashion accessories. And myRoomBud has been featured on major websites and cable TV shows. myRoomBud is a profitable, private company built and run by the kids who started it.

Roobit the Frog · Slops the Pig · RoomBette La French Maid · Roor the Tiger · Mooba the Cow

FooFoo the WereRabbit · Zeb the Zebra · Lucky the Ladybug

FIGURE 16-7: Some RoomBuds

Roomba Personalities with myRoomBud Is Alive Dashboard

The myRoomBud's dad and advisor, Greg Smith, used his programming expertise to add Roomba personalities to go along with a RoomBud costume. The myRoomBud Is Alive Dashboard (MiAD), shown in Figure 16-8, is a free program you can download that gives you a point-and-click way to control your Roomba. Besides providing the standard movement capabilities you're familiar with for a computer-controlled Roomba, MiAD has a set of buttons that will trigger the various personalities, like roar like a tiger or hop like a frog.

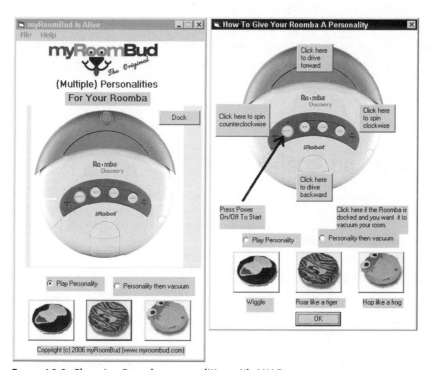

FIGURE 16-8: Changing Roomba personalities with MiAD

Build Your Own Costume

If you're handy with a needle and thread, you can build your own Roomba costume. Liz Goodman has posted a great step-by-step guide on Instructables about how to make a furry mouse costume out of fake fur and pink felt. The pattern shown can be assembled with a sewing machine or hand-stitched. Liz's site is:

```
http://www.instructables.com/id/ESYFHPW859EP286VS4/
```

The pattern is pretty simple and even if you have no sewing experience you can make it. You can also modify it to match some of the new algorithms and behaviors you've created from previous chapters. For example, make a bug costume. With pipe cleaners for antennae and shimmery insect-like fabric, it would be the perfect companion to the roach brain you created in Chapter 13.

You could also use the myRoomBud costumes but design your own behaviors for your Roomba. Get the ladybug costume and have your Roomba flit around your houseplants. (A green filter in front a light sensor lets you detect plants just like NASA satellites do.) Another idea is to send the frog to the bug zapper on the porch at night to get dinner. For both of these ideas you would need to modify the Roach code from Chapter 13 to make it go toward light rather than away from it. If your bug zapper isn't very bright, you could add a microphone and tune the code to listen for the zaps it makes.

Alternatively, create a little butler outfit for it and put lights or remote control code emitters on the fridge. Program Roomba to go find the light or emitter to fetch you a drink from the fridge. A simple servo-driven gripper could be added to the top of the robot and controlled by a Basic Stamp or Arduino.

Roomba APIs and Applications

The RoombaComm API you've worked with in this book's projects is just one example of how to communicate with the Roomba, and not a very advanced one at that. You've seen from the microcontroller code and the various incarnations of "roombacmd" on Linux that any language can be used to build a Roomba controlling library.

RoombaFX/Roomba Terminal

```
http://sourceforge.net/projects/roombafx
http://sourceforge.net/projects/roomba-term
```

RoombaFX is a C# framework by Kevin Gabbert for Roomba interfacing. If you're a Microsoft .NET user, then RoombaFX is for you. RoombaFX provides both high-level and low-level access to Roomba through well-designed C# objects.

Roomba Terminal is an application written in Visual Basic .NET. Also part of Roomba Terminal is Roomba Monitor, an application to display various Roomba statistics and serve as a tool to spy on the Roomba and run macros.

Figure 16-9 shows the Roomba Terminal main screen, allowing you to drive Roomba and read its sensors. Figure 16-10 shows the Sensors screen of Roomba Terminal and some of the more advanced actions that can be performed.

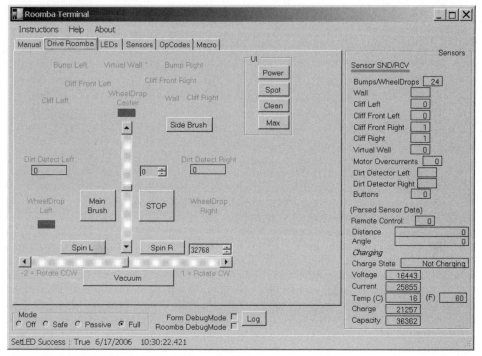

FIGURE 16-9: Roomba Terminal, using RoombaFX

SCI Tester

http://www.roombadevtools.com/productcart/pc/content_software.asp

Also for Windows, RoombaDevTools.com has created a simple Roomba SCI tester application with Visual Basic 6 source code (see Figure 16-11). It's called SCI Tester because it was created back when the ROI was called the SCI. It works with any of the Roomba interface products that RoombaDevTools sells, as well as the serial tether or Bluetooth adapter presented in Chapters 3 and 4. You can use the SCI Tester as an independent software verification of your interface hardware, if you're unsure of code you write.

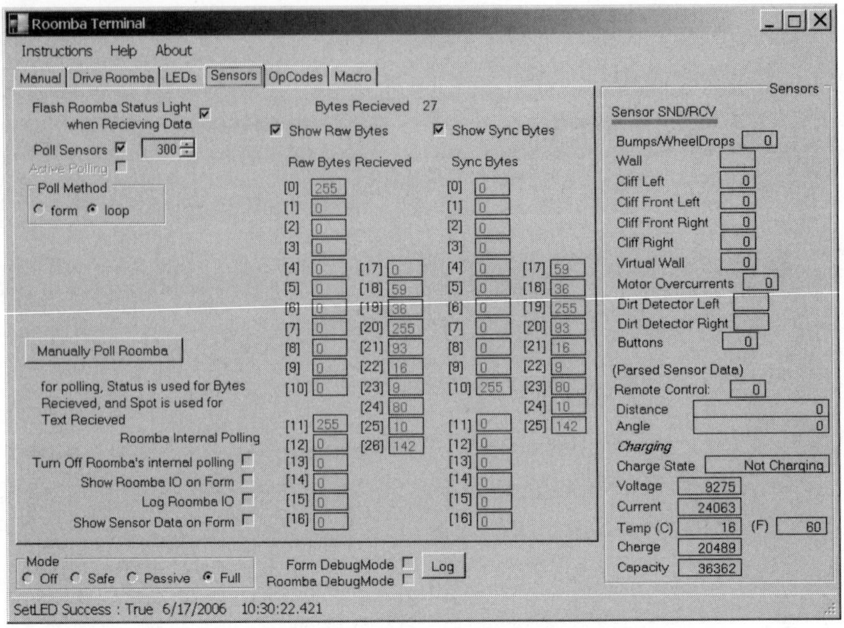

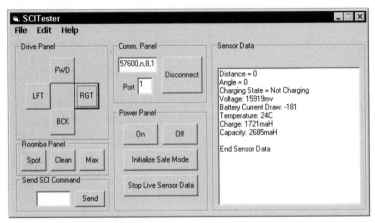

FIGURE 16-10: Sensor handling with Roomba Terminal and RoombaFX

If you program primarily on Windows, Visual Basic is a great way to create Windows programs quickly. You can create buttons, controls, and windows with code not much more complex than PBASIC. Check out the source code to the SCI Tester to see how easy it is to make a Windows program with it. You'll find that the source doesn't try to be a complex Roomba API, but is rather more like the roombacmd tools you created in Chapters 14 and 15. Microsoft Visual Studio, the program you use to create Visual Basic programs, normally costs several hundred dollars, but Microsoft has recently released an Express version for free that can create simple programs.

FIGURE 16-11: SCI Tester running

Warranty-Voiding Hacks

This book has purposefully steered away from projects that definitely void the warranty of your Roomba. Interfacing through the ROI probably doesn't void the warranty, but the following hacks definitely do.

Tip If the following hacks interest you, you should get a second Roomba to do them on. These hacks can destroy a Roomba, and you don't want to take your nice, expensive Roomba and void its warranty. You can find older used Roomba models on eBay. And factory-refurbished ones with a significant discount can be found on Amazon.com and other retailers on the Internet.

Line-Following Roomba

```
http://www.northridgerepair.com/plog/
```

Ben Miller discovered a way to turn the four cliff sensors on the front of Roomba into line sensors. He found that by placing a calibrated resistance across the photodiode of a cliff sensor, he could make the cliff sensor differentiate between carpet and black tape. Figure 16-12 shows the essence of his hack. The resistors are attached across the photodiode part of the cliff sensor and can be made reversible so you get normal functionality back.

Each sensor has to be calibrated by tuning the variable resistor until it triggers when it passes over the tape. Then, by programming the robot through the ROI (Ben used a Basic Stamp), you can poll the cliff sensors and watch for when Roomba crosses the black line.

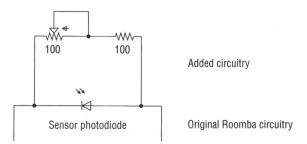

FIGURE 16-12: Schematic of added components to each cliff sensor

Vacuum Motor Connector Hacking

The vacuum motor of Roomba exists in the removable dustbin. This means power must be routed to that motor and if you remove the dustbin and look at the right side of Roomba, you'll notice two electrical connectors (see Figure 16-13). This connector provides power directly from the Roomba battery but switched through the vacuum motor bit in the ROI MOTORS command. The connector is even labeled plus/minus so you know the voltage polarity. Any manner of device that can be powered off +16 VDC can be hooked up to this port and switched on and off under software control. The vacuum motor draws a few hundred milliamps, so you can draw at least that much current from the port.

FIGURE 16-13: Vacuum motor connector, a powered output port

Battery Upgrading

http://www.roombareview.com/hack/battery.shtml

If you have a Roomba battery pack that stops working for some reason, normally you'd be resigned to getting a new one. However, you also could fix it and upgrade it. Originally documented by Craig Capizzi of RoombaReview.com, the battery pack is a simple linear array of 12 sub-C sized NiMh cells. The screws on the battery pack have strange triangular heads, but a Torx T6 seems to work okay too. The top of the pack is tacked down with glue but otherwise isn't too hard to take apart. Figure 16-14 shows what the battery pack looks like with the top off.

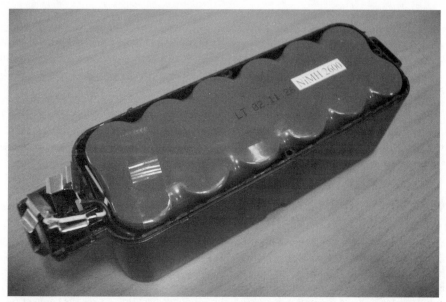

FIGURE 16-14: Battery pack partially apart

Once you pull the cells out of the plastic holder, you can see that they're all connected in a line (see Figure 16-15). There are several online battery retailers, but you may have the best luck with hobby stores. High-performance RC cars use 7.2V battery packs, which are just six of these sub-C cells in a series. The RC cars need enormous amounts of current, so you can easily find cells with a 3800 mAh rating.

FIGURE 16-15: Battery pack internals spread out

There are two mystery parts inside the battery pack. The first is a little flat rectangle connected in series with the cells. This is a positive temperature coefficient (PTC) resistor that acts as a sort of resettable fuse. If a short happens across the battery terminals, the cells heat up as they dump current. This heats the PTC resistor, increasing its resistance and limiting the current output. It's a great failsafe device. The other mystery part is the temperature sensor, usually located in the middle of the pack. In Figure 16-15 it's the small black protuberance in the center.

Note   Watch out! Even apparently discharged battery packs can contain a lot of charge and cause a nasty shock. The Roomba battery pack is designed to prevent rapid discharge, but if you start taking it apart you'll bypass that feature. Be careful.

Remote Control Hacking

The remote control that comes with Roomba is a standard circuit internally, but outputs custom infrared remote codes. You can use a learning universal remote to record the Roomba remote's commands and then take apart the remote and use it as another way of computer-controlling the robot. Figure 16-16 shows what the remote looks like taken apart. Each button circuit consists of two circuit traces interleaved and a rubber button containing a conductive coating on its bottom surface. When the button is pressed, the conductive coating makes the connection between the two interleaved traces.

FIGURE 16-16: Roomba remote taken apart

To hack this, find alternative ways of making that connection. One way is to bridge the connection electrically. Soldering wires to the interleaved traces is difficult because of the coating used, but there are metal test points on the traces that wires can be soldered to. Figure 16-17 shows wires soldered to the two traces that make up the Forward button. When you have these wires, you can run them out to an external circuit. You could hook them to a relay and control the relay with a computer's parallel port or microcontroller circuit. Repeat with the other buttons and you have a rudimentary way to remotely control the Roomba, albeit without sensors.

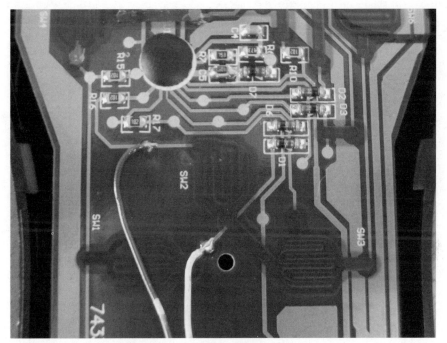

FIGURE 16-17: Soldering wires across the Forward button

If you've replaced the brain of your Roomba and made it autonomous, then the remote codes become just another sensor input and the remote control becomes a low-cost unidirectional communication mechanism. If you have two Roomba robots, you could have one emit remote commands to the other, thereby creating a way for the two to talk to each other.

Virtual Wall Hacking

Like the remote control, the Virtual Wall is a standard IR remote transmitter outputting a custom code. Unlike the remote, it only outputs a single code, which you can read with the virtual sensor byte. Figure 16-18 shows the virtual wall taken apart. There's a lot of unused space available for sticking in additional circuitry. The virtual wall has a directional IR transmitter on its front and what looks like the same hyperbolic lens and IR receiver the Roomba has on its top. It may also be a transmitter however, transmitting different codes.

FIGURE **16-18:** Virtual Wall taken apart, not much there

To just tell if the lens on top is for a transmitter or receiver, the easiest way is to point a digital camera at it and see if it lights up. The near-infrared used by remote control IR transmitters are invisible to the human eye but visible to electronic sensors. Figure 16-19 shows a virtual wall turned on. You can see light from the directional emitter on the front yet nothing from the lens on the top. The lens on top is likely only used by the Scheduler variant of Roomba to program when virtual walls are turned on and off.

Having a nicely machined powered enclosure with both an IR receiver and transmitter is a great hacker toy. You could uninstall and entirely replace the virtual wall circuitry with something that responds to arbitrary remote control codes and emits Roomba remote control codes. Or take advantage of the directionality of the emitter and mount the virtual wall unit on Roomba like a gun and have laser tag battles between Roombas. If you're concerned about power, there are chips by Maxim and others can convert the 2–3V from batteries into regulated 5 VDC.

FIGURE **16-19: Checking which parts are emitters and receivers**

Summary

Roomba owners have done a dizzying number of things with their Roombas. This list of existing hacks in this chapter is far from complete, and new ones are being created all the time. From the projects in the book and the ones on the Net, you may wonder if there's anything new to do with Roomba. However, any programmable system is infinitely deep in terms of what can be done with it. The Roomba, because of its extensibility, is even more so. Hacks using multiple Roombas or alternative sensors have barely been attempted. The ROI protocol that allows so many of these projects has only been available for a short time. The projects people have accomplished so far have focused mainly on just getting that to work. Now that low-level communication has been established, it's time to start doing some really innovative things. Roomba is a great low-cost platform for robotics exploration. What was once available to only academics is now for sale in every department store. Try out some of these hacks, invent some new ones, and let others know by posting your findings on one of the web sites mentioned in this chapter.

Soldering and Safety Basics

in this appendix

☑ Basic tools you need

☑ Safety considerations

☑ Soldering components

☑ Soldering wires

☑ Static safety

I f you've never built a circuit before, don't worry. The techniques are straightforward and can be picked up in an afternoon. Most circuits in this book can be sketched out on a solderless breadboard, but if you want something that can withstand vibrations and heavy use, then soldering your circuit is the way to go. A soldered circuit can last for decades.

At its most basic, soldering is the technique of joining metallic parts together with some sort of molten metal. The molten metal fuses the two parts together, almost like glue. This molten metal is called solder and is an alloy with a low melting point compared to most metals. The tool used to apply solder is a soldering iron: a pencil-shaped device that can apply high heat to very specific locations. A typical soldering iron can get up to 800ºF at its tip, yet be cool enough on its handle for you to hold it in your hand.

This appendix demonstrates some basic techniques for building and soldering circuits by hand. It doesn't cover the more complex techniques like soldering printed-circuit boards or surface-mount devices, although the techniques presented here can carry over to them.

Note When you feel like you are ready to tackle more complex soldering projects, an excellent resource is the ever-growing set of tutorials from SparkFun at `www.sparkfun.com/commerce/hdr .php?p=tutorials`. Another great set of soldering resources from do-it-yourselfers can be found on Instructables at `www .instructables.com/tag/keyword:soldering/`.

Tools Needed

If you're starting from scratch and want to build a basic electronic circuit assembly lab, get the following tools from Jameco:

- Weller WP25 soldering iron, Jameco part number 170587

- Soldering iron holder, Jameco part number 192153

- Solder, lead 0.031″, 60/40, rosin flux core, Jameco part number 141795

- Third-hand tool, Jameco part number 26690

- Diagonal cutters, Jameco part number 146712

- Needle nose pliers, Jameco part number 217891

- Digital Multimeter, Jameco part number 220767

- Desoldering braid, 0.075″, Jameco part number 124118

- Hookup wire, 22 gauge, 100 ft, Jameco part number 36792

- Heat-shrink tubing, 0.12″ diameter, 10 ft, Jameco part number 184721

The total cost of these items is a little over $100, and with them you can build and test just about any circuit you can imagine. You can also fix many simple issues with electronic devices in your home.

Soldering Iron

Soldering irons are rated by wattage. A higher wattage iron isn't hotter, but has more heat available to speed up heating up large items. For most hobbyist projects, a 25W to 35W iron is all that's needed. Do not use a solder gun. Those are 100W huge devices meant for large electrical or plumbing work, not small electronic work. And in general, avoid cordless irons as they can't provide enough power and have tips not usable for electronics.

Another factor in soldering iron shopping is the tip type. Make sure the tip is replaceable. The tip that comes with the iron will be fine when you're starting out, but as you get more experienced you'll want to get a finer tip for working on smaller things. Replacing tips on most irons is pretty easy, but you do have to let them cool down first. Certain higher-end irons allow you to change the temperature of the tip either by a dial on the base or replacing the tip.

Buying a Soldering Iron

Don't get the cheap $5 soldering irons you can find in electronics and tool supply shops. The tips destroy boards and they only last a few months. Spend the money on a good iron. Weller makes a decent 25W iron, the WP25, that's about $33. The Weller WTCPS solder station shown in Figure A-1 is a 42W iron and goes for around $130. A good iron lasts a long time. The iron shown in Figure A-1 has been in use for almost 20 years, with the only maintenance being tip replacement.

If you get an iron that doesn't come with a stand, be sure to order one. They cost only a few dollars and prevent a lot of mishaps. It's easy for a hot iron to slip off a table and onto the floor (or onto your lap). If you can, find one with a sponge holder because keeping the soldering iron tip clean is essential to getting good solder connections.

Soldering Iron Care

A few pointers to extend the life of your soldering iron and make it work at its best:

- Always keep the tip clean. Use a damp sponge (not dripping) and wipe off the tip before and after use, although the tip is hot enough to melt solder.

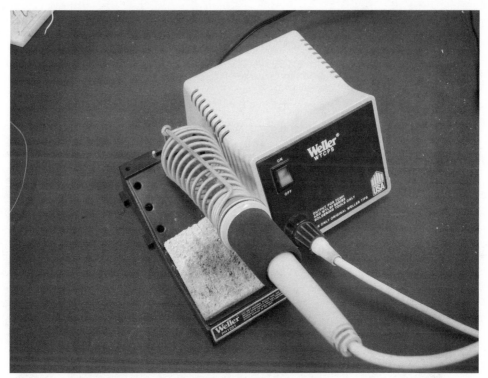

FIGURE A-1: Weller soldering iron with holder and cleaning sponge

- Never file or grind the tip. The tips of modern soldering irons are pre-tinned and pre-shaped. Using a file on them destroys this surface and may harm the temperature sensor (if it has one).

- Always keep the soldering iron in the holder when not in use. Laying the soldering iron down on the table not only invites painful accidents; it also can melt nearby unintended substances, which can be particularly troubling if they actually melt to the tip.

- If you do inadvertently melt something to the soldering iron, carefully wipe it off on the cleaning sponge while the iron is hot. If that doesn't clean it off, replace the tip.

- Only turn it on when you need it. Irons aren't meant to be on for more than a few hours. Don't leave it on overnight.

Solder

There are many types of solder. The three main aspects of various types of solder are: lead or lead-free, acid or rosin flux core, and diameter. For most electronics work, get lead solder with rosin core. Acid flux solder should never be used for electronics work. The diameter is less critical, but the 0.031″ variety as shown in Figure A-2 seems to work quite well for most work. A one-pound spool of it costs about $13, and it never seems to run out. The particular type of lead solder is usually 60/40: 60% tin and 40% lead. The combination of these two metals in those percentages creates a low-melting-point alloy that acts almost like a totally different metal. In the figure, the 66/44 represents the type of rosin flux in the solder and not the ratio of metals.

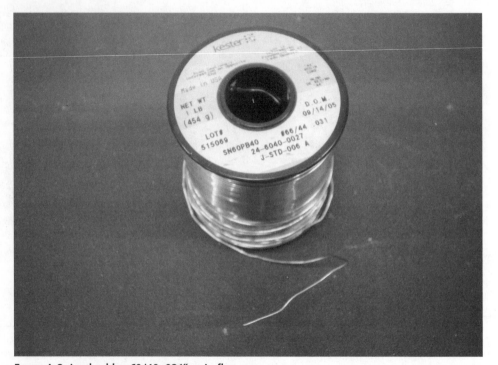

FIGURE A-2: Lead solder, 60/40 .031″ rosin flux core

If you were to dissect solder, you'd find that it's hollow and filled with flux. This flux core is a reducing agent that cleans the metal parts to be soldered of oxidation and other impurities. For electronics rosin flux is used because acid flux would eat away at the parts. Rosin flux is a sticky organic brown substance that you'll see on your circuit boards when you solder. You can use alcohol or specially formulated flux remover to get rid of it if you want.

With the recent Restriction of Hazardous Substances (RoHS) directive issued by the European Union in 2003, lead-based solder will no longer be used for commercial electronics production. Instead, new lead-free solder is substituted. This is good because lead is poisonous if ingested and if handled for long durations. However, these lead-free solders haven't been completely figured out yet. They have higher melting points, can lead to mechanically weaker junctions, and can disassociate into their component metals over time. Normal lead-based solder will be available for a long time and will work with components for a long time. It's proven and works well.

Third-Hand/Helping-Hand

Electronic assembly was originally created by aliens with extra sets of arms. Not really, but trying to hold a soldering iron, solder, wire, and board all at once makes you think that surely this process wasn't invented by humans. Fortunately, there is a large assortment of helping-hand or third-hand tools that can assist you. Figure A-3 shows one type that has two spring-loaded alligator clips on arms and a magnifying glass (which no one seems to use).

The alligator clips can be abrasive on boards and wires so to make them a bit gentler, slide small bits of rubber tubing (such as heat-shrink tubing) on each jaw of the clips.

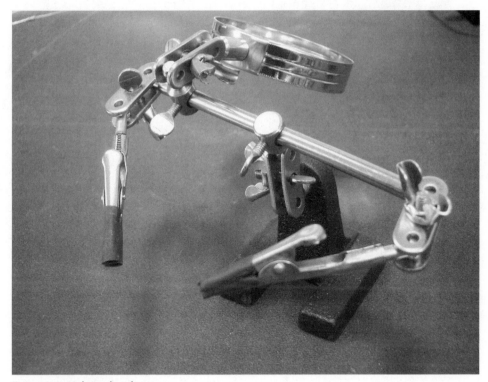

FIGURE A-3: Helping hands

Cutters and Pliers

A good pair of diagonal cutters (or flush-cut cutters) is essential to trimming part leads. The diagonal part means the cutting edges are at an angle to the handles, enabling you to more easily make a flush cut. A decent pair of diagonal cutters is about $6.

Needle-nose pliers are useful for placing parts, bending component leads, holding leads while soldering, and basically doing all those things your fingers are too big for. They run about $16 dollars for a decent pair. Figure A-4 shows both cutters and pliers. They last a long time. The ones in the figure are over a decade old.

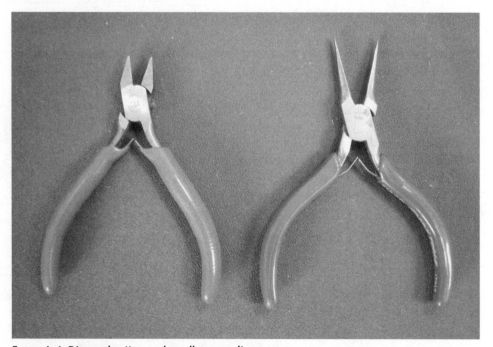

FIGURE A-4: Diagonal cutters and needle-nose pliers

Digital Multimeter

A digital multimeter is a Star Trek tricorder in many ways. It can measure DC and AC voltage and current, resistance, capacitance, and frequency. It can also test diodes or transistors. Figure A-5 shows a 20-year-old digital multimeter that's still going strong. It originally was several hundred dollars. You can get essentially the same accuracy today with additional functions for about $35. Being able to verify the value of resistors and capacitors is very handy with these devices, but mostly you'll use it in continuity mode to check if two points in a circuit are connected.

Other Tools and Supplies

There are a few other things that are handy to have around. A spool of hookup wire is useful for making jumpers and otherwise hooking up stuff. Get the solid wire, not the stranded, for jumpers and other short cable runs. The solid wire is stiff and holds its shape nicely. A 100-foot roll costs about $5.50. Get a few colors.

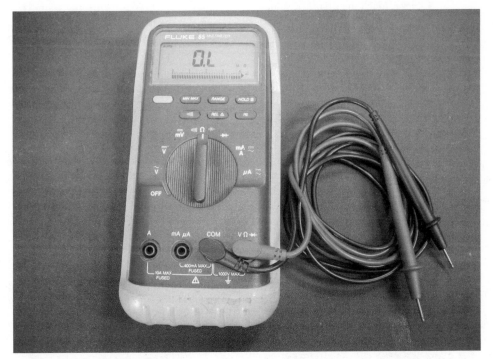

FIGURE A-5: Fluke digital multimeter

When joining two wires, heat-shrink tubing is a nice way to both insulate and mechanically strengthen the solder joint. Heat-shrink tubing comes in a wide variety of diameters and shrinks to about 50 percent of its original inside diameter when heated (AKA a 2:1 shrink ratio). A 10-foot length of it costs about $1.20 (it's sold by the original diameter size). Get a few different diameters, or buy a pre-built kit of selected lengths and diameters.

For other insulation tasks, electrical tape is useful. It's a bit messy and becomes unstuck over time. You shouldn't use it if you can use heat-shrink tubing.

A hot glue gun from a craft store is perfect for tacking down jacks and wires. A few strategic dabs of hot glue can greatly improve the mechanical stability of tall components. Since often the only thing holding something up is its solder joint, if it's going to be flexed a lot, add some hot glue to stabilize it. Hot glue is also an insulator that is useful when you need custom insulation and heat-shrink tubing isn't possible.

Safety Considerations

Building circuits uses dangerous chemicals and devices that can burn or otherwise harm you.

- Know at all times where your soldering iron is. The tip gets up to 800°F and can burn your skin so fast it cauterizes and deadens nerve endings so you won't feel it at first.

- Never touch the tip of the soldering iron unless you know it's cool. One way to check its temperature is to hold your hand above its tip, several inches away. If you can still feel warmth, it's too hot to touch. Even so, once the air above the tip feels cool, you probably should still give it a few more minutes to cool off. It is better to be a little patient than to end up with an unnecessary scar.

- Wear safety goggles. Molten solder can splash, especially when coming into contact with cool things like large metal surfaces or the cleaning sponge.

- Keep your work area well ventilated. The rosin flux and other fumes from soldering are not meant to be inhaled.

- Wash your hands thoroughly after handling solder and any components. Lead is toxic with prolonged exposure. Many components have lead on them, but also a lot of them are just dirty.

- Never solder on a powered circuit. The soldering iron is grounded for static safety reasons, but that presents a shock hazard if the circuit is connected. Always disconnect all power before soldering.

You *will* burn your fingers a little now and again. It's part of the initiation of learning how to solder. Sometimes you have to hold a part in place with your fingers while you solder it. But try to minimize burning yourself by using the proper tools and procedures as described in this appendix.

Soldering Components

Soldering two parts together not only creates a connection that can conduct electricity; it also makes a mechanical bond that can be quite strong when done correctly.

To solder two things together, you must first *tin* each part with a thin layer of solder. Tinning is the process of applying a thin layer of solder to a metal part. The part turns silver colored where the solder has been applied but otherwise shouldn't look different than before. That is, it shouldn't be blobby with solder.

Solder naturally likes flowing along metal, so tinning parts isn't very hard. Many components already come pre-tinned to make them easier to solder. Usually you'll want to add just a little more solder when making solder joints.

The recipe used when soldering is:

1. Tin the first piece.

2. Tin the second piece.

3. Position the two pieces together.

4. Heat the two pieces simultaneously with the soldering iron.

The important part of that recipe is that you're not adding more solder when making the actual joint: The pieces should have enough solder that when heated the solder on each flows together to form one bond.

When soldering parts to a board, this recipe is somewhat truncated because the two pieces (board and component) are already touching. Then the application of solder must be controlled so you don't add too much. Figure A-6 shows three different cases of a component lead and board; you want the right-most one. The meniscus of the solder should be concave, sloping up the sides of the part and down to the board like a ramp. If you have too much solder like the middle case, the solder blob will often hide air or dirt and create a bad connection. This is called a *cold* solder joint. It may work initially, but a small flex of the board might make the electrical connection across the joint go intermittent.

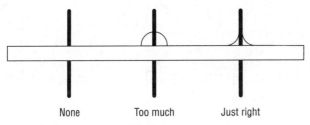

None   Too much   Just right

FIGURE A-6: How much solder to use

Placing the Component

Figure A-7 shows a new component being placed on a board. To start out, clamp the board down in the third-hand tool to stabilize it; then insert the component. When inserted, splay its leads out to hold the part in. If the part leads are too short to do this, you can hold the part from the bottom with a finger.

Applying Solder

As in Figure A-8, with the part in the middle, bring the solder and soldering iron together quickly. When the solder melts, it will naturally try to go onto the soldering iron. By putting the part in between, you'll get the solder onto the part. Be sure to be close to the board so the melting solder touches the board's copper pad too.

The result will look like Figure A-9. Add just enough solder so it appears to crawl up the part.

Note   All parts are heat sensitive and will break if heated for too long. Apply the soldering iron only long enough to melt the solder and have it flow properly.

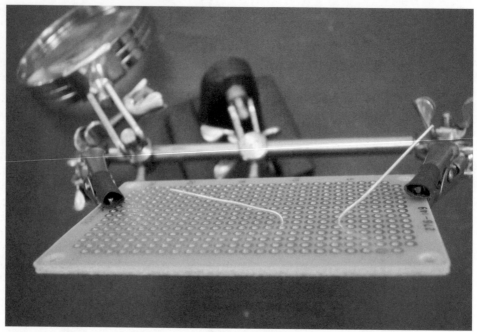

FIGURE A-7: Bend component leads if possible to hold them

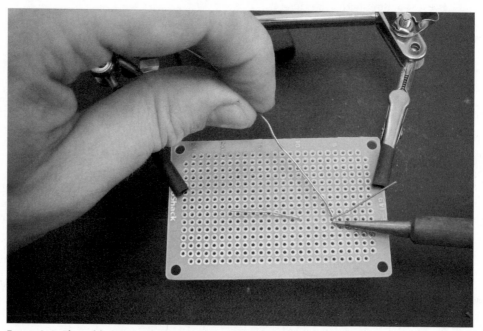

FIGURE A-8: The soldering component, iron on one side, solder on the other

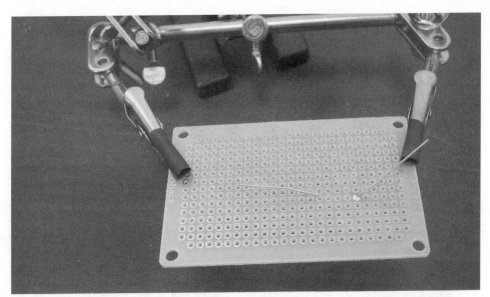

FIGURE A-9: The finished solder joint: shiny and not blobby

Making a Bad Solder Joint

Figure A-10 shows the other lead of the part soldered, but badly, creating a blobby dull solder joint. If this cold solder joint works at all, it will fail soon with no warning.

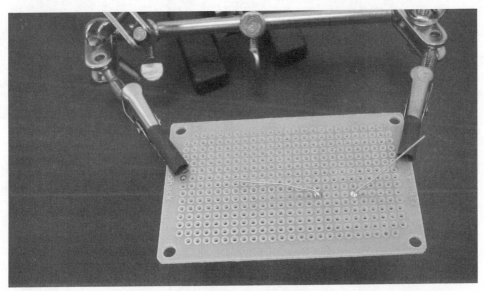

FIGURE A-10: The left side is a bad solder joint: dull and blobby.

Fixing Mistakes

Because solder likes metal so much, getting it off of parts can be a problem. There are many *desoldering* tools available but one of the easiest to use is the desoldering braid (or wick, as shown in Figure A-11). Desoldering braid works by creating a metallic surface that solder likes even better than your part. It's a fine mesh of copper that has a very high surface area. Since solder likes to cover metallic surfaces, applying desolder wick is almost like sucking the solder off a part.

FIGURE A-11: Desoldering braid

Figure A-12 shows how to apply the desoldering wick to the bad solder joint in Figure A-10. Notice it's between the joint and the soldering iron. The wick will turn silver colored and get bigger as it soaks up solder. Figure A-13 shows the result. All the extra solder is gone and now a proper solder joint can be made.

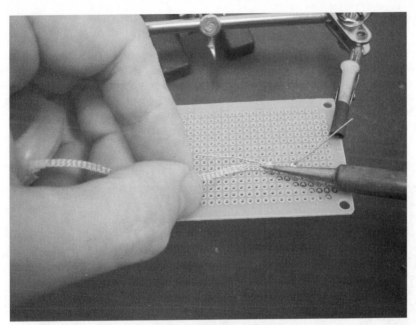

FIGURE A-12: Using desoldering braid to remove solder

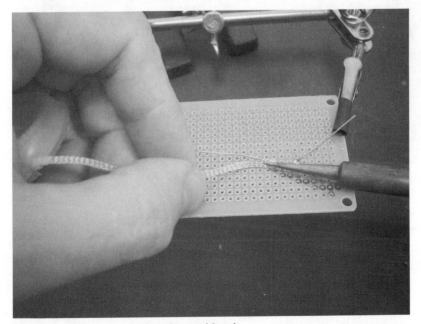

FIGURE A-13: Clean and ready to be resoldered

Testing

When you're happy with your solder joints, pull out the multimeter and verify the connection is good. Figure A-14 shows the previously bad solder joint being tested. The resistance should be less than an ohm. If it's any greater, something is probably wrong with the joint and you should re-heat it at the least, or desolder it and try again. If you're soldering many connections, it's usually faster to solder all the connections and then check them all.

When you're happy with your connection, you can snip off the extraneous leads close to the board with the cutters close, as in Figure A-14.

FIGURE A-14: Testing continuity with multimeter

Making Connections

Using these prototyping boards means running wires between parts yourself. The snipped-off portion of the leads make great jumpers, as shown in Figures A-15 and A-16. Using the needle-nose pliers, bend one of the leads into a U-shape to match the distance you need it to cover. Poke it through the holes from the other side of the board like a component and bend it slightly to keep it in place. To connect it to an existing component, bend its lead over like in Figure A-16. Then, solder both ends of the jumper down.

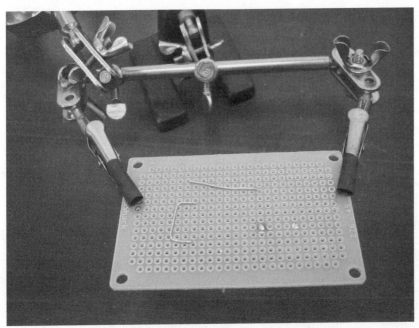

FIGURE A-15: Cut leads, one bent for use as a jumper

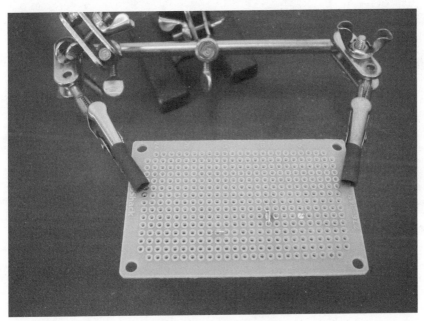

FIGURE A-16: Jumper inserted and bent to make connection

Soldering Wires

You'll often need to splice a wire or solder a new wire onto an existing cable. You can get really professional results quickly by using the third-hand tool and heat-shrink tubing. To start out, strip about 1/4″ from the ends of the two wires you want to solder together and place them in the arms of the third-hand tool.

Tinning Wires

With the wires in the third-hand tool, tin each wire by putting the solder on one side of the wire and the soldering iron on the other side and bring the two together to meet at the wire (see Figure A-17). You just need a thin layer of solder, and it will be barely visible.

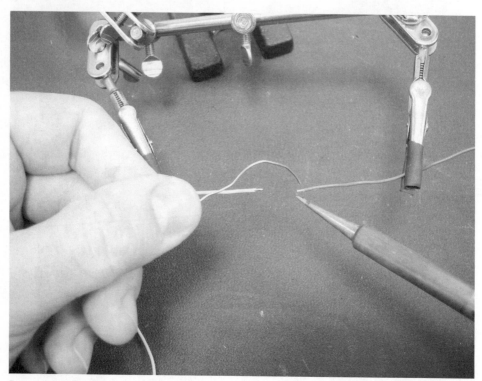

FIGURE A-17: Tinning one of the wires

Making the Solder Joint

If you want to use heat-shrink tubing, cut off an inch length and slide it onto one of the wires, but out of the way for now. Move the two wires together so they touch. With the soldering iron, lightly touch the two wires. Watch the solder on each flow together to form the joint as shown in Figure A-18. You may find having a tiny bit of fresh solder on the iron helps this process, but otherwise you shouldn't need solder at this point. The solder joint is now done and should be quite strong.

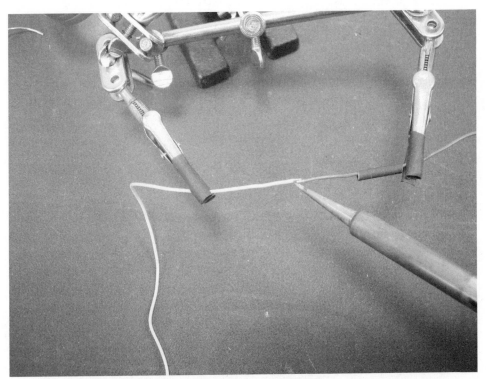

FIGURE A-18: Making the connection: notice solder is not needed.

Adding Insulation with Heat-Shrink Tubing

Slide the piece of heat-shrink tubing so it covers the solder joint and heat it so it shrinks down and forms a tight fit around the joint. This not only insulates the joint from the elements and electrical shorts. It also adds some mechanical strength to the area around the joint. You can use a special heat-shrink gun or a lighter, but the most readily available source of heat is your iron, so using the lower temperature side of the tip, wave the iron around the heat shrink tubing and watch it shrink, as in Figure A-19. You'll have to move all around the tubing to make sure all of it shrinks. Figure A-20 shows the finished result.

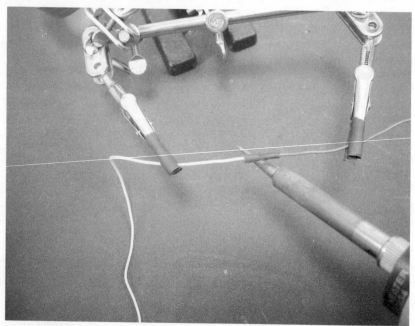

FIGURE A-19: Heating the heat-shrink tubing with the side of the iron tip

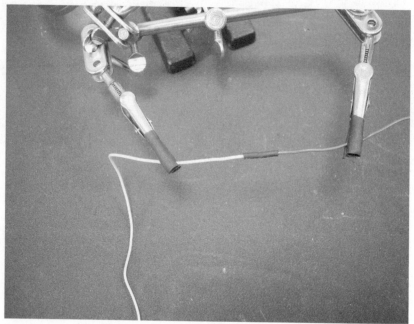

FIGURE A-20: The finished product, mechanically secure and electrically protected

Handling Static-Sensitive Components

Static electricity is the bane of the electronics hacker. The zap you get on a doorknob after shuffling your feet on the carpet is a heavy-duty example of the static that can completely destroy a sensitive electronic device. All IC chips and similar devices are static sensitive and you can damage internal parts of chips without knowing there's even static electricity around.

There's a lot of technology to help dissipate static electricity, but the most important thing is how you handle the devices you're working with. Static electricity likes to travel to ground, so if you have a grounded pipe or metal computer case, touch it periodically to drain any static build-up. Figure A-21 shows a grounding strap you wear on your wrist and plug into the grounded hole of a standard AC socket. They sell for a few bucks at Jameco. These are useful when you're moving around a lot but want to be sure you can't zap an expensive component. The WiMicro in the picture is over $100, so a little bit of prevention could save you a lot of money, not to mention hours of debugging a dead part.

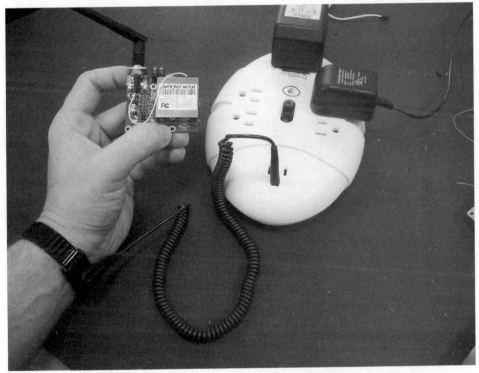

FIGURE A-21: Grounding strap for yourself, plugged into ground socket

Figure A-22 shows a standard anti-static bag used as a holder for an expensive part. These bags are made of metalized plastic and are meant to keep all pins of a part at the same voltage potential. What's dangerous about static electricity is that it can put a several-thousand-volt spike between two pins on a part. The current is very low, but the delicate internals of an IC cannot withstand such a high voltage. Figure A-23 shows ICs pushed into anti-static foam, which serves a similar purpose of keeping all the parts at the same voltage level.

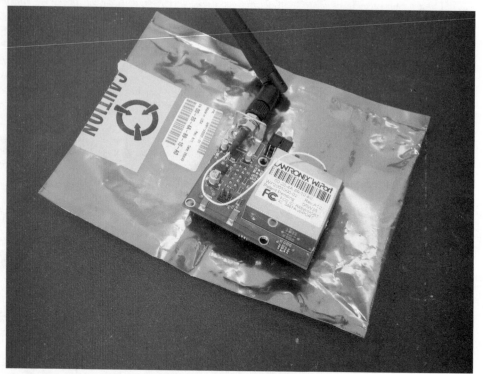

FIGURE A-22: Anti-static bag with sensitive component on top

If you solder directly to ICs (which is not recommended unless you have to), make sure you use a grounded soldering iron. All the better soldering irons (like the Weller ones) are grounded.

If you live in a humid climate, static prevention issues become less of an issue, but if you live in the desert, take extra care. Once you get into the habit of not shifting too much in your chair and occasionally touching the grounded chassis of your computer or work light, you don't have to consciously worry about static issues.

FIGURE A-23: Anti-static foam

Summary

Soldering is fun. On the one hand, you get to play with molten metal and a dangerous device that can burn holes in things. On the other hand, you get to construct a working circuit using your bare hands. You can now build all of the circuits in this book and 90 percent of the circuits out there. The main difference with the more advanced circuits is the smaller parts.

Electrical Diagram Schematics

When you first stumble upon a circuit schematic it looks like a bunch of mysterious squiggly lines and weird little curves, not unlike hieroglyphics. Schematic symbols are indeed a language unto their own, but it's a relatively recent language, derived from a hundred-year-old way of writing wiring diagrams for telegraphs and scientific experiments. Only in the last 50 years or so has it become standard enough for anyone to understand, used and refined by hackers like you who were trying to figure out unambiguous ways to share their hacks with friends and colleagues. The various symbols in schematics are based very much on the physical devices they represent. The way one draws a schematic both influences and is influenced by the physical layout of component parts and wires.

It's possible to draw a schematic that mimics closely the physical instantiation of a circuit. The first circuit schematics were sketches just like this, but as time passed people discovered better ways to translate a circuit to paper. Some of the changes involved short-cuts similar to contractions in English: no need to write the whole thing down if everyone knows what you mean. Other changes were more conceptual like the addition of pronouns: you can say "this" instead of "an appendix about schematics" when talking about *this* appendix.

Because electrical schematic drawings are a lot like language, everyone who draws a schematic has his or her own style and idioms. Getting used to the idioms of different groups can take a few minutes. For example, European hackers have a different style than American ones. But thanks to Internet communication and common programs to draw schematics, the various accents used across the globe are becoming more unified.

If you're interested in drawing your own schematics, you can use any drawing program, preferably a vector-based one. People who draw many schematics use a schematic capture program. One of the best ones for hobbyist use is Eagle by Cadsoft, available at `http://cadsoft.de/`. It's available for Linux, Mac OS X, and Windows, and is free for non-commercial use. A completely open-source toolkit is the gEDA project at `www.geda.seul.org/`. It has many converts but is a little harder to use. A very easy-to-use system for Windows is PCB123 software. PCB123 will make a board from your schematic, at a very reasonable rate, but you have to use their software. The professionals use either ProTel (now Altium) or OrCAD. These comprehensive tools have huge part libraries and can even simulate your schematic. They also have a professional price tag to match their capabilities.

Conventions

When drawing schematics you should follow a few conventions. These aren't hard rules and will be broken in the interest of making a schematic easier to understand.

- Signal flow goes from top-left to bottom-right.

- Positive voltages are on top; negative voltages are at the bottom.

- Each component is labeled with the part's value, for example, 220 for 220 ohms, or 7805 for the 7805 voltage regulator.

- Each component is labeled with a unique identifier to distinguish it from other parts of the same type, for example R2 (for resistor #2) or IC4 (for integrated circuit #4). Sometimes integrated circuit chips (ICs) are labeled with identifiers starting with U (for U #4) instead of IC.

- No diagonal wires, only up-down and left-right.

- Minimize the crossing of wires.

To read a schematic, note the above conventions and just dive in. You'll often find little sub-circuits you understand, like the power supply and LED light circuits that keep popping up in this book. When you understand a sub-circuit, you can focus on the other parts of the circuit you don't understand. Most circuits are built by connecting sub-circuits and they should be fairly intuitive when you understand the basics presented in this appendix.

If you come across a symbol you don't recognize, don't worry. It's probably a new symbol created with a special purpose (or perhaps a symbol from a different idiom, as mentioned earlier in this appendix). Usually you can tell by context and similarity with previous symbols as to what it does. Otherwise there will be some explanatory text to go with the circuit diagram.

Similarly, when you're drawing a schematic and you don't have a symbol for a part, make one up based on what you think it should look like. Then, if it's not obvious, label your new part so others understand. It is generally a good idea to familiarize yourself with some of the commonly used idioms. That way, you can make sure that you don't create a symbol that already exists, and you can also make sure that you make a symbol that will clearly imply what type of component it is, when possible. For example, a resistor should look like a resistor. By creating your symbols this way, you can make it easier for other people to understand your schematics later.

It's All About the Connections

Schematics describe the connectivity between components. They are wiring diagrams. It's not important where the parts are placed on the page but rather how the parts are connected to each other. There is no one right way of drawing a schematic; in fact, there's an infinite number of ways. You can see throughout the book that I predominantly use the U.S. convention as drawn by the Eagle schematic capture software. Sometimes other conventions are used to match the style in which a circuit is normally seen. For example, the Basic Stamp circuits use a slightly different idiom for power and ground to match the Basic Stamp documentation. The particular layout used is due to convention or author preference.

Figure B-1 shows three different ways of drawing a flashlight circuit made out of a battery, resistor, and LED lamp.

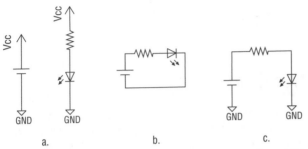

a.   b.   c.

FIGURE B-1: Different but equivalent schematics to light an LED lamp

Graph Theory

In a way, schematic diagrams are a lot like subway and train maps. Subway maps show the connectivity between stations, but misrepresent the distance between stations. The geographic layout between stations isn't as important as showing the connections between them. Both schematics and subway maps are examples of *graphs*, a mathematical concept that describes a set of objects (called nodes or vertices) and their connections (called edges or lines). The study of graphs is called *graph theory*, a field of study that, besides electronics, is critical in Internet search engines (connectivity of web pages), information storage and retrieval (connectivity of data), telephone and Internet routing (connectivity of a telephone network), and many other fields.

Wires

Schematics are made up of two types of pieces: components and wires. Wires join component parts together and are represented as simple lines. Sometimes a schematic cannot be drawn without having one line cross another. In such a case, the lines should just be drawn on top of one another, as in the left-most example in Figure B-2. If the two wires *should* connect, then a small dot is placed on the intersection to represent the connection. This representation likely grew out of the reality that a real connection would be accomplished by a small dot of solder. Usually you see intersecting wires depicted like the right-most example in Figure B-2, where one wire seems to grab on to an existing one.

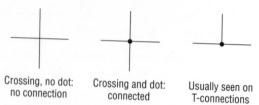

Crossing, no dot: Crossing and dot: Usually seen on
   no connection   connected   T-connections

FIGURE B-2: Wires and their connections

Power and Ground Symbols

A great shortcut to avoiding drawing lots of wires is the use of labeled arrow symbols. Generally, arrows indicate a wire with a signal going off the page or connected elsewhere on the page. The very common cases for using labeled arrows are for the ground and power signals in a circuit.

Ground is an important concept in circuits, as all other voltages and signals in circuits are measured in reference to the ground wire. The name *ground* comes from the first circuits where one wire was literally pushed into the earth. Figure B-3 shows a variety of different ground symbols. It's always an arrow pointing down and labeled with GND or Gnd, or Vss. Vss is the more general way of saying negative supply voltage, but that almost always means zero volts, that is, ground.

When building a circuit, all ground symbols are connected together.

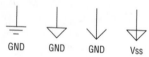

GND   GND   GND   Vss

FIGURE B-3: Common symbols for ground

Similar to the ground symbol is the power symbol. Figure B-4 shows a few of the most common symbols for power. Sometimes the explicit voltage being used is shown (+5V), but usually the general label for positive supply voltage is used. Vcc and Vdd both mean positive supply voltage.

+5V   Vcc   Vdd

FIGURE B-4: Common variations
for power or positive voltage

The Vdd and Vss labels come from the MOSFET transistor that enabled high-density integrated circuits. A MOSFET has a drain (positive pin) and a source (negative pin). Vdd meant the voltage for all drain pins, and Vss meant the same for all source pins. The Vcc label comes from the earlier BJT transistor type that had collector and emitter pins instead of drain and source. As you might expect, there is a Vee label to go along with the Vcc, but today it's more common to use Gnd instead.

A circuit may have multiple voltage sources, each distinct from one another. For example, in the Roomba adapter schematics you see Vpwr or +16VDC to indicate the power from Roomba and Vcc+ or +5VDC to indicate the regulated power coming from the 7805 voltage regulator.

When building a circuit, all power symbols with the same value are connected together.

Basic Components

When you have power and ground, you can start hooking up components between the two to do things. Simple components like resistors and capacitors are considered *passive* since they do not require a source of energy to perform their task. Passive components usually have two leads (also known as pins or terminals). They are the simplest parts physically but often have the most interesting symbols. In contrast, active components like integrated circuits (ICs) require power and have complex internal functionality, but are represented by simple rectangles bristling with short lines indicating their connection pins.

Resistors

Resistors are the most basic of components. They are commonly used to limit the amount of current or act as part of a filter circuit. Figure B-5 shows the symbols for several different types of resistors. This back-and-forth squiggle common to all the symbols is representative of the resistance that a resistor provides: It's harder to move down a curvy road than a straight one. (As you can see, these symbols are made by regular people looking for good analogies.)

FIGURE B-5: Types of resistors: fixed, variable, potentiometer, photocell, thermistor

The left-most symbol is for the standard fixed resistor; its resistance value doesn't change. Fixed resistors are often used to restrict the amount of current to other components, like the resistor that's part of the LED sub-circuits in this book. Without the resistor, the LED would draw too much current and burn out. Two different types of variable resistor are the next two symbols. Most knobs on electronic devices are variable resistors. The second-to-last symbol is for a photocell: a light-sensitive resistor. These act just like normal variable resistors, but instead of a knob, the amount of light hitting them changes their resistance. The last symbol is for a thermistor, a resistor that changes its resistance value based on the temperature. Thermistors are sometimes used in thermostats for heaters and air conditioners. New types of variable resistors are created all the time (bend-sensitive, force-sensitive, and so on), and so new symbols are also created.

Capacitors

Capacitors store small amounts of electricity and are useful as parts of filters or to smooth out power supply fluctuations. The amount of electricity a capacitor can store is its charge capacity, thus its name. Figure B-6 shows three different symbols for capacitors. The symbol comes from the fact that capacitors were first made using two metal plates next to each other. The middle symbol isn't used as much as it used to be because of its similarity to the battery symbol. The last symbol is for a polarized capacitor, like the electrolytic capacitors used in power supplies. A polarized capacitor needs to be oriented with its positive terminal attached to the more positive part of the circuit than its negative terminal. Otherwise the capacitor won't work and might fail.

FIGURE B-6: Capacitor symbols: regular, old-style regular, and polarized

Tip Many components are polarized like this and will indicate their polarity both physically and in their schematic symbol. Be alert when a symbol has an arrow or a plus sign. Polarized parts wired backward are one of the leading mistakes made when building circuits.

Diodes

Diodes only let current flow in one direction. The most common use is to turn the alternating current of AC from a wall socket into the single direction current of DC needed by most gadgets. A diode added before a battery connector protects the circuit in case the batteries are inserted backward. The arrow of the diode indicates the direction current is allowed to flow in the diode. Figure B-7 shows the symbols for a few different types of diodes. The left symbol is for a regular diode. The middle symbol is for an LED (light-emitting diode), a common part of any electronic device. The act of current flowing through an LED lights it up. The right-most symbol is a photodiode. A photodiode will generate current when light falls on it. Photodiodes are used as the receiver in all your devices that have infrared remote controls. In the diode symbol, sometimes the arrow is solid and sometimes it's just an outline. There's no difference between the two representations.

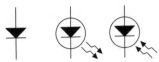

FIGURE B-7: Diodes: regular, LED, and photodiode

Other Components

The preceding sections describe the most common components you'll run into when building projects. Figure B-8 shows some other parts you may also see.

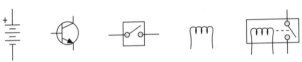

FIGURE B-8: Battery, transistor, switch, inductor, and relay

The first symbol is for a battery. It has a positive and negative terminal, as you'd expect. A single short-dash/long-dash pair originally indicated a single cell of a battery (approximately 1.5V). A stacked set of cells becomes a battery with the voltage indicated by the number of cells. This has fallen out of practice and now a general battery symbol is often shown with a voltage value given next to it.

The next symbol is a transistor. It's used as either an amplifier or an electrically controlled switch.

The middle symbol is for a switch or button. Sometimes switches will have multiple contacts operated by a single push, represented as two switch symbols joined together with a dotted line.

The next-to-last symbol is an inductor or coil. An inductor is sort of like a capacitor, but instead of storing charge, it stores magnetic fields. The electromagnets you might have played with in high school science classes are a special type of inductor.

The final symbol is for a relay. It's a compound symbol made up of an inductor (the electromagnet) and a switch. When current flows through the electromagnet part of the relay, it creates a magnetic field to pull down the switch contacts. Relays are great for turning on and off things that require more power than your circuit can provide, like motors.

With the above basic components you can read just about any circuit written before 1960. There are a lot of fun circuits to build with that toolkit: alarm systems, telephones, audio amplifiers, clocks, and even rudimentary computers. But even the most rudimentary computer has hundreds of transistors. Imagine being required to draw (and read!) a hundred transistors. Some manner of summarization was needed for these more complicated parts.

Integrated Circuits and Other Complex Components

When electrical components were being created and discovered, it seemed appropriate to draw specialized symbols for each one. As individual components became more complex internally, it became harder to describe symbolically what a component did. The most complex devices came to be represented as simple boxes. An intermediary form is the relay above: a compound device with a box drawn around it to show it's a single unit. The internals of that box don't need to be shown as long as the terminals exiting are labeled appropriately. It will still be understood to be a relay.

This compartmentalization is important with integrated circuits (ICs) like voltage regulators, which contain many resistors and transistors, or microcontrollers like the Basic Stamp or Arduino's ATMega8, which contain millions of transistors. Figure B-9 shows the schematic symbols for those three ICs. Instead of showing the internals as with the relay, only the box is shown with the meaning of each of the terminals (pins) labeled.

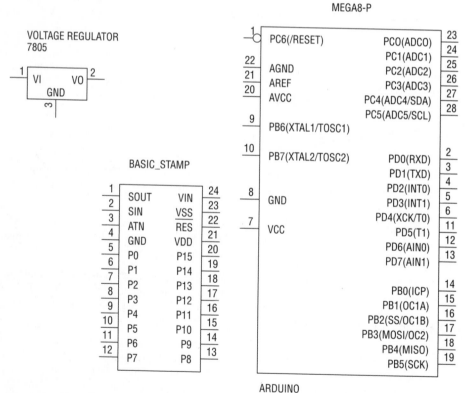

FIGURE B-9: Some ICs: voltage regulator, Basic Stamp, and ATMega8 used in Arduino

Summary

You can start understanding schematics in an afternoon. Everyone keeps references around to remind them what the various parts are. Use these references to help you understand the schematic. Often the references contain other similar schematics that help you out. If you'd like to build your own circuits and draw your own schematics, find some parts you'd like to use and search on the Internet for schematics that use them. Alternatively, find one of the many circuit databases on the Internet and find a circuit you want to build. One pretty good site to browse is www.discovercircuits.com/, which has over 23,000 circuits to look at. Two others are http://web-ee.com/ and www.epanorama.net/, both of which also have tutorials, datasheets, and forums to talk about electronics.

iRobot Roomba Open Interface (ROI) Specification

The following is the complete iRobot Roomba Open Interface (ROI) specification. It is also available as a downloadable PDF from http://irobot.com/developers, which will always have the most recent version of the specification. I'd like to thank iRobot for giving permission to reprint the specification here and for being so hacker-friendly in general. This book would not be possible without their willingness to let users find new uses for the Roomba.

You will undoubtedly run into references to the Roomba Serial Command Interface (SCI). This was the original name of the ROI. There is no difference between the ROI and the SCI; only the name has changed. The original SCI name was likely an internal name used by iRobot engineers that escaped into public use. Unfortunately the name Serial Command Interface can be confusing in hacker contexts where so many things speak through serial communications and many microcontrollers have a Serial Communications Interface (SCI). The new ROI name positively identifies it as pertaining to Roomba, creates an abbreviation not common in hacker circles, and emphasizes the fact that Roomba is a hackable system.

Because the ROI is simply SCI renamed, there are no ROI versus SCI compatibility issues to worry about. The SCI was in use and unchanged for a year, a part of hundreds of thousands of Roombas. If you have a Roomba that mentions SCI, all software and hardware designed for a Roomba ROI will work, and vice versa. If you already have an SCI specification PDF (originally available from http://irobot.com/hacker), you'll notice it's the same as the specification printed in this appendix.

This does not mean that iRobot will not change the ROI in the future. Just as it changed the SCI, over time iRobot might decide to update the ROI. But even if it updates the Roomba communication port, it almost certainly will remain backward compatible with the current ROI. As you can tell from the specification, there is plenty of room for additional commands and functionality. The name change does signify the current functionality is tied to Roomba and thus may not work for other products iRobot makes such as Scooba.

Chapter 2 expands upon the following specification, covering parts of the ROI that may not be obvious or may only be evident through use. However, the specification should always be your starting point. It is the official authority. Check the iRobot developer URL mentioned above and `http://roombahacking.com/docs/` for future updates and discussions about the specification.

iRobot® Roomba®
Open Interface (ROI)
Specification

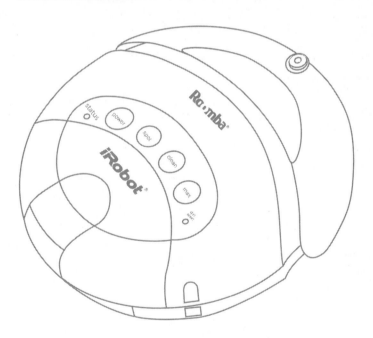

www.irobot.com

iRobot® Roomba® Open Interface (ROI) Specification

Roomba Open Interface Overview

Versions of iRobot® Roomba® Vacuuming Robot manufactured after October, 2005 contain an electronic and software interface that allows you to control or modify Roomba's behavior and remotely monitor its sensors. This interface is called the iRobot Roomba Open Interface or iRobot ROI.

iRobot ROI is a serial protocol that allows users to control a Roomba through its external serial port (Mini-DIN connector). The ROI includes commands to control all of Roomba's actuators (motors, LEDs, and speaker) and also to request sensor data from all of Roomba's sensors. Using the ROI, users can add functionality to the normal Roomba behavior or they can create completely new operating instructions for Roomba.

Physical Connections

To use the ROI, a processor capable of generating serial commands such as a PC or a microcontroller must be connected to the external Mini-DIN connector on Roomba. The Mini-DIN connector provides two way serial communication at TTL Levels as well as a Device Detect input line that can be used to wake Roomba from sleep. The connector also provides an unregulated direct connection to Roomba's battery which users can use to power their ROI applications. The connector is located in the rear right side of Roomba beneath a snap-away plastic guard.

ROOMBA'S EXTERNAL SERIAL PORT
MINI-DIN CONNECTOR PINOUT

This diagram shows the pin-out of the top view of the female connector in Roomba. Note that pins 5, 6, and 7 are towards the outside circumference of Roomba.

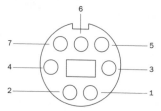

Pin	Name	Description
1	Vpwr	Roomba battery + (unregulated)
2	Vpwr	Roomba battery + (unregulated)
3	RXD	0 – 5V Serial input to Roomba
4	TXD	0 – 5V Serial output from Roomba
5	DD	Device Detect input (active low) – used to wake Roomba from sleep
6	GND	Roomba battery ground
7	GND	Roomba battery ground

The RXD, TXD, and Device Detect pins use 0 – 5V logic, so a level shifter such as a MAX232 chip will be needed to communicate with a Roomba from a PC, which uses rs232 levels.

Serial Port Settings

Baud: 57600 or 19200 (see below)
Data bits: 8
Parity: None
Stop bits: 1
Flow control: None

By default, Roomba communicates at 57600 baud. If you are using a microcontroller that does not support 57600 baud, there are two ways to force Roomba to switch to 19200:

METHOD 1:

When manually powering on Roomba, hold down the power button. After 5 seconds, Roomba will start beeping. After 10 seconds, Roomba will play a tune of descending pitches. Roomba will now communicate at 19200 baud until the battery is removed and reinserted (or the battery voltage falls below the minimum required for processor operation) or the baud rate is explicitly changed via the ROI.

METHOD 2:

You can use the Device Detect to change Roomba's baud rate. After you have awakened Roomba (using Device Detect or by some other method) wait 2 seconds and then pulse the Device Detect low three times. Each pulse should last between 50 and 500 milliseconds. Roomba will now communicate at 19200 baud until the battery is removed and reinserted (or the battery voltage falls below the minimum required for processor operation) or the baud rate is explicitly changed via the ROI.

Here is a Python code fragment that illustrates this method (Device Detect is connected to the PC's RTS line via a level shifter):

```
ser = serial.Serial(0, baudrate=19200,
timeout=0.1)

ser.open()

# wake up robot
ser.setRTS (0)
time.sleep (0.1)
ser.setRTS (1)
time.sleep (2)

# pulse device-detect three times
for i in range (3):
        ser.setRTS (0)
        time.sleep (0.25)
        ser.setRTS (1)
        time.sleep (0.25)
```

Roomba Open Interface Modes

The Roomba ROI has four operating modes: off, passive, safe, and full. On a battery change or other loss of power, the ROI will be turned off. When it is off, the ROI will listen at the default baud bps for an ROI Start command. Once it receives the Start command, the ROI will be enabled in passive mode. In passive mode, users can do the following:

- Request and receive sensor data using the Sensors command
- Execute virtual button pushes to start and stop cleaning cycles (Power, Spot, Clean, and Max commands)
- Define a song (but not play one)
- Set force-seeking-dock mode

Users cannot control any of Roomba's actuators when in passive mode, but Roomba will continue to behave normally, including performing cleaning cycles, charging, etc. When in passive mode, users can then send the Control command to put the robot into safe mode.

In safe mode, the users have full control of the robot, except for the following safety-related conditions:

- Detection of a cliff while moving forward (or moving backward with a small turning radius)
- Detection of wheel drop (on any wheel)
- Charger plugged in and powered

When one of the conditions listed above occurs, the robot stops all motors and reverts to passive mode.

For complete control of the robot, users must send the Full command while in safe mode to put the ROI into full mode. Full mode shuts off the cliff and wheel-drop safety features. (The robot will still not run with a powered charger plugged in.) This mode gives users unrestricted control of the robot's actuators. To put the ROI back into safe mode, users can send the Safe command.

If no commands are sent to the ROI when it is in safe or full mode, Roomba will wait with all motors off and will not respond to button presses or other sensor input.

To go back to passive mode from safe or full mode, users can send any one of the four virtual button commands (Power, Spot, Clean, or Max). These button commands are equivalent to the corresponding button press in normal Roomba behavior. For instance, the Spot command will start a spot cleaning cycle.

Allow 20 milliseconds between sending commands that change the ROI mode.

Roomba Open Interface Commands

Listed below are the commands that users send to the ROI over to the serial port in order to control Roomba. Each command is specified by a one-byte opcode. Some commands must also be followed by data bytes. The meaning of the data bytes for each command are specified with the commands below. The serial byte sequence for each command is also shown with each separate byte enclosed in brackets. Roomba will not respond to any ROI commands when it is asleep. Users can wake up Roomba by setting the state of the Device Detect pin low for 500ms. The Device Detect line is on Roomba external Mini-DIN connector.

Start   Command opcode: 128   Number of data bytes: 0

Starts the ROI. The Start command must be sent before any other ROI commands. This command puts the ROI in passive mode.

Serial sequence: [128]

Baud   Command opcode: 129   Number of data bytes: 1

Sets the baud rate in bits per second (bps) at which ROI commands and data are sent according to the baud code sent in the data byte. The default baud rate at power up is 57600 bps. (See Serial Port Settings, above.) Once the baud rate is changed, it will persist until Roomba is power cycled by removing the battery (or until the battery voltage falls below the minimum required for processor operation). You must wait 100ms after sending this command before sending additional commands at the new baud rate. The ROI must be in passive, safe, or full mode to accept this command. This command puts the ROI in passive mode.

Serial sequence: [129] [Baud Code]

Baud data byte 1: Baud Code (0 – 11)

Baud code	Baud rate in bps
0	300
1	600
2	1200
3	2400
4	4800
5	9600
6	14400
7	19200
8	28800
9	38400
10	57600
11	115200

Control Command opcode: 130   Number of data bytes: 0

Enables user control of Roomba. This command must be sent after the start command and before any control commands are sent to the ROI. The ROI must be in passive mode to accept this command. This command puts the ROI in safe mode.

Serial sequence: [130]

iRobot® Roomba® Open Interface (ROI) Specification

Safe   Command opcode: 131   Number of data bytes: 0

This command puts the ROI in safe mode. The ROI must be in full mode to accept this command.

Note: In order to go from passive mode to safe mode, use the Control command.

Serial sequence: [131]

Full   Command opcode: 132   Number of data bytes: 0

Enables unrestricted control of Roomba through the ROI and turns off the safety features. The ROI must be in safe mode to accept this command. This command puts the ROI in full mode.

Serial sequence: [132]

Power   Command opcode: 133   Number of data bytes: 0

Puts Roomba to sleep, the same as a normal "power" button press. The Device Detect line must be held low for 500 ms to wake up Roomba from sleep. The ROI must be in safe or full mode to accept this command. This command puts the ROI in passive mode.

Serial sequence: [133]

Spot   Command opcode: 134   Number of data bytes: 0

Starts a spot cleaning cycle, the same as a normal "spot" button press. The ROI must be in safe or full mode to accept this command. This command puts the ROI in passive mode.

Serial sequence: [134]

Clean   Command opcode: 135   Number of data bytes: 0

Starts a normal cleaning cycle, the same as a normal "clean" button press. The ROI must be in safe or full mode to accept this command. This command puts the ROI in passive mode.

Serial sequence: [135]

Max   Command opcode: 136   Number of data bytes: 0

Starts a maximum time cleaning cycle, the same as a normal "max" button press. The ROI must be in safe or full mode to accept this command. This command puts the ROI in passive mode.

Serial sequence: [136]

Drive   Command opcode: 137   Number of data bytes: 4

Controls Roomba's drive wheels. The command takes four data bytes, which are interpreted as two 16 bit signed values using twos-complement. The first two bytes specify the average velocity of the drive wheels in millimeters per second (mm/s), with the high byte sent first. The next two bytes specify the radius, in millimeters, at which Roomba should turn. The longer radii make Roomba drive straighter; shorter radii make it turn more. A Drive command with a positive velocity and a positive radius will make Roomba drive forward while turning toward the left. A negative radius will make it turn toward the right. Special cases for the radius make Roomba turn in place or drive straight, as specified below. The ROI must be in safe or full mode to accept this command. This command does change the mode.

Note: The robot system and its environment impose restrictions that may prevent the robot from accurately carrying out some drive commands. For example, it may not be possible to drive at full speed in an arc with a large radius of curvature.

Serial sequence: [137] [Velocity high byte] [Velocity low byte]
[Radius high byte] [Radius low byte]

Drive data bytes 1 and 2: Velocity (-500 – 500 mm/s)

Drive data bytes 3 and 4: Radius (-2000 – 2000 mm)
Special cases: Straight = 32768 = hex 8000
Turn in place clockwise = -1
Turn in place counter-clockwise = 1

Example:

To drive in reverse at a velocity of -200 mm/s while turning at a radius of 500mm, you would send the serial byte sequence [137] [255] [56] [1] [244].

Velocity = -200 = hex FF38 = [hex FF] [hex 38] = [255] [56]

Radius = 500 = hex 01F4 = [hex 01] [hex F4] = [1] [244]

Motors   Command opcode: 138   Number of data bytes: 1

Controls Roomba's cleaning motors. The state of each motor is specified by one bit in the data byte. The ROI must be in safe or full mode to accept this command. This command does not change the mode.

Serial sequence: [138] [Motor Bits]

Motors data byte 1: Motor Bits (0 – 7)

0 = off, 1 = on

Bit	7	6	5	4	3	2	1	0
Motor	n/a	n/a	n/a	n/a	n/a	Main Brush	Vacuum	Side Brush

Example:

To turn on only the vacuum motor, send the serial byte sequence [138] [2].

Leds   Command opcode: 139   Number of data bytes: 3

Controls Roomba's LEDs. The state of each of the spot, clean, max, and dirt detect LEDs is specified by one bit in the first data byte. The color of the status LED is specified by two bits in the first data byte. The power LED is specified by two data bytes, one for the color and one for the intensity. The ROI must be in safe or full mode to accept this command. This command does not change the mode.

Serial sequence: [139] [Led Bits] [Power Color] [Power Intensity]

Leds data byte 1: Led Bits (0 – 63)

Dirt Detect uses a blue LED: 0 = off, 1 = on

Spot, Clean, and Max use green LEDs: 0 = off, 1 = on

Status uses a bicolor (red/green) LED: 00 = off, 01 = red, 10 = green, 11 = amber

Bit	7	6	5	4	3	2	1	0
LED	n/a	n/a	Status (2 bits)		Spot	Clean	Max	Dirt Detect

Power uses a bicolor (red/green) LED whose intensity and color can be controlled with 8-bit resolution.

Leds data byte 2: Power Color (0 – 255)

0 = green, 255 = red. Intermediate values are intermediate colors.

Leds data byte 3: Power Intensity (0 – 255)

0 = off, 255 = full intensity. Intermediate values are intermediate intensities.

Example:

To turn on the dirt detect and spot LEDs, make the status LED red, and to light the power LED green at half intensity, send the serial byte sequence [139] [25] [0] [128]

Song   Command opcode: 140   Number of data bytes: 2N + 2, where N is the number of notes in the song

Specifies a song to the ROI to be played later. Each song is associated with a song number which the Play command uses to select the song to play. Users can specify up to 16 songs with up to 16 notes per song. Each note is specified by a note number using MIDI note definitions and a duration specified in fractions of a second. The number of data bytes varies depending on the length of the song specified. A one note song is specified by four data bytes. For each additional note, two data bytes must be added. The ROI must be in passive, safe, or full mode to accept this command. This command does not change the mode.

Serial sequence: [140] [Song Number] [Song Length] [Note Number 1] [Note Duration 1] [Note Number 2] [Note Duration 2] etc.

Song data byte 1: Song Number (0 – 15)
Specifies the number of the song being specified. If you send a second Song command with the same song number, the old song will be overwritten.

Song data byte 2: Song Length (1 – 16)
Specifies the length of the song in terms of the number of notes.

Song data bytes 3, 5, 7, etc.: Note Number (31 – 127)
Specifies the pitch of the note to be played in terms of the MIDI note numbering scheme. The lowest note that Roomba can play is note number 31. See the note number table for specific notes. Any note number outside of the range of 31 to 127 will be interpreted as a rest note and no sound will be played during this note duration.

Song data bytes 4, 6, 8, etc.: Note Duration (0 – 255)
Specifies the duration of the note in increments of 1/64 of a second. Therefore, half-second long note will have a duration value of 32.

Note Number Table for Song Command (with Frequency in Hz)

Number	Note	Frequency	Number	Note	Frequency
31	G	49.0	80	G#	830.6
32	G#	51.0	81	A	880.0
33	A	55.0	82	A#	932.3
34	A#	58.3	83	B	987.8
35	B	61.7	84	C	1046.5
36	C	65.4	85	C#	1108.7
37	C#	69.3	86	D	1174.7
38	D	73.4	87	D#	1244.5
39	D#	77.8	88	E	1318.5
40	E	82.4	89	F	1396.9
41	F	87.3	90	F#	1480.0
42	F#	92.5	91	G	1568.0
43	G	98.0	92	G#	1661.2
44	G#	103.8	93	A	1760.0
45	A	110.0	94	A#	1864.7
46	A#	116.5	95	B	1975.5
47	B	123.5	96	C	2093.0
48	C	130.8	97	C#	2217.5
49	C#	138.6	98	D	2349.3
50	D	146.8	99	D#	2489.0
51	D#	155.6	100	E	2637.0
52	E	164.8	101	F	2793.8
53	F	174.6	102	F#	2960.0
54	F#	185.0	103	G	3136.0
55	G	196.0	104	G#	3322.4
56	G#	207.7	105	A	3520.0
57	A	220.0	106	A#	3729.3
58	A#	233.1	107	B	3951.1
59	B	246.9	108	C	4186.0
60	C	261.6	109	C#	4434.9
61	C#	277.2	110	D	4698.6
62	D	293.7	111	D#	4978.0
63	D#	311.1	112	E	5274.0
64	E	329.6	113	F	5587.7
65	F	349.2	114	F#	5919.9
66	F#	370.0	115	G	6271.9
67	G	392.0	116	G#	6644.9
68	G#	415.3	117	A	7040.0
69	A	440.0	118	A#	7458.6
70	A#	466.2	119	B	7902.1
71	B	493.9	120	C	8372.0
72	C	523.3	121	C#	8869.8
73	C#	554.4	122	D	9397.3
74	D	587.3	123	D#	9956.1
75	D#	622.3	124	E	10548.1
76	E	659.3	125	F	11175.3
77	F	698.5	126	F#	11839.8
78	F#	740.0	127	G	12543.9
79	G	784.0			

Play   Command opcode: 141   Number of data bytes: 1

Plays one of 16 songs, as specified by an earlier Song command. If the requested song has not been specified yet, the Play command does nothing. The ROI must be in safe or full mode to accept this command. This command does not change the mode.

Serial sequence: [141] [Song Number]

Play data byte 1: Song Number (0 – 15)
Specifies the number of the song to be played. This must match the song number of a song previously specified by a Song command.

Sensors Command opcode: 142   Number of data bytes: 1

Requests the ROI to send a packet of sensor data bytes. The user can select one of four different sensor packets. The sensor data packets are explained in more detail in the next section. The ROI must be in passive, safe, or full mode to accept this command. This command does not change the mode.

Serial sequence: [142] [Packet Code]

Sensors data byte 1: Packet Code (0 – 3)
Specifies which of the four sensor data packets should be sent back by the ROI. A value of 0 specifies a packet with all of the sensor data. Values of 1 through 3 specify specific subsets of the sensor data.

Force-Seeking-Dock Command opcode: 143 Number of data bytes: 0

Turns on force-seeking-dock mode, which causes the robot to immediately attempt to dock during its cleaning cycle if it encounters the docking beams from the Home Base. (Note, however, that if the robot was not active in a clean, spot or max cycle it will not attempt to execute the docking.) Normally the robot attempts to dock only if the cleaning cycle has completed or the battery is nearing depletion. This command can be sent anytime, but the mode will be cancelled if the robot turns off, begins charging, or is commanded into ROI safe or full modes.

Serial sequence: [143]

Roomba Open Interface Sensor Packets

The robot will send back one of four different sensor data packets in response to a Sensor command, depending on the value of the packet code data byte. The data bytes are specified below in the order in which they will be sent. A packet code value of 0 sends all of the data bytes. A value of 1 through 3 sends a subset of the sensor data. Some of the sensor data values are 16 bit values. These values are sent as two bytes, high byte first.

Sensor Packet Sizes

Packet code	Packet Size
0	26 bytes
1	10 bytes
2	6 bytes
3	10 bytes

Bumps Wheeldrops

Packet subset: 1

Range: 0 - 31

Data type: 1 byte, unsigned

The state of the bump (0 = no bump, 1 = bump) and wheeldrop sensors (0 = wheel up, 1 = wheel dropped) are sent as individual bits.

Bit	7	6	5	4	3	2	1	0
Sensor	n/a	n/a	n/a	Wheeldrop			Bump	
				Caster	Left	Right	Left	Right

Note: Some robots do not report the three wheel drops separately. Instead, if any of the three wheels drops, all three wheel-drop bits will be set. You can tell which kind of robot you have by examining the serial number inside the battery compartment. Wheel drops are separate only if there is an "E" in the serial number.

Wall

Packet subset: 1

Range: 0 – 1

Data type: 1 byte, unsigned

The state of the wall sensor is sent as a 1 bit value (0 = no wall, 1 = wall seen).

Cliff Left

Packet subset: 1

Range: 0 – 1

Data type: 1 byte, unsigned

The state of the cliff sensor on the left side of Roomba is sent as a 1 bit value (0 = no cliff, 1 = cliff).

Cliff Front Left

Packet subset: 1

Range: 0 – 1

Data type: 1 byte, unsigned

The state of the cliff sensor on the front left side of Roomba is sent as a 1 bit value (0 = no cliff, 1 = cliff).

Cliff Front Right

Packet subset: 1

Range: 0 – 1

Data type: 1 byte, unsigned

The state of the cliff sensor on the front right side of Roomba is sent as a 1 bit value (0 = no cliff, 1 = cliff)

Cliff Right

Packet subset: 1

Range: 0 – 1

Data type: 1 byte, unsigned

The state of the cliff sensor on the right side of Roomba is sent as a 1 bit value (0 = no cliff, 1 = cliff)

Virtual Wall

Packet subset: 1

Range: 0 – 1

Data type: 1 byte, unsigned

The state of the virtual wall detector is sent as a 1 bit value (0 = no virtual wall detected, 1 = virtual wall detected)

Motor Overcurrents

Packet subset: 1

Range: 0 – 31

The state of the five motors' overcurrent sensors are sent as individual bits (0 = no overcurrent, 1 = overcurrent).

Bit	7	6	5	4	3	2	1	0
Motor	n/a	n/a	n/a	Drive Left	Drive Right	Main Brush	Vacuum	Side Brush

Dirt Detector Left

Packet subset: 1

Range: 0 - 255

Data type: 1 byte, unsigned

The current dirt detection level of the left side dirt detector is sent as a one byte value. A value of 0 indicates no dirt is detected. Higher values indicate higher levels of dirt detected.

Dirt Detector Right

Packet subset: 1

Range: 0 – 255

Data type: 1 byte, unsigned

The current dirt detection level of the right side dirt detector is sent as a one byte value. A value of 0 indicates no dirt is detected. Higher values indicate higher levels of dirt detected.

Note: Some robots don't have a right dirt detector. You can tell by removing the brushes. The dirt detectors are metallic disks. For robots with no right dirt detector this byte is always 0.

Remote Control Command

Packet subset: 2

Range: 0 – 255 (with some values unused)

Data type: 1 byte, unsigned

The command number of the remote control command currently being received by Roomba. A value of 255 indicates that no remote control command is being received. See Roomba remote control documentation for a description of the command values.

Buttons

Packet subset: 2

Range: 0 – 15

Data type: 1 byte, unsigned

The state of the four Roomba buttons are sent as individual bits (0 = button not pressed, 1 = button pressed).

Bit	7	6	5	4	3	2	1	0
Button	n/a	n/a	n/a	n/a	Power	Spot	Clean	Max

Distance

Packet subset: 2

Range: -32768 – 32767

Data type: 2 bytes, signed

The distance that Roomba has traveled in millimeters since the distance it was last requested. This is the same as the sum of the distance traveled by both wheels divided by two. Positive values indicate travel in the forward direction; negative in the reverse direction. If the value is not polled frequently enough, it will be capped at its minimum or maximum.

Angle

Packet subset: 2

Range: -32768 – 32767

Data type: 2 bytes, signed

The angle that Roomba has turned through since the angle was last requested. The angle is expressed as the difference in the distance traveled by Roomba's two wheels in millimeters, specifically the right wheel distance minus the left wheel distance, divided by two. This makes counter-clockwise angles positive and clockwise angles negative. This can be used to directly calculate the angle that Roomba has turned through since the last request. Since the distance between Roomba's wheels is 258mm, the equations for calculating the angles in familiar units are:

Angle in radians = (2 * difference) / 258

Angle in degrees = (360 * difference) / (258 * Pi).

If the value is not polled frequently enough, it will be capped at its minimum or maximum.

Note: Reported angle and distance may not be accurate. Roomba measures these by detecting its wheel revolutions. If for example, the wheels slip on the floor, the reported angle of distance will be greater than the actual angle or distance.

Charging State

Packet subset: 3

Range: 0 – 5

Data type: 1 byte, unsigned

A code indicating the current charging state of Roomba.

Code	Charging State
0	Not Charging
1	Charging Recovery
2	Charging
3	Trickle Charging
4	Waiting
5	Charging Error

Voltage

Packet subset: 3

Range: 0 – 65535

Data type: 2 bytes, unsigned

The voltage of Roomba's battery in millivolts (mV).

Current

Packet subset: 3

Range: -32768 – 32767

Data type: 2 bytes, signed

The current in milliamps (mA) flowing into or out of Roomba's battery. Negative currents indicate current is flowing out of the battery, as during normal running. Positive currents indicate current is flowing into the battery, as during charging.

Temperature

Packet subset: 3

Range: -128 – 127

Data type: 1 byte, signed

The temperature of Roomba's battery in degrees Celsius.

Charge

Packet subset: 3

Range: 0 – 65535

Data type: 2 bytes, unsigned

The current charge of Roomba's battery in milliamp-hours (mAh). The charge value decreases as the battery is depleted during running and increases when the battery is charged.

Capacity

Packet subset: 3

Range: 0 – 65535

Data type: 2 bytes, unsigned

The estimated charge capacity of Roomba's battery. When the Charge value reaches the Capacity value, the battery is fully charged.

Roomba Open Interface Commands Quick Reference

Command	Opcode	Data Byte 1	Data Byte 2	Data Byte 3	Data Byte 4	Etc.
Start	128					
Baud	129	Baud Code (0 – 11)				
Control	130					
Safe	131					
Full	132					
Power	133					
Spot	134					
Clean	135					
Max	136					
Drive	137	Velocity (-500 – 500)		Radius (-2000 – 2000)		
Motors	138	Motor Bits (0 – 7)				
Leds	139	Led Bits (0 – 63)	Power Color (0 – 255)	Power Intensity (0 – 255)		
Song	140	Song Number (0 – 15)	Song Length (0 – 15)	Note Number 1 (31 – 127)	Note Duration 1 (0 – 255)	Note Number 2, etc.
Play	141	Song Number (0 – 15)				
Sensors	142	Packet Code (0 – 3)				
Force-Seeking-Dock	143					

Baud data byte 1: Baud Code (0 – 9)

Baud code	Baud rate in bps
0	300
1	600
2	1200
3	2400
4	4800
5	9600
6	14400
7	19200
8	28800
9	38400
10	57600
11	115200

Motors data byte 1: Motor Bits

0 = off, 1 = on

Bit	7	6	5	4	3	2	1	0
Motor	n/a	n/a	n/a	n/a	n/a	Main Brush	Vacuum	Side Brush

Leds data byte 1: Led Bits (0 – 63)

Dirt Detect uses a blue LED: 0 = off, 1 = on

Spot, Clean, and Max use green LEDs: 0 = off, 1 = on

Status uses a bicolor (red/green) LED: 00 = off, 01 = red, 10 = green, 11 = amber

Bit	7	6	5	4	3	2	1	0
LED	n/a	n/a	Status (2 bits)		Spot	Clean	Max	Dirt Detect

Power uses a bicolor (red/green) LED whose intensity and color can be controlled with 8-bit resolution.

Leds data byte 2: Power Color (0 – 255)

0 = green, 255 = red. Intermediate values are intermediate colors.

Leds data byte 3: Power Intensity (0 – 255)

0 = off, 255 = full intensity. Intermediate values are intermediate intensities.

Roomba Open Interface Sensors Quick Reference

Packet Code	Packet Size
0	26 bytes
1	10 bytes
2	6 bytes
3	10 bytes

Name	Groups	Bytes	Value Range	Units
Bumps Wheeldrops	0, 1	1	0 – 31	
Wall	0, 1	1	0 – 1	
Cliff Left	0, 1	1	0 – 1	
Cliff Front Left	0, 1	1	0 – 1	
Cliff Front Right	0, 1	1	0 – 1	
Cliff Right	0, 1	1	0 – 1	
Virtual Wall	0, 1	1	0 – 1	
Motor Overcurrents	0, 1	1	0 – 31	
Dirt Detector - Left	0, 1	1	0 – 255	
Dirt Detector - Right	0, 1	1	0 – 255	
Remote Opcode	0, 2	1	0 – 255	
Buttons	0, 2	1	0 – 15	
Distance	0, 2	2*	-32768 – 32767	mm
Angle	0, 2	2*	-32768 – 32767	mm
Charging State	0, 3	1	0 – 5	
Voltage	0, 3	2*	0 – 65535	mV
Current	0, 3	2*	-32768 – 32767	mA
Temperature	0, 3	1	-128 – 127	degrees C
Charge	0, 3	2*	0 – 65535	mAh
Capacity	0, 3	2*	0 – 65535	mAh

* For 2 byte sensor values, high byte is sent first, followed by low byte.

Bumps Wheeldrops

Bit	7	6	5	4	3	2	1	0
Sensor	n/a	n/a	n/a	Wheeldrop			Bump Left	Bump Right
				Caster	Left	Right		

Motor Overcurrents

Bit	7	6	5	4	3	2	1	0
Motor	n/a	n/a	n/a	Drive Left	Drive Right	Main Brush	Vacuum	Side Brush

Buttons

Bit	7	6	5	4	3	2	1	0
Button	n/a	n/a	n/a	n/a	Power	Spot	Clean	Max

Charging State Codes

Code	Charging State
0	Not Charging
1	Charging Recovery
2	Charging
3	Trickle Charging
4	Waiting
5	Charging Error

Index

Note to the Reader: Throughout this index **boldfaced** page numbers indicate primary discussions of a topic. *Italicized* page numbers indicate illustrations.

A

AARON art, 168
accuracy of distance and angle measurements, **125**, *125*
acid flux solder, 385
actionPerformed method, 143
actions
  driving. *See* driving actions
  translating MIDI notes into, **163–164**
ActionScript language, 134
active components, 409
Adams, Bryan, 367–368
Adams, Ricci, 154
addEventListener method, 118
addKeyListener method, 105
addresses. *See* IP addresses
alarm clock, **200–202**
Altium software, 405
ampere-hours, 14
analog sensors, **190**
analogWrite method, 291
AND operator
  in bit operations, 31
  for sensor data, 120
Angle, Colin, 4
ANGLE command, 190
angle sensors
  as mouse pointers, 191
  packets for, 424
angles
  computing, 37
  measuring, **124–125**, *125*
  for mouse simulation, 190
  rotating, **101–102**
angular speed, 101
annulus, 90, *91*
antennae for theremins, 194
anti-static bags, 402, *402*
anti-static foam, 402
applications
  RoombaFX framework for, **372–373**, *373–374*
  SCI tester, 373–374
Arduino microcontrollers, **276–277**
  for bump turns, **288–290**
  coding and running modes in, **286–288**, *287*
  environment for, **279–281**, *280*
  hooking up to Roomba, **282**
  for mobile mood light, **290–294**, *291–292*
  overview, **277–279**, *278*

prototyping shield for, **284–286**, *285–286*, 293
  starting, **281–282**, *282*
arp command, 219
art, **167**
  brushes for, **170–174**
  canvas for, **174–175**, *175–176*
  parts and tools for, **168–169**, *169*
  by robots, 168
  spirals. *See* spirals
  testing modifications for, **176–177**, *177*
at method, **98–100**
ATMega8 microcontroller
  in projects, *278*, 279
  in schematic diagrams, 412, *412*
  serial ports in, 287
auto-refreshing webcam images, 342–343, *343*
autonomous operations, sensors for, **122–123**
autonomous Roombas, **365–366**
  Erdos, **370**
  Gumstix boards, **367–368**, *367–368*
  iPaq PDAs, **368–369**, *369*
  Mind Control, **366**, *366–367*
available power, **14**
AVR Butterfly board, 282
AVR-Libc library, 281
AVRLib library, 281
AWARE robotic intelligence, 10

B

bad solder joints, **393**, *393*
Banzi, Massimo, 279
Barragan, Hernando, 279
BASIC language interpreter, 261
Basic Printing Profile (BPP), 68
Basic Stamp microcontrollers, **261**
  Basic Stamp 2, **262–263**
  for bump turns, **268–272**
  environment for, **264–265**, *264–266*
  hooking up to Roomba, **266–268**, *267*
  for robot roach, **273–275**, *273–274*
  in schematic diagrams, 412, *412*
batteries and battery packs, **14**
  Basic Stamp 2, 262, 266
  in schematic diagrams, 411, *411*
  SitePlayer Telnet, 221
  upgrading, **376–378**, *377*
  vision systems, 358–359
  WL-HDD device, **325–328**, *326–328*
  WRTSL54GS, **337**, *337–338*
  XPort, 221
battery charge value, 38
battery current value, 38
BAUD command and baud rates
  Arduino, 288
  Basic Stamp 2, 268
  Bluetooth interface, 82

opcodes and data bytes for, 27
  overview, 29
  specification for, 419
  WiMicro, **245**
beepers, piezo, **151–153**, *152*
behavior-based robotics architecture, 4
belkin_sa driver, 317
bin release, 307
bits
  for LEDs, 32
  for sensors, **120**
  setting and clearing, **31**
blu2i module, 71
BlueSMiRF boards, 66. *See also* Bluetooth interface
Bluetooth interface, **65–66**, *66*
  benefits, **66**
  building, **71–76**, *72–77*
  cable for, **72**, **74–75**, *75*
  circuit for, **70–71**
  configuring, **82–83**
  connection tests for, **75–76**, *76*
  enclosures for, **76**, *77*
  for Erdos, 370
  operation of, **67**
  pairing with, **78–80**, *79–80*
  parts and tools for, **68–69**, *69*, **72–73**, *73*
  power classes for, **67**
  profiles for, **68**
  soldering for, **73–75**, *74–75*
  testing, **83**
  virtual serial ports for, **80**, *81*
  voltage checks for, **74**
  working with, **83**
Bluetooth Setup Assistant, **78–80**, *79–80*
Board of Education, 263, *263*
  for robot roach, 273, *274*
  setup for, 265, *266*
  wiring, 267, *267*
Bonjour networking, 214
boot_wait bootloader, 299, 305
bootloaders
  for Arduino, 281
  for single board computers, 299
BPP (Basic Printing Profile), 68
braid, desoldering, 394, *394–395*
brain replacement, **257**
  Arduino for. *See* Arduino microcontrollers
  Basic Stamp for. *See* Basic Stamp microcontrollers
  microcontrollers vs. microprocessors in, **257–258**, *259*
  parts and tools for, **258–260**
  solderless breadboards for, **260–261**, *260*
  in vision systems, **334–346**
breadboards, **260–261**, *260*
bricking routers, 299
Brooks, Rodney, 4

brushes
    attaching, **171–174**, *171–174*
    types of, **170**
bump sensors
    for mouse simulation, 191
    operation, **15**
    optical interrupters, **112**, *112*
bump turns
    Arduino for, **288–290**
    Basic Stamp for, **268–272**
bumpLeft method, 120
bumpRight method, 120
bumps wheeldrops sensor packets, 423
BumpTurn.bs2 program, **270–272**, 275
BumpTurn.Java program, 122–123
BusyBox program, 302
button sensors
    micro-switches for, **113**, *114*
    packets for, 424
buttons
    commands for, **36–37**, *36–37*
    MyGUI for, **142–144**, *144*
    in schematic diagrams, 411, *411*
    on WRTSL54GS, **363**
bytes
    bits in, 31
    for ROI commands, **27–28**
    for sensor data, **121**

C

C language and libraries
    for Linux, 303
    for vision systems, **346–351**
cables, **21–23**, *21, 23*
    Bluetooth interface, **72**, **74–75**, *75*
    drive motor unit, 89, *90*
    serial interface tether, **42–43**, **49–50**, *51*,
        54, *55*
    SitePlayer Telnet adapter, 211–213, *213*
    vision systems, **353–357**, *354–357*
calculators, graphing, 182
cameras in vision systems, 333, **339–340**, *340*
    drivers for, **341**
    power consumption by, 359
    for taking pictures, **341–342**
    for viewing pictures, **342–343**, *343*
canvases, laying out, **174–175**, *175–176*
capacitive touch sensors, 16
capacitors
    for MAX232 transceivers, 48
    in schematic diagrams, **410**, *410*
    voltage ratings for, 46
    in voltage regulators, 45
capacity sensor packets, 424
Capizzi, Craig, 376
carpets, current variations from, 14
carrier-sense multiple access with collision
        avoidance (CSMA/CA)
        technique, 233
carrier-sense multiple access with collision
        detection (CSMA/CD)
        technique, 232
cell phone sync cable hack, **42–43**

CF (Compact Flash) board, 298
CGI (Common Gateway Interface), **351**, *352*
chalk, 170, 175
channelLoop method, 198–199
charge sensor packets, 424
charging state sensor packets, 424
circles, unit, 180, *180*
Circuit Cellar magazine, 258
clamps for attaching brushes, **171–174**,
        172–174
classes for Bluetooth power, **67**
Clean bit, 32
Clean button for mouse simulation, 191
CLEAN command
    modes for, 26
    opcodes and data bytes for, 28
    overview, 29
    sending, 98
    specification for, 420
cleaning motors, **30**
clearing bits, **31**
cliff front left sensor packets, 423
cliff front right sensor packets, 423
cliff left sensor packets, 423
cliff right sensor packets, 423
cliff sensors
    for line-following Roombas, 375
    optical object detectors, **113**, *113*
    for pitch control, **198–199**
    for theremin simulation, 195, *195*
    wheel-drop, 16
cliffFrontLeft method, 120
clock, alarm, **200–202**
Cobox Micro, 222, 236
cockroach, robot, **273–275**, *273–274*
code stick programmers, 366, *366–367*
code structure, **84**
coding mode in Arduino, **286–288**, *287*
Cohen, Harold, 168
coils in schematic diagrams, 411, *411*
cold solder joints, 391, *391*
collision avoidance, 232–233
color of Power LEDs, 32
command center for vision systems,
        360–362, *361*
command line
    for driving actions, **97**
    for OpenWrt, **324–325**
commands
    in ROI specification, 419–422, 425
    structure of, **27–28**
CommAPI, 85
Common Gateway Interface (CGI), **351**, *352*
Compact Flash (CF) board, 298
compatibility of Processing, **135**
compilers vs. interpreters, **276–277**, *277*
compiling
    Java programs, 96
    roombacmd program, **349–350**
components
    motors, **15**
    power, **14**
    sensors, **15–16**
    soldering, 391, *392*
    underside, **13**, *13*

computeRoombaLocation method, 148, 190
computeSensors method, 117–118
computing
    angles and distances, 37
    position, **147–148**
configuring
    Bluetooth interface, **82–83**
    OpenWrt, **309–313**, *309–313*
    Telnet, **245**
    vision systems, **352–353**
    WiMicro boards, **243–246**, *243*
    XPort, **223–225**, *224*
connect method
    RoombaComm, 62, **85–87**
    RoombaCommTCPClient, 226–227
connections, 396, *397*
    Bluetooth interface, **75–76**, *76*
    ROI specification for, 418
    in schematic diagrams, **406–407**, *407*
    serial interface tether, **54–56**, *57*
connectors, ROI, **21–24**, *21–24*
constants in RoombaComm, 98
control, ROI for, **20**
CONTROL command
    modes for, 25–26
    opcodes and data bytes for, 27
    overview, 29
    specification for, 419
control method, 62, 84, 87
conventions for schematic diagrams, **406**
converting
    note names to MIDI note numbers,
        154, *155*
    radius/velocity to left/right speeds, **94–96**,
        94–95
core MIDI, **164–166**
cost of Processing, **134–135**
costumes
    building, **372**
    RoomBud, **370**, *371*
CP2103 chip, 317
cpuinfo command, 320
createSong method, 156, 160
Creative Instant webcam, 340, *340*
cross-platform compatibility of Processing, **135**
crystals, **152–153**
CSMA/CA (carrier-sense multiple access
        with collision avoidance)
        technique, 233
CSMA/CD (carrier-sense multiple access
        with collision detection)
        technique, 232
Cuartielles, David, 279
current
    through LEDs, 46–47
    monitoring, 38
    in Ohm's Law, 47
    variations in, 14
    in vision systems, **358–360**, *359–360*
current sensor packets, 424
currentTimeMillis method, 127
curves
    moving in, **102–104**, *103*
    parametric, **178–181**, *179–180*
cutters, **388**, *388*